# Chemistry matters

Richard Hart

Head of Chemistry,
Churchdown Comprehensive School, Gloucestershire

Oxford University Press

Oxford University Press,
Walton Street, Oxford OX2 6DP

Oxford London Glasgow
New York Toronto Melbourne Auckland
Kuala Lumpur Singapore Hong Kong Tokyo
Delhi Bombay Calcutta Madras Karachi
Nairobi Dar es Salaam Cape Town

and associated companies in
Beirut Berlin Ibadan Mexico City Nicosia



ISBN 0 19 914050 2

First published 1978
reprinted 1979, 1980, 1981, 1982, 1983

Front cover: used nuclear fuel rods
(see pages 56–7)

Rear cover: Bikini Island 1946

Phototypeset in Great Britain by
Filmtype Services Limited, Scarborough

Printed in Great Britain
at the University Press, Oxford
by Eric Buckley
Printer to the University

# Acknowledgements


## Photographs

A–Z Botanical 238
AERE Harwell 123
Associated Press Ltd 240 top
Paul Brierley 246
British Leyland Motor Corporation 7 133 bottom
British Petroleum Co. Ltd 276 top right top left bottom 277
British Steel Corporation (Brian Richards) 256
S. J. Brown 225
Ronald Chapman 168 bottom
Courage Beers 182 top
Courtaulds Ltd 291
De Beers 88
Colin Dexter 195
EMI Ltd Central Research Laboratory 294 bottom right
EMI Electronics Ltd 295 top
GPG Products 289
Mary Evans Picture Library 16 top right 17 top left 72 top
Dr T. Faulkner 55
Fox Photos Ltd 145 bottom 154 top
GEC 89 bottom
Glass Manufacturers Association 169 top right
Houseman (Burnham) Ltd 132 top
ICI Ltd 168 top right
ICI Ltd Plastics Division 286
Ilford Ltd 73
Indusfoto 41
Keystone Press Agency 58 60
Metal Box Ltd 31 bottom
National Coal Board 273
Picturepoint Ltd 151 261
Pilkington Brothers Ltd 169 bottom
Popperfoto 42 top right 74 bottom 102 top 103 right 114 top 115 top 155 top 156 bottom 282 top
Press Association Ltd 208 top
Michael Putland 294 top
Radio Times Hulton Picture Library 194 267 bottom
Royal Geographical Society 144 top right 145 top left
Shell UK Ltd 241 224 top
Strand Studios 279
Syndication International 30 top 43 top 111 266 top
Thermit Welding (GB) Ltd 252 right left
John Topham Picture Library 116 top
UKAEA 59 top front cover
Watney Mann and Truman Brewers Ltd 183
ZETA backcover

## Extra time material

**Introduction** Adapted from 'The Chemistry of a Mini' by C. E. Jones, from *Science and the Car* (Wheaton Press, 1975), by permission of the author.

**Chapter 3** 'It never rains . . .' extracted from 'When the Country's Taps Run Dry' by Dr. Tom Margerison, which appeared in the *Daily Telegraph Magazine* (No. 607), by permission of the Daily Telegraph.

'Liquid Assets Down the Drain' extracted from an article in *The Guardian* of 12 August 1976 by Lesley Adamson, by permission of the Guardian Newspapers Ltd.

**Chapter 4** Second article condensed from *Black Rain* by Masuji Ibuse, translated by John Bester, (Copyright 1969) originally published in English by Kodansha International Ltd. By permission of Kodansha International Ltd., and Martin Secker & Warburg Ltd.

**Chapter 6** Adapted from 'The Diamond Maker' by Ivor Smullen, first printed in Mayfair, reprinted by permission of Fisk Publishing Co. Ltd on behalf of the author.

**Chapter 8** Adapted from articles from *The Daily Herald*, December 1952 and October 1953 by permission of Syndication International Ltd.

**Chapter 9** Adapted from an article first published in the *Observer Colour Supplement.*

**Chapter 10** Extracted from *High Adventure* by Sir Edmund Hillary, by permission of Hodder & Stoughton Ltd.

**Chapter 11** Extracted from *The Hindenburg* by M. Mooney (Rupert Hart-Davis Ltd.) by permission of Granada Publishing Ltd.

**Chapter 14** Adapted from 'Chemical Warfare – the Initial Horror' by Dr. Robert Jones, first printed in *New Scientist London* (No. 945 Vol. 66) the weekly review of Science & Technology, by permission of New Science Publications.

**Chapter 18** Concorde quotation from 'Supersonic Profile' from *Shell Aviation News* (No. 433 1976).

'Planes of the Future' extracted from an article in the *New Scientist* (No. 519 1977).

'Samurai!' from *Ascent of Man* by J. Bronowski, by permission of BBC Publications.

**Chapter 19** 'The Petrol Tree' adapted from an article from *The Guardian*, 1 February 1977, by permission of the Guardian Newspapers Ltd.

# Contents

# Introduction

This book is about chemistry, and it can be used for both reference and reading. It does not contain experiments for you to do. Your teacher will give you these. I hope you will use the book to help you write your notes, to help you with revision, and above all, I hope you will enjoy reading it.

Each chapter deals with an important topic in chemistry. Sometimes, well-known experiments that you may have seen or done, are described. At the end of each chapter, there is a summary which tells you what you should have learned about that topic. There is also a section called 'Extra Time'. These passages may show you how chemistry matters both outside of school, as well as inside it.

This book has a contents page and an index. Use them both. If you want to revise or study a large topic, look on the contents page to find the chapter which deals with it. If you want to look up a specific fact, like a definition or a formula, look for the word in the index. Next to it you will find one or more numbers. These are the pages on which information about that word appears. Turn to those pages and look for the facts that you want.

Chemistry is a very important subject. It is involved in almost everything in our daily lives from the clothes we wear to the air we breathe, and the food we eat. The article that follows should give you some idea of how important chemistry really is.

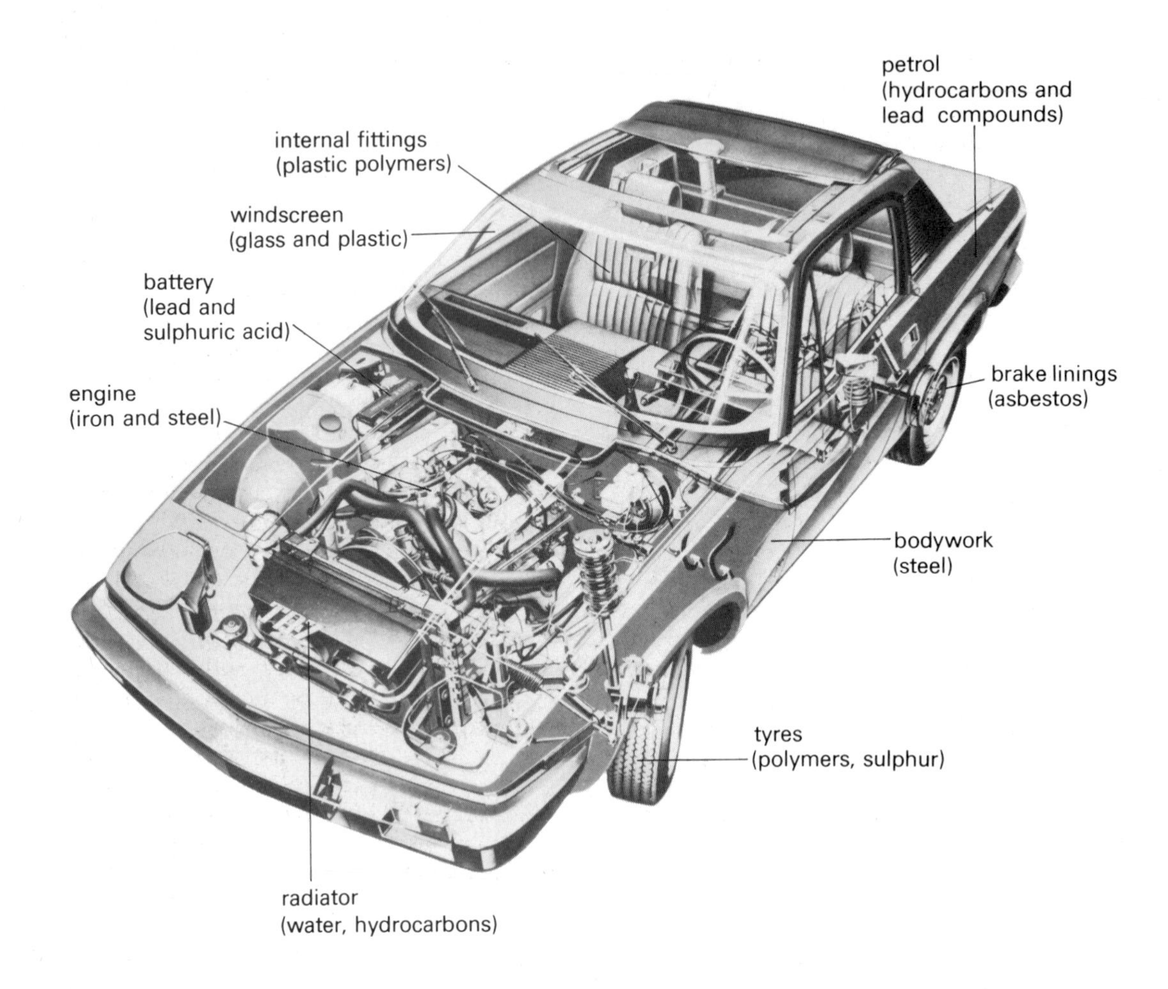


# Chemistry on four wheels

*Chemists turn raw materials – mainly iron ore and crude oil – into iron, steel, paint, plastic, glass, lubricants, and fuel. In doing so, they use hundreds, if not thousands, of different chemical reactions.*

**Out along the motorway at 70 miles per hour . . .**

The whole car, including the engine and the chassis, the tyres and the bodywork, and the petrol that it uses to drive it along – are all produced by man from natural materials. The chemical structure of the car is illustrated in the diagram.

**The engine and the bodywork**

The engine is made of cast iron. It is made by pouring molten iron into moulds; as it cools it solidifies. The engine comes out exactly the same shape as the mould. By making it this way it makes the engine very strong indeed. On the other hand, cast iron is not malleable, and therefore it is not suitable for use in making the bodywork of the car. Cast iron cannot be bent or easily formed into the shapes that are needed. But if iron is purified and small amounts of carbon and other elements are added to it, it becomes steel which is softer and easy to bend. Other parts of the car such as the radiator grille and the bumpers are also made of steel, but they are plated with chromium which does not rust and which has an attractive shine.

**Paint**
Cars are painted to help prevent rust. Before the paint is applied, the car body is first thoroughly cleaned with a solvent and sodium hydroxide solution. It is then treated with a coating of zinc phosphate. This helps to stop rust and also provides a suitable surface for the first coat of special paint. The car body is then baked in an oven so that the paint layer becomes very hard. After this, up to five further layers of other paints may be applied.

**The glass and fittings**
The chemicals used to make the internal fittings of the car are made from crude petroleum oil. Many of these are called polymers. One polymer called Bakelite is used for the distributor cover, and polyurethane foam is used for the seat cushions. Polyvinyl chloride (PVC) is used for the seat covers, and another polymer called polyethene is used for many of the plastic fittings such as the windscreen washer bottle. Polystyrene is used to make the housing of the steering wheel.

The windows are made of a special safety glass that does not shatter when struck hard. This glass is made of sodium silicate which is heated to make it tougher. The windows are sometimes made up of several layers of glass and plastic which make it extra tough and even less likely to cause injury if it is smashed in an accident.

**The car battery**
The car battery consists of six cells which give the total voltage of twelve volts. The electrodes are grids made of lead. The electrolyte is dilute sulphuric acid. When in use the battery has to be kept topped up with distilled water.

**Ore and oil come together**
In the manufacture of the complete car, the iron and other metal ores have been treated and processed in a variety of different ways to make them useful. The crude oil is converted into all kinds of different plastics and lubricants which are used both in the structure of the car, and to make sure that it runs smoothly. The oil also provides the main driving force for the car, which is of course petrol.

**Chemical reaction on four wheels**
The petrol used in a car is obtained from crude petroleum after fractional distillation. Petrol burns easily, sometimes explosively, to produce carbon dioxide and water. It also produces some carbon monoxide, which is highly poisonous. Because of this, car engines must not be run in enclosed spaces such as garages, where the gas is not able to disperse.

In the cast iron engine there are four cylinders containing pistons which are lubricated by the engine oil. A mixture of petrol and air is drawn into each cylinder as the piston descends. The inlet valve which let the petrol and air mixture in, closes. The piston returns up the cylinder, compressing the fuel. An electric spark from a sparking plug ignites the mixture which burns very quickly to produce a large volume of gases which expand and force the piston down. If the fuel burns too rapidly the piston may vibrate rather than move smoothly. This vibration can cause damage to the engine. It is called knocking, and it can be reduced by the addition of lead tetraethyl to the petrol. A disadvantage results, however, from the use of this chemical, because lead tetraethyl burns in the engine to form lead oxide and this eventually comes out in the exhaust fumes as lead vapour. Lead is very poisonous.

**Cool it, man!**
The engine is cooled by a water radiator. Without the radiator the engine would overheat and the pistons might weld to the cylinder casing. During the winter, anti-freeze is added to the water because otherwise it might freeze, and doing so expand and crack the radiator. Anti-freeze is a chemical called ethane-1,2-diol which dissolves in the water and lowers its freezing point to a safe level.

**Chemical reactions make it go**
Hundreds, if not thousands, of different chemical reactions are used to convert the variety of raw materials into the end product, which is the car able to travel at high speed. Quite a number of these chemical reactions have been mentioned in this section. There has been a lot to think about all in one go, but you will find many of the aspects of the chemistry dealt with in this section, in different parts of the book.

# 1 Atoms and molecules

## 1.1 Names

In this chapter, you will meet, and use the terms *atom*, *element*, *molecule* and *compound*. You probably know these terms and use them already, but remember that each has a very precise meaning. We talk about an atom of sodium, but a molecule of sugar. We say that sodium is an element, but that sugar is a compound. The following sections explain what these terms mean.

## 1.2 Atoms

Atoms can be thought of as very small particles from which all other substances are built up. They are so small that no one has ever seen them. Nevertheless, we believe they exist, because with them we can explain the results of many experiments. Scientists have built up an 'identikit' picture of the atom.

**1** Atoms are very small spheres. They can be of different sizes, but an average atom has a diameter of about half a thousand millionth of a metre. This can be expressed as:

$$\frac{1}{2\,000\,000\,000} \text{ or } 0{\cdot}5 \times 10^{-9}\text{ m or } 0{\cdot}5 \text{ nanometre (nm)}.$$

This is incredibly small. You would need to place two thousand million atoms side by side for them to stretch one metre.

**2** Until this century, scientists thought that atoms could not be split. (The word atom meant 'unsplittable' in ancient Greek.) However, it has been shown that atoms can be broken into smaller parts. Even so, it is still useful most of the time to think of atoms as tiny spheres.

**3** Everything is made of atoms. Sometimes atoms join together into pairs or larger groups. They are then called molecules. You can read more about them in the next section.

In conclusion: all substances on Earth and in space are made of atoms or molecules. Atoms can be thought of as tiny spheres that cannot easily be broken up.

1.3
## Elements

Substances made from only one sort of atom are called *elements*. There are about one hundred different elements, and each contains a different sort of atom.

Carbon is an element. It contains only carbon atoms. All carbon atoms behave in the same way.

Copper is another element. It contains only copper atoms. All copper atoms behave in the same way. Copper atoms are different from carbon atoms. They have different masses, sizes and chemical properties. Copper and carbon are two different elements.

For convenience, each element is given a *symbol*. You will find a complete list of elements and their symbols at the back of the book, but figure 1 contains some of the more common elements and their symbols.

| **element** | **symbol** | **element** | **symbol** |
|---|---|---|---|
| aluminium | Al | magnesium | Mg |
| calcium | Ca | nitrogen | N |
| carbon | C | mercury | Hg |
| chlorine | Cl | oxygen | O |
| copper | Cu | silver | Ag |
| gold | Au | sulphur | S |
| hydrogen | H | tin | Sn |
| iodine | I | uranium | U |
| iron | Fe | zinc | Zn |
| lead | Pb | | |

**Figure 1**
*Common elements and their symbols.*

The symbol often comes from the first one or two letters of the element's name – either its modern name, or an old one. For example, gold has the symbol Au, taken from *Auram*, the Latin word for the sun.

1.4
## Molecules

Sometimes, an element consists of *molecules*. Molecules are groups of atoms joined together. The symbol for hydrogen gas is $H_2$. This tells us that hydrogen gas is made of two hydrogen atoms joined together. These are called hydrogen molecules. If a substance is made up from groups of more than one atom joined together, we give it a *formula*. This tells us what atoms are present in each group. A molecule of chlorine is made from two identical chlorine atoms:

$Cl_2$ is the formula for chlorine gas.

A molecule of oxygen gas is made from two identical atoms of oxygen:

$O_2$ is the formula for oxygen gas.

## 1.5 Compounds

Water is a *compound*. It contains molecules. The atoms in the molecules of water are not all the same sort.

Compounds are substances made up of molecules whose atoms are not all the same sort.

The formula for water is $H_2O$. This means that a molecule of water contains two atoms of hydrogen and one atom of oxygen, joined together in a compound.

The formula for carbon dioxide is $CO_2$. This tells us that it is a compound made of one carbon atom and two oxygen atoms, making a molecule of carbon dioxide.

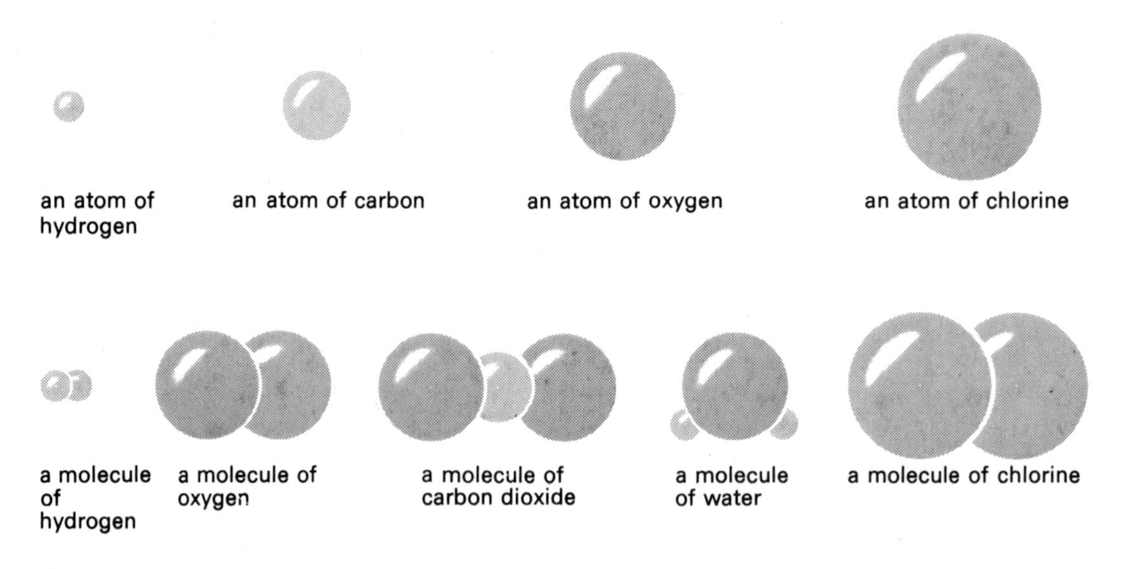


**Figure 2** *Atoms and molecules in elements and compounds.*

For each of the molecules in figure 2, the formula tells us what elements are present, and how many atoms of them there are in each molecule.

## 1.6 Reasons for believing in atoms and molecules

Atoms cannot be seen, even with the most powerful microscope. Some very large molecules such as proteins can be seen as blurred outlines with an electron microscope. Another instrument called a field ion microscope can produce pictures showing the arrangement of atoms in metals – but neither of these machines can give us pictures of atoms that we can study.

Scientists believe in atoms and molecules, because there is so much evidence, all of it pointing to the same thing. Detectives use a variety of clues in the same way to find a criminal, even though it may be that no-one actually saw him commit the crime. Like detectives, we may be wrong . . . The next six paragraphs give you some of the evidence. What do you think?

**Crystals have a regular shape.** Many substances occur as crystals. Naphthalene, used in moth-balls, forms crystals. Its crystals have straight edges and flat surfaces. This can be explained by suggesting that naphthalene is made of molecules. When naphthalene crystallises from a solution, the molecules arrange themselves in regular rows, to give the crystal shape. Other crystals of other substances have different shapes, but they are always regular and do not vary. In each case the molecules are packing together in a regular pattern.

**Crystals dissolve.** When a lump of sugar is dropped into water, it dissolves. We cannot see the sugar but we know it is still there, because the water tastes sweet. The sugar crystals have disappeared, but the water remains, so the water must have broken the crystals down in some way.

This can be explained by suggesting that sugar crystals contain molecules arranged in a regular order. Water is a liquid which does not have a definite shape, so the molecules cannot be in a regular order. If we could see the molecules before dissolving, they might appear as in Figure 3**a**.

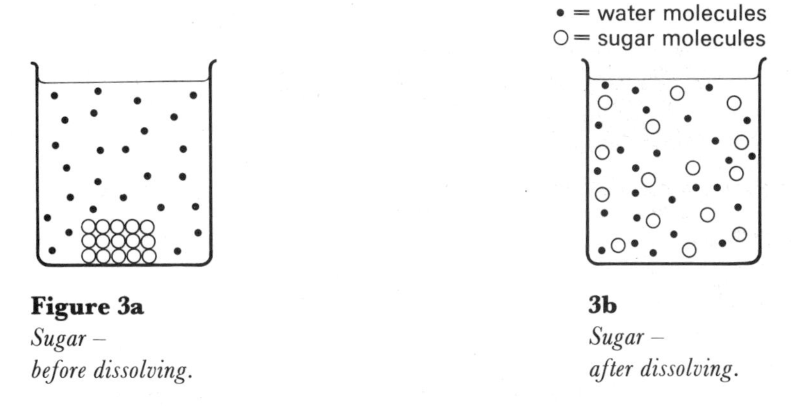


**Figure 3a** *Sugar – before dissolving.*

**3b** *Sugar – after dissolving.*

As the sugar dissolves, the molecules break away from each other and mix in with the water molecules. The result is a solution. A solution is a mixture of solid molecules and liquid molecules. It is shown in Figure 3**b**.

This experiment looks more convincing if you use a coloured substance such as copper sulphate. (See figure 4.) The copper sulphate mixes with the water to make a clear blue solution.

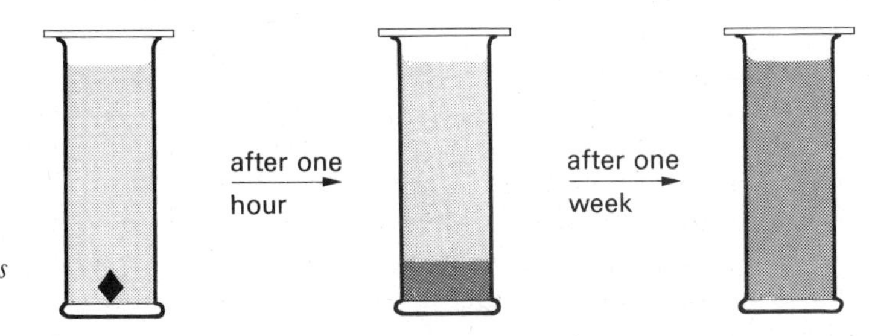


**Figure 4** *A copper sulphate crystal slowly dissolves in water.*

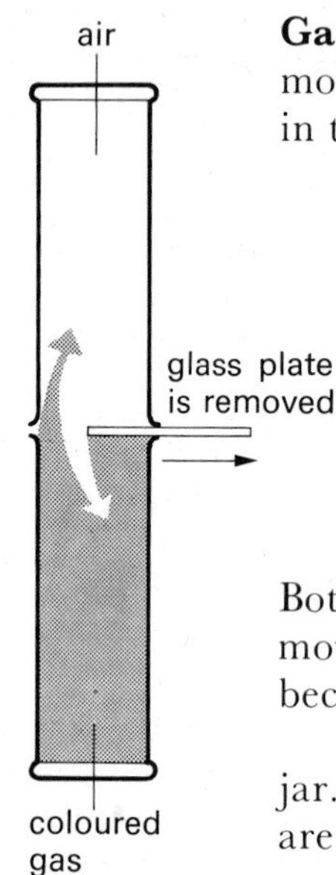


**Figure 5**
*Gases mix quickly.*

**Gases mix easily.** When a solid dissolves in a liquid, the molecules slowly mix with one another. If two gases are put together in the same container they mix very quickly. (See figure 5.)

Both gases consist of molecules with no regular arrangement. They move about at random. You cannot see the individual molecules because they are too small.

When the glass plate is removed, the colour rises into the other jar. After a few minutes, the gases have mixed completely. There are big spaces between the molecules, and the molecules are in constant motion, so it would be very surprising if they did not mix.

**Brownian Motion.** In 1827, Robert Brown, a botanist, was looking at some grains of pollen floating on water when he noticed that they were moving in an erratic way. He had seen the pollen grains with the aid of a microscope, but not the much smaller water molecules. These were bumping into the pollen grains causing them to move jumpily about. This was later called *Brownian Motion.* It can be seen whenever small particles of dust or smoke, or even tiny crystals are suspended in a liquid or a gas. Usually, they have to be illuminated with a bright light and viewed through a microscope.

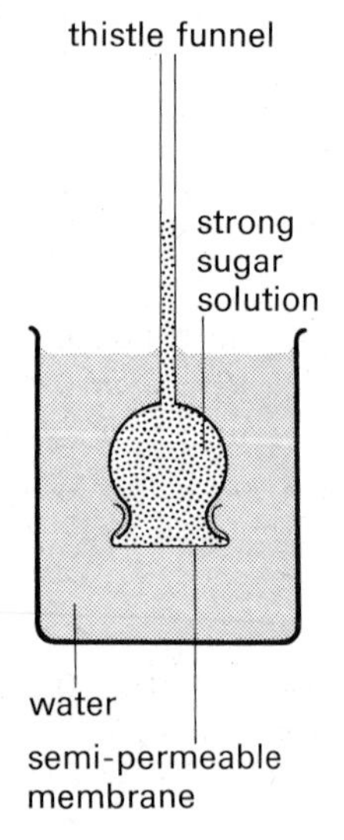


**Figure 6**
*A laboratory demonstration of osmosis.*

**Osmosis.** Plant cells need to contain just the right amount of water if the plant is to thrive. If a plant is put into salty water, water goes out of the plant cell into the salty water, so the plant wilts. The water passes through the plant walls, but the salt cannot. This process is called *osmosis.* It is difficult to explain osmosis without using the idea of molecules.

It can be demonstrated in the laboratory by the experiment illustrated in figure 6. The thistle funnel contains a strong sugar solution. The beaker contains pure water. A material called a semi-permeable membrane is stretched over the end of the thistle funnel. It has very small holes in it which allow water molecules to pass through it, but not sugar molecules. The water flows through the membrane, but the sugar does not, so the level in the thistle funnel stem rises. It stops when the weight of liquid balances the pressure of the osmosis, or the membrane breaks. The bladder or guts of a pig (from your local butcher) can be used to make a membrane. Cellophane or parchment could also be used.

**Solids, liquids and gases exist.** The very fact that our whole world is made up of solids, liquids and gases, fits the idea that everything is made up from atoms and molecules. Each of the three groups have easily recognisable properties.

Solids keep their shape. They are not easily squashed or stretched, but if they are, they keep the same volume. This is because the atoms or molecules in solids are packed closely together in a regular way and are held together by strong forces which attract the molecules to each other. The individual molecules do not move about, but they do vibrate. The more energy they have (in other words, the hotter they get), the more they vibrate. At a certain temperature, they vibrate so much that their regular arrangement breaks down. At this point, the solid melts and turns into a liquid. The temperature at which this happens is called the *melting point* of the substance.

Liquids take the shape of the vessel they are put into. They cannot be compressed or expanded by squeezing or stretching them. The molecules in a liquid move about but they are still quite close together. They are attracted to each other by small forces.

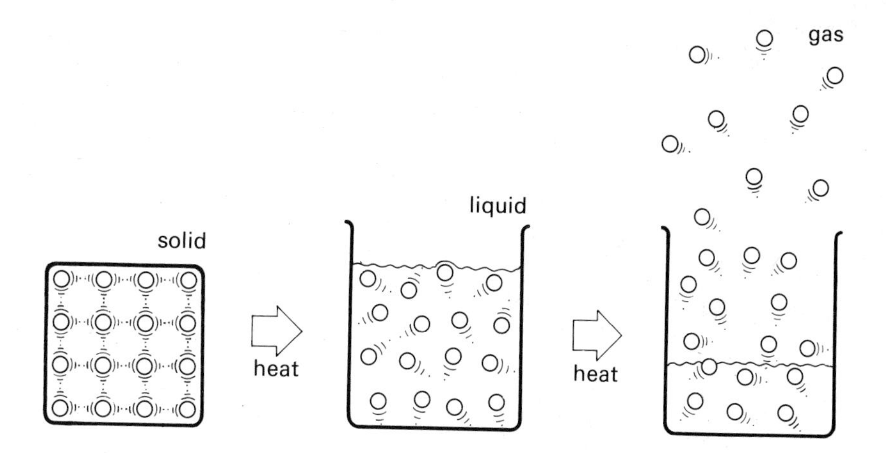


**Figure 7**
*Molecules in solids, liquids and gases.*

If a liquid is made hotter and hotter, the molecules move about faster and faster. Eventually, at a temperature called the *boiling point*, the molecules have enough energy to leave the liquid and form a gas. The molecules in a gas are not attracted to each other. They move about very quickly and independently. There are large spaces between the molecules so gases can be easily compressed and expanded.

This idea about the way molecules behave in solids, liquids and gases is called the *Kinetic Theory*. This is shown in figure 7.

**Summary.** All the experiments and properties which we have seen over the last six paragraphs are best explained using the idea of molecules. We have not proved that molecules really exist, but it is very difficult to explain all these observations without them.

## 1.7 Gas or vapour?

We talk about oxygen gas, but water vapour. What's the difference? Both words are used, but there is a strict scientific difference between them. In simple terms, the word *vapour* is used when a substance has formed into a gas even though it has not reached its boiling point. The smell of ethanol comes from ethanol vapour, and air contains water vapour.

## 1.8 Heating and cooling curves

If you measured the temperature of a block of ice as it was slowly heated until it melted, and then measured the temperature of the water until it boiled, you could plot a *heating curve* graph. The results would appear similar to those in figure 8.

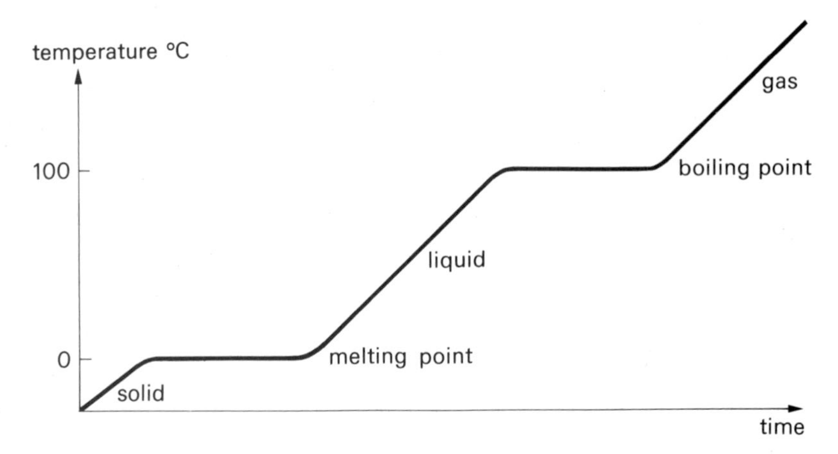


**Figure 8**
*A heating curve for ice.*

The gradual temperature increase stops at the melting point until all the ice has melted. Similarly, the temperature increase stops at the boiling point until all the water has turned to steam.

The same shape of curve is seen for any solid, to liquid, to gas change. If a gas is slowly allowed to cool, the gradual temperature decrease would also stop at the 'boiling' and 'freezing' temperatures. This would be a *cooling curve*.

## 1.9 Melting points and boiling points

The apparatus used to measure the melting point of a solid is shown in figure 9**a**.
The solid is put into a very narrow tube called a melting point tube, and this is attached to a thermometer with a rubber band. The thermometer and tube are immersed in paraffin oil and this is gently warmed up until the solid melts. At this point, the temperature is quickly noted.

To measure the boiling point of a liquid, it must be carefully boiled. A thermometer is held in the vapour, just above the liquid. (See figure 9**b**.) If the liquid is pure, the temperature registered on the thermometer is its boiling point.

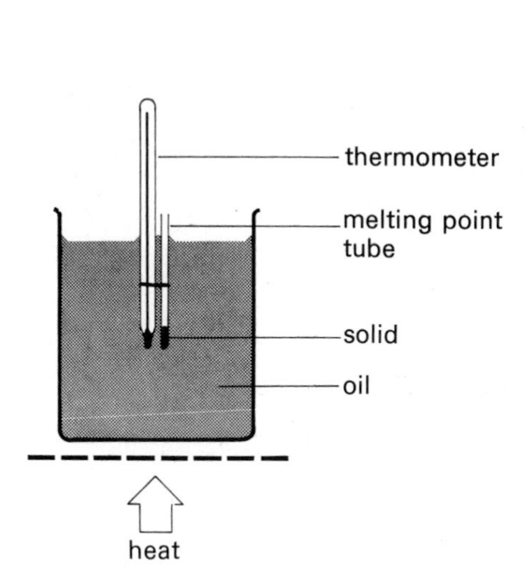


**Figure 9a**
*Melting point apparatus.*

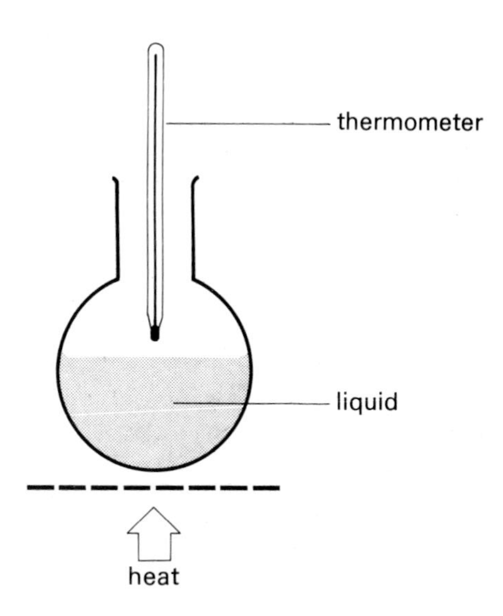


**9b**
*Boiling point apparatus.*

**Summary**

At the end of this chapter you should be able to:

**1** Explain what atoms and molecules are.

**2** Explain the meaning of the terms element and compound, giving examples of each.

**3** Find the symbols for elements in the table at the end of the book.

**4** Explain the structure of crystals, and describe the processes of dissolving, mixing gases, osmosis and Brownian Motion in terms of atoms and molecules.

**5** Explain how molecules are arranged and how they move in solids, liquids and gases.

**6** Explain what is meant by the terms melting point and boiling point.

**7** Measure the melting point of a solid and the boiling point of a liquid.

# It's a gas!

# The story of Robert Boyle and Jacques Charles

*Robert Boyle, who lived nearly three hundred years ago, experimented with gases. He measured how their volumes were affected by different pressures. About a hundred years later the Frenchman Jacques Charles found out even more. The article that follows explains their results in a way that has never been used before . . .*

### Oven and bicycle pump

Despite what you may read in Physics text books, I have reason to believe that Robert Boyle, the seventeenth century scientist did his experiments in this way. He had a bike pump. And an oven. The oven was perfectly normal – eye-level grill, four rings, smart enamelled finish, all the trimmings, but the pump was something special. Instead of the connector at the end, there was a pressure gauge.

### Volume halved, pressure doubled

Robert Boyle fixed his pump so that it was full of air and he put it in the oven so that it stayed at the same temperature. That was important. He then pushed the pump handle in so that the volume of the air in the pump was halved. To his great surprise, he found that the reading on the pressure gauge had doubled. (See the diagrams below.) In other words, the number found by multiplying the pressure by the volume, was the same after his experiment as it was before it.

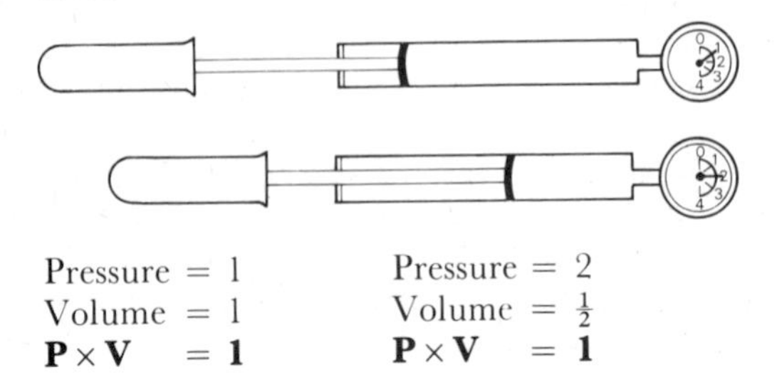


In this way he discovered what we now call Boyle's Law.

**'For a fixed mass of gas at constant temperature, the product of pressure and volume stays the same.'**

In other words, $P \times V$ at the start of an experiment is equal to $P \times V$ at the end of the experiment, so long as no gas escapes, and the temperature stays the same.

We can explain why Boyle's Law works using the kinetic theory you read about in the chapter. Imagine a gas in a container such as Robert Boyle's bike pump. (Look at the next diagram.)

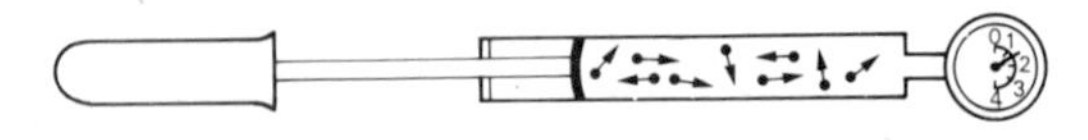

### Air molecules hit walls

The molecules of the gas inside the pump are in constant motion, bumping into one another and into the walls of the pump. The force of the molecules hitting the walls of the pump is the pressure of the gas. When you reduce the volume of the gas, the molecules have less space in which to move about, so they collide with each other and with the walls of the pump more often. In other words, as the volume is reduced, the pressure is increased.

## Oven and pump found in junk shop

Professor Jacques Charles lived about 100 years later, so I reckon he must have found Robert Boyle's bike pump and oven in a junk shop. Recognizing their true value immediately, he began a series of experiments that were very similar to Robert Boyle's. He fixed the amount of air in the pump so that the handle was half in, at normal room temperature. He then put it in the oven, and began to heat it up . . .

## Temperature doubled – so did volume

He saw that as the air in the pump got hotter, the handle was gradually pushed out, even though the reading on the gauge remained the same. When he let it cool, it gradually went back in again.

V went up when T went up;
V went down when T went down;

So $\frac{V}{T}$ stayed the same –

i.e. $\frac{V}{T}$ = a constant value.

The volume always kept in step with the temperature. This was Charles' discovery. The scientific statement of this 'keeping in step' is:

**'For a fixed mass of gas at constant pressure, the volume of the gas is always proportional to its temperature.'**

This is Charles' Law. His results can be shown in a graph.

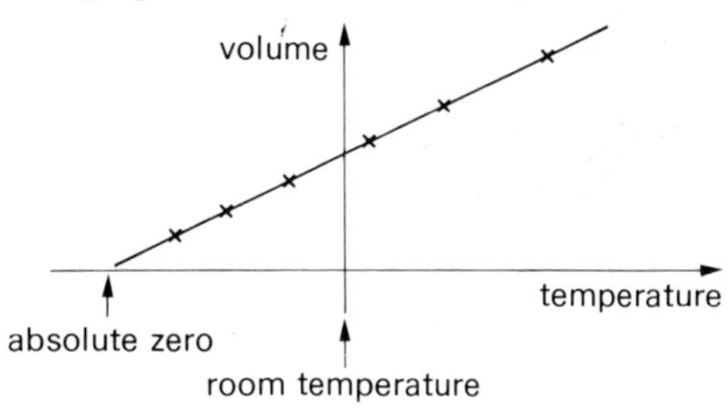


## Absolutely freezing

The graph shows that if the temperature is made to go *lower and lower*, the volume gets *smaller and smaller*. The temperature at which the volume would get so small as to be zero is called *absolute zero* temperature. This is the coldest temperature which can ever be reached! But just as it is impossible to make the volume compress to nothing, it is impossible to reach absolute zero temperature.

When scientists realized that there was a 'coldest temperature possible', a new temperature scale was invented called the *Kelvin scale*, where absolute zero has the value of zero Kelvins, or 0 K. (written *without* a ° sign):

0 K = absolute zero. This was found to be −273 °C.
So 273 K = 0 °C,
and 373 K = 100 °C.

Any centigrade temperature can be turned into a Kelvin temperature by adding 273, and in any calculations involving gases, Kelvin temperatures must be used.

## Walls move back to stop pressure increase

Like Boyle's Law, Charles' Law can also be explained in terms of the kinetic theory. At a certain temperature, the molecules of a gas move about at a certain speed. The pressure of the gas is the force with which the molecules hit the walls of the vessel they are contained in. If the temperature is increased, the molecules move faster and hit the walls more often. This would tend to cause an increase in pressure. But the pressure is kept the same by allowing the molecules to use up their extra energy in pushing the walls back. In other words, the volume increases, but the pressure stays the same. The same sort of thing happens if the temperature is lowered. The volume must decrease.

## Questions

**1** The following list gives the symbols or formulae of a number of substances. Use the list to write down:

**a** those which are elements,
**b** those which are compounds,
**c** those which contain atoms but not molecules, and
**d** those which contain molecules.

| | |
|---|---|
| magnesium Mg | iron Fe |
| water $H_2O$ | hydrogen gas $H_2$ |
| oxygen gas $O_2$ | sulphur dioxide $SO_2$ |
| calcium Ca | lead Pb |
| ammonia $NH_3$ | carbon dioxide $CO_2$ |
| manganese Mn | lead(IV) bromide $PbBr_4$ |

**2** The following list contains the names of some compounds and their formulae. In each case, say what elements are in the compound and say how many of each type of atom is found in one molecule of that compound. Use the table of elements at the back of the book to help you.

| | |
|---|---|
| copper(II) sulphate $CuSO_4$ | iron(III) oxide $Fe_2O_3$ |
| zinc oxide ZnO | vanadium(V) oxide $V_2O_5$ |
| hydrochloric acid HCl | calcium carbonate $CaCO_3$ |
| sodium hydroxide NaOH | ammonium sulphate $(NH_4)_2SO_4$ |
| sulphuric acid $H_2SO_4$ | aluminium sulphate $Al_2(SO_4)_3$ |

**3** The following list contains the names of some compounds and the numbers of each type of atom in a molecule of the compound. Try to write the formula for the compound using the table of elements at the back of the book to help you.

copper oxide = 1 atom of copper + 1 atom of oxygen

iron(II) carbonate = 1 atom of iron + 1 atom of carbon + 3 atoms of oxygen

phosphorus(V) chloride = 1 atom of phosphorus + 5 atoms of chlorine

nitric(V) acid = 1 atom of hydrogen + 1 atom of nitrogen + 3 atoms of oxygen

potassium hydroxide = 1 atom of potassium + 1 atom of oxygen + 1 atom of hydrogen

red lead oxide = 3 atoms of lead + 4 atoms of oxygen

barium sulphate = 1 atom of barium + 1 atom of sulphur + 4 atoms of oxygen.

iron(III) iodide = 1 atom of iron + 3 atoms of iodine

sodium phosphate = 3 atoms of sodium + 1 atom of phosphorus + 4 atoms of oxygen

calcium hydrogen carbonate = 1 atom of calcium + 2 atoms of hydrogen + 2 atoms of carbon + 6 atoms of oxygen

**4** Flask A contains a coloured gas. Flask B has had all the air pumped out of it. When the tap is open, the gas will flow from flask A to flask B. How quickly do you think this will happen? Explain your answer.

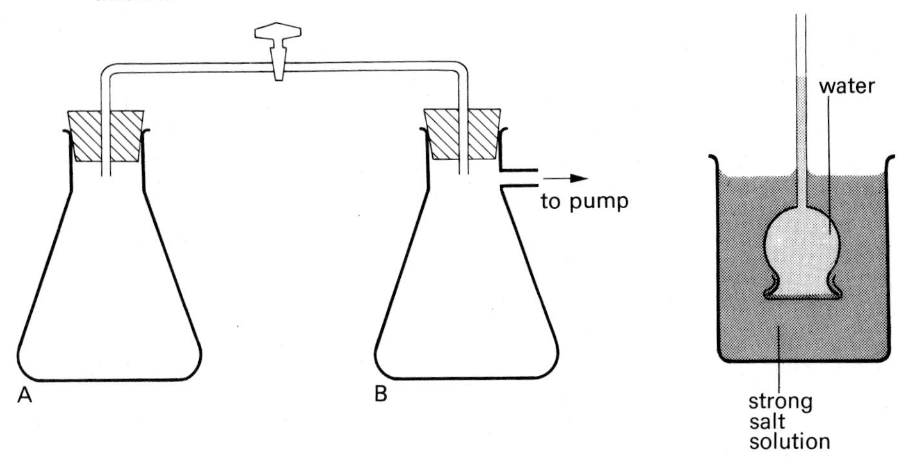


**5** The beaker contains a strong solution of salt. The thistle funnel contains ordinary water, and its end is sealed with a semi-permeable membrane. What will happen when the apparatus is assembled as shown?

**6** Make as long a list as you can of the ways in which solids differ from liquids.

**7** Look at this graph and answer the questions that follow.

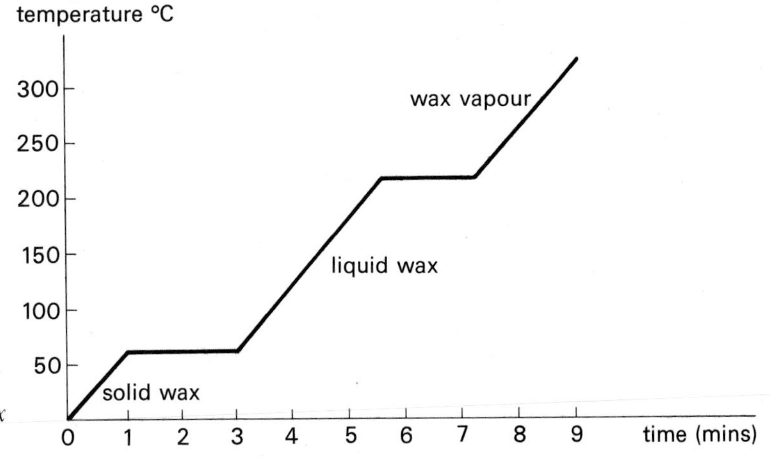


*Heating curve for wax*

**a** What is the melting point of the wax?

**b** From the time it starts to melt, how long does it take until all the wax has turned into a liquid?

**c** If wax gas is cooled, at what temperature will it start to condense?

**d** Draw diagrams to show how the molecules in wax might be arranged in the solid, liquid and gas states.

# 2 Pure substances and mixtures

## 2.1 Pure substances and mixtures

We must begin by distinguishing between *pure substances* and *mixtures*. Pure substances contain only one type of thing. Elements (like gold, carbon, oxygen) are pure substances because they are made up of atoms or molecules each one of which is identical. A compound (like pure water) is a pure substance because each of the molecules in the compound are the same.

Mixtures contain more than one sort of thing. A cup of tea, for example, is a mixture containing several things including water and sugar. The air we breathe is a mixture of several different gases. Figure 1 shows some of the mixtures that can be made from solids, liquids and gases.

| | **Solid** | **Liquid** | **Gas** |
|---|---|---|---|
| **Solid** | Alloys, such as brass and bronze. Salt and sand. Soil. | | |
| **Liquid** | Solutions such as brine. Jellies. Milk. | Emulsions such as hair oil or salad cream. Solutions such as wine. | |
| **Gas** | Smoke. Pumice stone. A sponge. | Mists and sprays. Foams. | The air. |

**Figure 1**
*Some mixtures.*

As you can see, there are many different types of mixture. You can't use the same method to separate each one. The filter on a car engine gets the dirt particles out of smoky air, but you can't get the alcohol out of wine using the same method. The following section explains some of the methods which are used to separate different sorts of mixtures.

## 2.2 Separating mixtures

**Dissolving, filtering and evaporating.** If a mixture consists of two solids, one of which will dissolve in a solvent, and the other of which will not, the solvent may be used to separate the mixture. Take, for example, salt and sand.

The mixture is added to water, warmed and stirred. The soluble part, salt, will dissolve in water, and the insoluble part, sand, will not.

The mixture is then filtered. The insoluble part remains in the filter paper and is called the *residue*. The residue may be washed and dried. The soluble part which is in solution, passes through the filter paper and is collected in the flask. This part is called the *filtrate*.

The filtrate is evaporated to drive off the water, re-forming the original solid, salt.

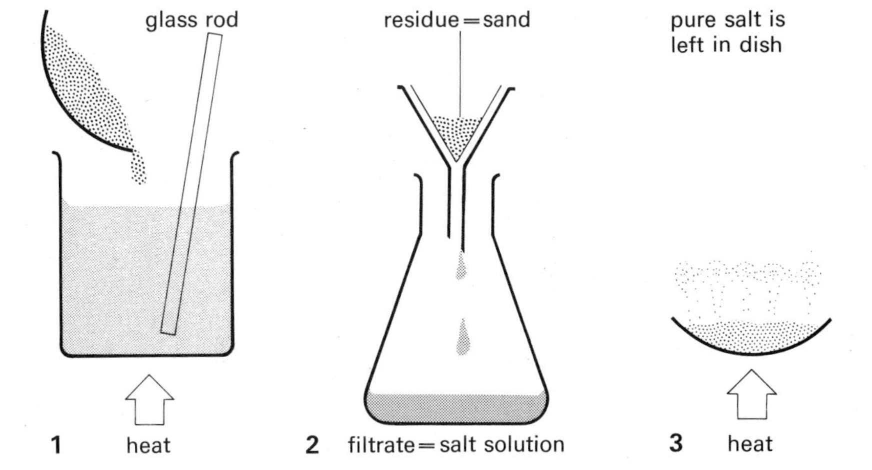


**Figure 2**
*Separating salt and sand.*

Some mixtures contain two substances neither of which will dissolve in water. In this case, another solvent must be used to separate the mixture. For example, a mixture of sulphur and sand may be separated by the same technique as in figure 2, but using methylbenzene instead of water. Sulphur dissolves in methylbenzene, but sand does not.

**Decanting and centrifuging.** Sometimes a mixture of an undissolved solid in a liquid (such as sand in water) can be separated by carefully pouring the liquid off leaving the solid behind. This is called *decanting*. (See figure 3**a**.) It is quicker than filtering, but not as good.

Sometimes, the solid may not sit at the bottom of the beaker, but may be suspended in the liquid. This is called a *suspension*. A good example of a suspension is dirty water. The dirt can be separated from the water by filtration, and that is how it is done at a water works, but a small sample of the mixture can be separated more quickly using a *centrifuge*. (See figure 3**b**.)

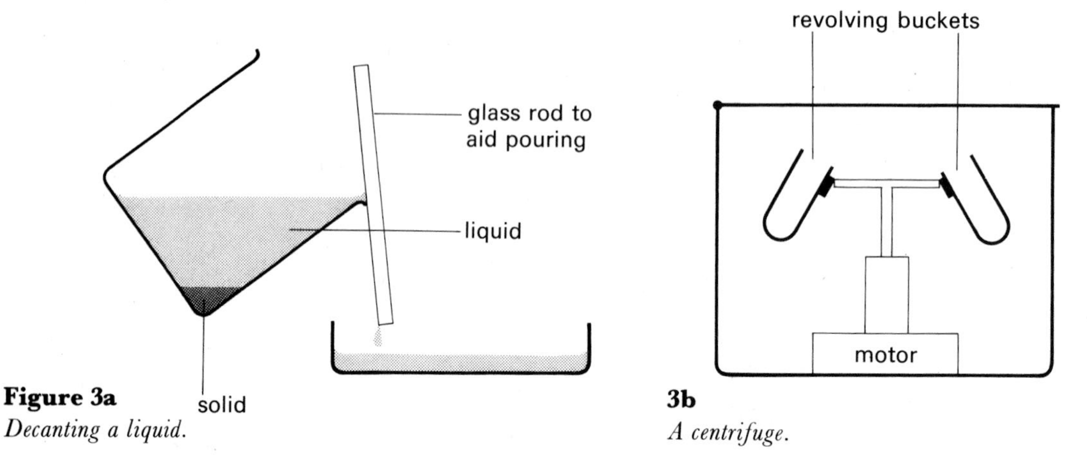


**Figure 3a**
*Decanting a liquid.*

**3b**
*A centrifuge.*

The electric motor revolves the spindle very quickly and the buckets containing the test tubes are spun around. The buckets tip up and the solid matter is flung to the bottom of the test tube. After spinning, the clean liquid can be removed with a pipette.

**Distillation.** Both solutions and mixtures of liquids may be separated by *distillation*. Figure 4 shows the apparatus used to obtain pure water from a solution of salt and water.

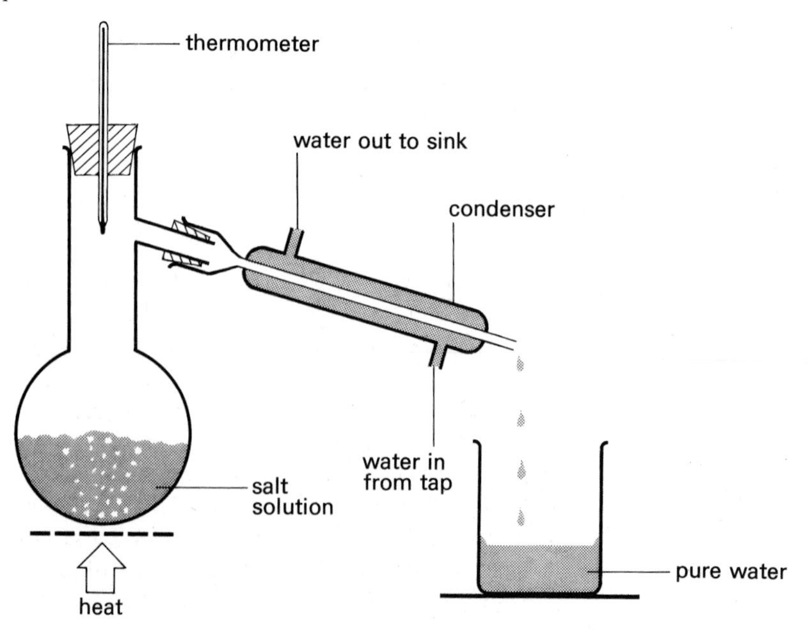


**Figure 4**
*Simple distillation.*

The salt solution in the flask is boiled. The water part boils and becomes steam. The steam passes into the condenser where it is turned back to water, which then drips into the collecting beaker. If the mixture consists of two liquids dissolved in each other, (these are called miscible liquids), a special type of distillation apparatus called a *fractionating column* must be used. Figure 5 shows the apparatus that might be used to separate a mixture of ethanol and water.

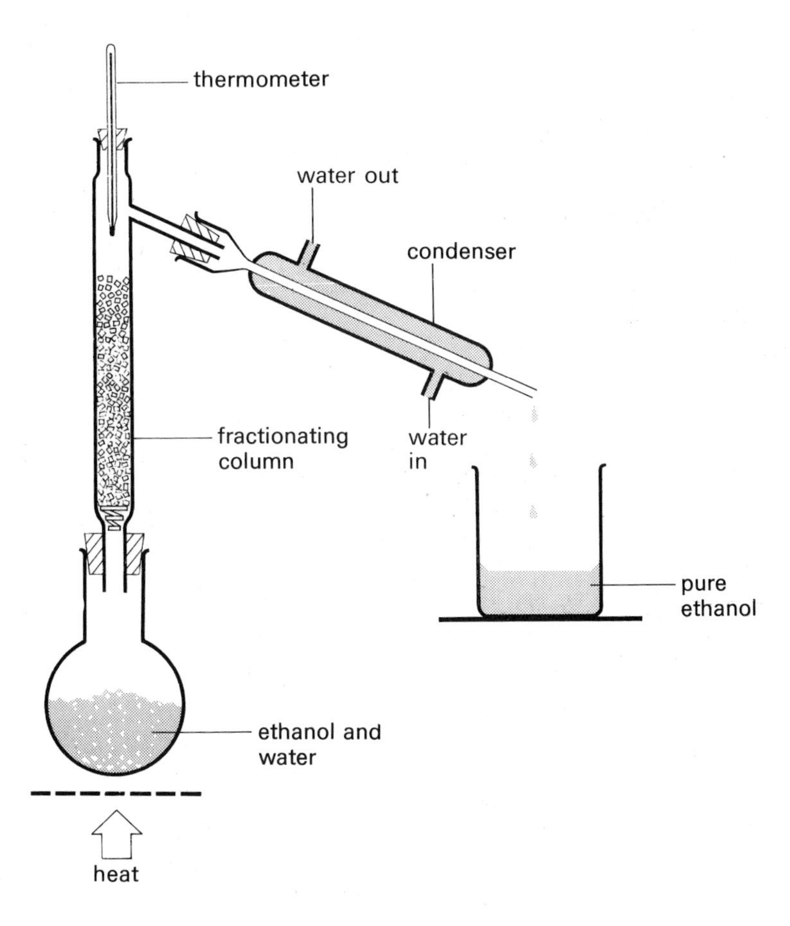


**Figure 5**
*Fractional distillation.*

This time, when the mixture is heated, both the ethanol and the water will start to evaporate. However, the boiling point of water is 100°C and that of ethanol is 78°C, so the ethanol will evaporate more easily. The fractionating column is a long tube packed with small glass beads which assist the separation of the two gases, by providing a large surface area for them to condense and evaporate from. Only the ethanol reaches the top of the column, so only pure ethanol forms in the condenser. Note that the thermometer is placed at the top of the fractionating column so that it registers the temperature of the substance which is distilling. This technique is called *fractional distillation* because more than one liquid is boiling.

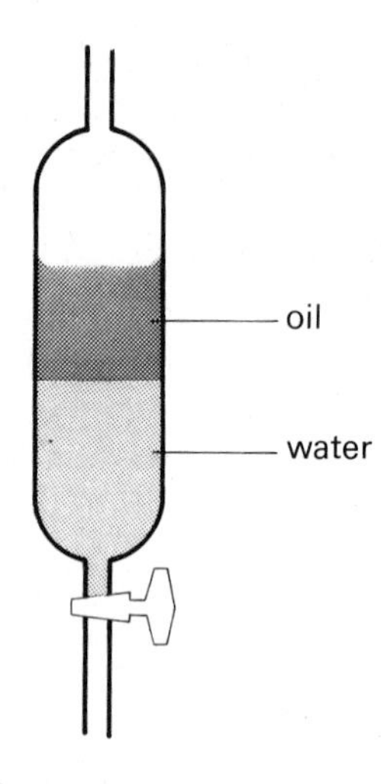


**Figure 6**
*A separating funnel.*

**The separating funnel.** Liquids that do not dissolve in each other are called *immiscible liquids*. When they are useful together, as for example in paint, hair oil, and salad dressing, they have to be shaken together to get the liquids to mix. This sort of mixture is called an *emulsion*.

In salad dressing, oil and water are mixed in the same bottle. They mix into an emulsion when they are shaken up, but soon separate into two layers. They can be separated using a separating funnel. (See figure 6.)
The tap is opened and the water is allowed to run out. The tap is closed before the oil reaches the bottom.

**Sublimation.** Some types of solid do not melt when they are heated. Instead, they change directly from a solid to a gas. When the gas is cooled, it does not condense into a liquid, but changes directly back to a solid. The two processes are called *dissociation* and *sublimation*. They are shown in figure 7.

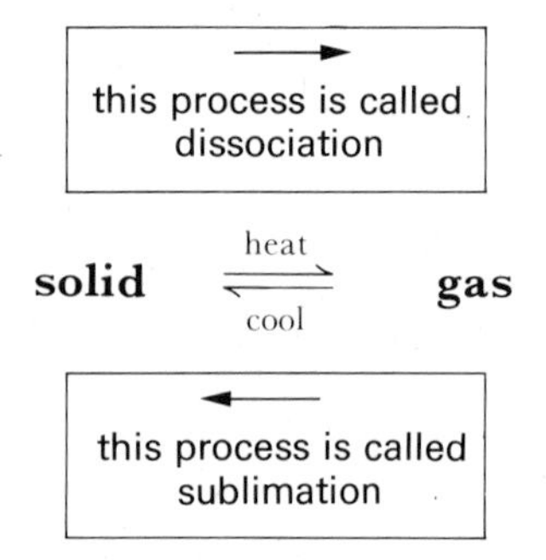


**Figure 7**
*Dissociation and sublimation.*

Very few substances sublime. Ammonium chloride and iodine are examples of substances that do. A mixture of sodium chloride and ammonium chloride may be separated by gently heating it so that the ammonium chloride turns into a gas leaving the sodium chloride behind. If a cold surface is held over the heated mixture, the ammonium chloride will sublime back to a solid. (See figure 8.)

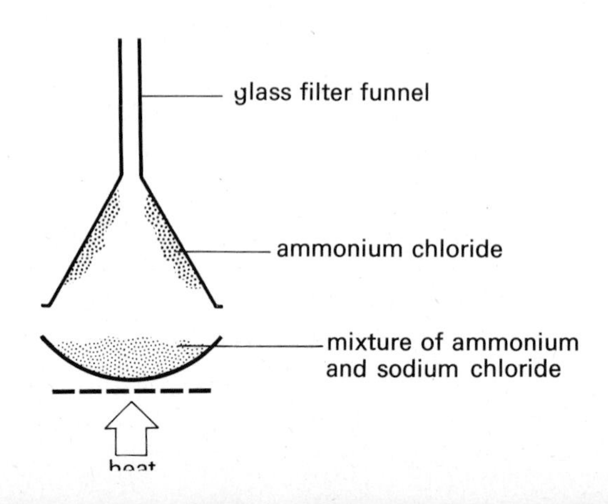


**Figure 8**
*Separating a mixture by sublimation.*

**Chromatography.** This technique was originally discovered when scientists were extracting coloured dyes from plants.

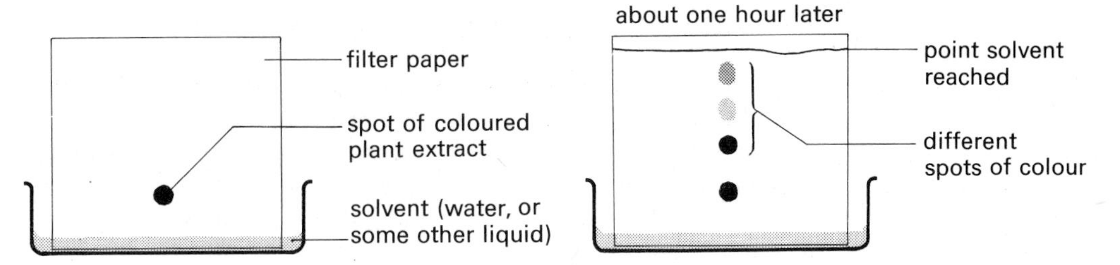


**Figure 9** *Separating coloured pigments by chromatography.*

A modern version of their experiment is shown in the diagram above. It consists of a coloured extract made from a mixture of different compounds. As the solvent soaks up the paper, the different coloured compounds follow it at different speeds, so they gradually become separate.

Chromatography provides a means of not only separating mixtures, but identifying what is in them. For example, a mixture of inks whose composition is unknown can be analysed by using inks whose colours are known alongside them. Figure 10 shows what might happen.

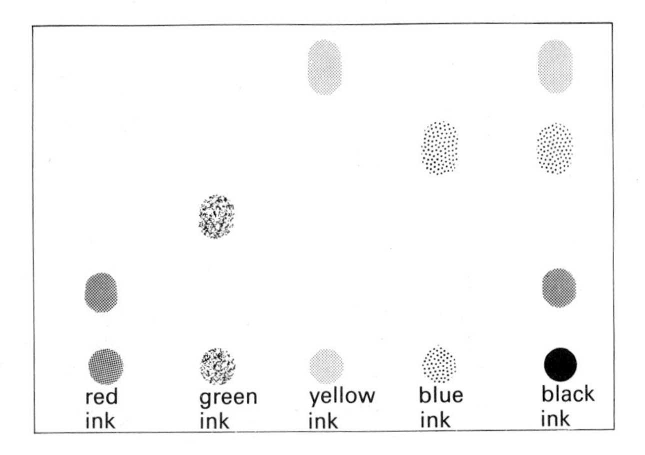


**Figure 10** *Identifying the components of an ink.*

The red, green, yellow and blue inks must be pure substances because they only give one spot of colour as they soak up the paper. The black ink, however, gives three spots, at the same height as the red, blue and yellow inks. Therefore, we may conclude that the black ink was made of red, blue and yellow ink.

The piece of paper with chromatography on it is called a *chromatogram*. If necessary the chromatogram can be cut up so that the different components can be soaked out into different solutions. With special chromatography equipment, minute amounts of substances can be detected. This makes it a very important tool – both for chemists, and for detectives . . .

## 2.3 Is it pure?

How can you tell if something is pure? You could use some of the techniques we have just discussed.

For example, you could add the substance to water: see if some of it will dissolve and some will not. You could heat it: see if some of it sublimes. If it were liquid, you could distil it: see if it will separate into two parts. If you know what the substance is, you could measure its melting or boiling point and see if it is the same as the value quoted in a reference book of chemical data. Impurities in a substance make the melting point lower than it should be and the boiling point higher than it should be.

## 2.4 Solubility of solids

If a substance will dissolve in a liquid we say that it is *soluble*. A soluble solid is called a *solute* and the liquid it dissolves in is called the *solvent*. Whatever the solute and solvent, they mix to give a *solution*:

a solute + a solvent ⟶ a solution.

Different substances dissolve in a solvent by varying amounts. They have different *solubilities*. To talk about the solubility of a substance, you must mention three things:

**1** The temperature of the solution.

**2** The name and the mass of the solute which is dissolved.

**3** The name and the mass of the solvent it is dissolved in.

For example: 'At 20 °C, the solubility of sodium chloride in water is 36 g of sodium chloride per 100 g of water.'

It is most important to include the temperature, because solutes have different solubilities at different temperatures.

**Saturated solutions.** When a solvent has dissolved as much solute as it possibly can at a given temperature, the solution is said to be *saturated*. The only way to be sure that as much solute as possible has dissolved, is for some to be left over at the bottom of the test tube.

**A saturated solution is one which contains as much solute as it can at a given temperature, and still has some undissolved solute left over.**

For example: at 40 °C, about 60 g of potassium nitrate dissolves in 100 g of water. This is a saturated solution.

If the temperature is changed, the solubility changes as well. For most solutes, their solubility increases if the temperature of the solution is changed. At 100°C, the solubility of potassium nitrate in water goes up to 245 g per 100 g of water. So 60 g of solute in 100 g of water at 100 °C would not be a saturated solution because more solute could dissolve. Similarly, if the temperature is lowered, the solubility decreases. At 10 °C, only 21 g of potassium nitrate can dissolve in 100 g of water, so if the saturated solution is cooled from 100 °C to 10 °C some of the solute will crystallise out from the solution. (See figure 11.)

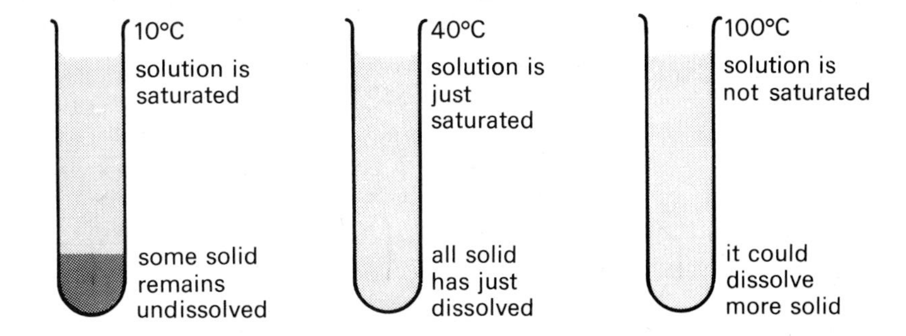


**Figure 11**
*Solubility changes with temperature.*

## 2.5 Solubility curves

These variations of solubility with temperature are best shown on a graph called a *solubility curve*. Figure 12**a** shows a solubility curve for potassium nitrate in water.

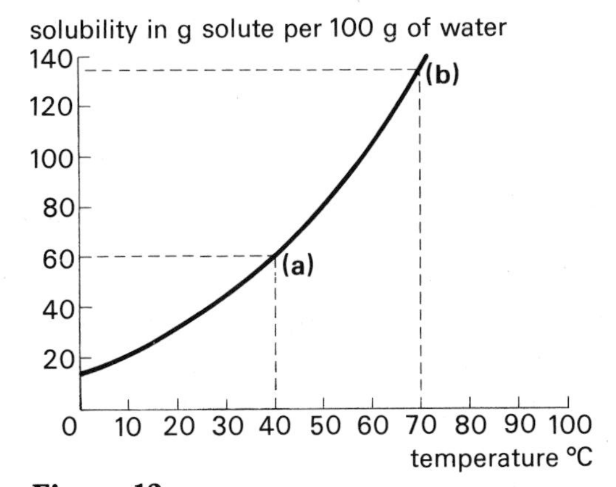


**Figure 12a**
*A solubility curve for potassium nitrate in water.*

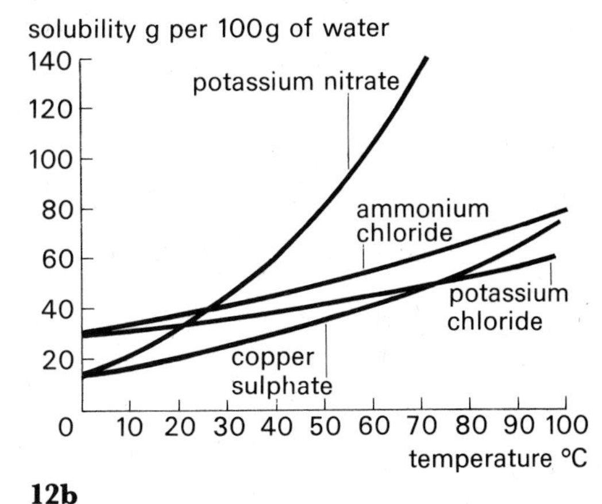


**12b**
*Comparing solubilities.*

From this graph it is possible to find the solubility of potassium nitrate at any temperature from 0 °C to 100 °C.
Look at example (**a**) on the graph.
At 40 °C, the solubility of potassium nitrate is 60 g per 100 g of water.
Look at example (**b**).
At 70 °C, the solubility of potassium nitrate is 135 g per 100 g of water.

**Solubility of different substances.** Different substances have different solubilities. All of these can be displayed on the same graph at once so the solubilities can be compared. (See figure 12**b**.)

For example: the graph tells you that below 26 °C, ammonium chloride is more soluble than potassium nitrate – but above 26 °C, it is less soluble.

## 2.6 Measuring solubility

Figure 13 shows the apparatus which is used to measure the solubility of a solute in water at a particular temperature. The procedure is as follows:

**1** The water bath is kept as close as possible to the temperature at which the solubility is to be measured. The water is constantly stirred. The flame of the bunsen burner is adjusted to keep the temperature of the water bath steady.

**2** Solute is added to the water in the test tube, and the solution is stirred. Solute is added until no more will dissolve, no matter how much the solution is stirred. The solution must then be saturated at that temperature.

**3** An evaporating basin is weighed.

**4** Some of the saturated solution is carefully taken from the test tube with a warm pipette, and it is transferred to the evaporating basin . . . It is weighed immediately.

**5** The solution in the basin is carefully evaporated until all the water has gone and only dry solute remains.

**6** It is weighed again.

To see how the result is calculated, here are some sample figures.

### The solubility of potassium nitrate.

| | |
|---|---|
| Temperature at which experiment was performed | . . . 40 °C |
| Mass of evaporating basin | . . . 30·0 g |
| Mass of basin + solution | . . . 46·4 g |
| Mass of basin + solute after evaporation | . . . 36·4 g |

**1** The mass of solute in the sample. This is obtained by subtraction:

| | |
|---|---|
| mass of basin and solute | 36·4 g |
| *minus* mass of basin | −30·0 g |
| = *mass of solute* | 6·4 g |

**2** The mass of water in the sample of solution. This is obtained by doing another subtraction:

| | |
|---|---|
| mass of basin and solution | 46·4 g |
| *minus* mass of basin and solute | −36·4 g |
| = *mass of water in the solution* | 10·0 g |

At 40 °C, 6·4 g of potassium nitrate dissolve in 10 g of water.

So the solubility of potassium nitrate is 64 g per 100 g of water at 40 °C.

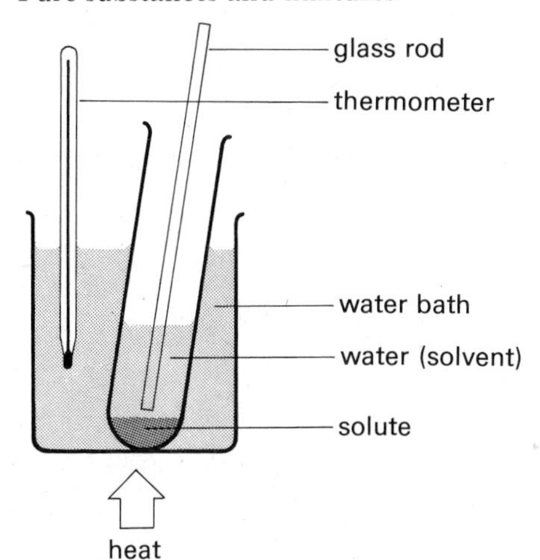


**Figure 13**
*Measuring the solubility of a solute in water.*

2.7
## Solubility of gases

Gases also dissolve in solvents. Fish use the oxygen that is dissolved in water; swimming baths have chlorine dissolved in the water to kill bacteria.

Most solids increase their solubility as the temperature of the solvent is raised. Gases, on the other hand, become less soluble.

Pop has the gas carbon dioxide dissolved in it. As the pop warms up in your mouth the gas 'undissolves', making carbon dioxide bubbles on your tongue. This is the 'fizz'.

Pressure also has an effect. Increased pressure will make more gas dissolve. Carbon dioxide is put into pop at high pressure. When you take the top off a bottle of pop, the pressure is released, the solubility of the gas is decreased, and bubbles of carbon dioxide come out of solution. The gas stays out, too, if you leave the top off!

## Summary

At the end of this chapter, you should be able to:

**1** State the differences between a pure substance and a mixture.

**2** Describe how to separate mixtures by means of: dissolving and filtering; evaporating; decanting; centrifuging; distilling; using a separating funnel; subliming, and by chromatography.

**3** Decide whether or not a substance is pure by means of tests.

**4** Explain what is meant by the term solubility, and describe the effect of temperature changes upon the solubility of a solute in a solvent.

**5** Explain what is meant by a saturated solution.

**6** Draw and interpret a solubility curve.

**7** Describe an experiment to measure the solubility of a substance at a stated temperature, and calculate its solubility from the results obtained.

**8** Describe the effect of temperature and pressure on the solubility of gases in solvents.

# A load of old rubbish

*Rubbish has always been a problem. If you had lived in London in the 17th century, you would have taken very great care when walking through the narrow city streets. People threw their rubbish (and not very pleasant rubbish at that) out of the window into the street below. There was nowhere else to put it. Today, there are people who take an interest in rubbish . . .*

**Binmen to the rescue**

Nowadays, we take it for granted that the 'binmen' will call once a week and take away all our rubbish. Every year, they remove eighteen million tonnes of paper, tea-leaves, old bottles, beer cans, the remains of yesterday's dinner, and such like . . . a load of old rubbish!

But is it really rubbish? It has come in very useful in the past for filling old quarries, making sites which can eventually be used for buildings. But more important than that, people have come to realise that some of the 'rubbish' can be separated, and reused. All of a sudden, the waste becomes useful raw material.

Take tin cans. Are you having baked beans for tea? Or a ring pull can of orangeade for your lunch? If you are, you will shortly be adding a few more cans to the total of 10 000 million which people in Britain throw away each year. Put end to end, they'd stretch round the world twenty times! The metal in them is worth £15 000 000 . . . What about paper? One tonne of paper costs about £350; and 3 trees are chopped down for each tonne. In Britain, we throw away 4 million tonnes of waste paper each year!

**Volunteer help to carry the can?**

But there are many technical problems in separating the waste. Rag and bone men, and their modern equivalent – people collecting old newspapers and milk bottle caps for charity – are one quite effective way of separating rubbish, as long as the donors are prepared to co-operate, by putting everything in neat bundles. But without voluntary help it is not economic to separate waste this way.

No-one has yet invented a full size machine which can separate normal domestic rubbish into separate piles of metal, glass, plastics and so on. There have been many attempts to develop such machines.

One of the most interesting plants for attempting this job has been built in Newcastle-upon-Tyne. Three firms – the Metal Box Company, the British Steel Corporation, and Batchelor-Robinson – provided the money to build the machine; Tyne and Wear County Council were happy to provide the rubbish!

The money to build the plant was provided by people in the 'metals' business, so

naturally enough, the plant concentrates on tin cans. The problems are:
how to separate the tins;
how to clean the tins;
how to convert the tins into separate piles of pure tin and steel.

**The tin-grabber**
The machine shown in the diagram does both the first and the second job at the same time.

The rubbish is dumped in at the right hand by the 'Hymac' grab and travels up the sloping conveyor belt. This is vibrating all the time to loosen up the rubbish. As it travels along to the left, it is vibrated even more – at the end, the rubbish falls down a series of chutes with strong magnets attached which pull the tins to one side, where they fall onto a separate conveyor. This is the main part of the process, and is the most difficult part to do well. The system in this machine is the most advanced in both Europe and America, because it can separate most of the iron and, at the same time, 'shake off' most of the sticking paper and plastic.
Everything besides the tins ends up dropping through the chute at the left-hand end of the machine. The tins travel back underneath the horizontal conveyor – the first separation has been made!

But things are never as easy as they seem. Tins are in fact made mainly from steel, with a very fine layer of tin on the inside and outside to prevent rusting. To get pure steel and pure tin, another more difficult separation has to be done.

Batchelor-Robinson developed the equipment to do this job. The tins are broken up and then dumped into a bath full of alkali. (Use the index!) As in the salt and sand experiment, the tin dissolves and the steel remains. Another separation has been done. The tin which has reacted with the alkali, is removed by a special type of separation process called 'electrolysis', which you can read about in chapter 17.

**Rubbish? No such thing!**
The Newcastle plant is rescuing about 85 million tins per year. The next job will be to build plants to make the paper waste available for re-use – another very necessary separation.

No-one has yet devised a machine which can separate all the components of domestic rubbish, but the effort which needs to be made will become worthwhile if the price of the materials involved rises too much.

This is exactly what happened during the Second World War. Materials became very scarce, so the Government appealed to householders not to throw away anything.

'Teapot spouts make jumping boots for paratroopers' ran one headline. 'Old saucepans into Spitfires', ran another. What next? How long will it be before we see the headlines 'New cars from old baked bean tins'? What a load of old . . . (?)

*The separating plant at Newcastle on Tyne.*

## Questions

**1** Which of these are mixtures, and which are pure substances? soil; copper sulphate; 18 carat gold; hydrogen; water; silver; air; sea water; lead; carbon dioxide.

**2** Describe how you would separate the following mixtures:

**a** powdered copper sulphate and powdered copper;

**b** a suspension of chalk in water;

**c** two miscible liquids, one of which boils at 60 °C and the other of which boils at 100 °C;

**d** powdered iodine and sand;

**e** tetrachloromethane and water (tetrachloromethane does not dissolve in water);

**f** pigments obtained from various crushed flower petals.

**3** What will happen if a saturated solution of potassium nitrate which contains 100 g of water is cooled from 70 °C to 30 °C? Use the solubility curve in figure 12**b** to help you. What will happen if you do the same thing to a saturated solution of potassium chloride?

**4** Study the solubility curve of potassium nitrate and lead nitrate shown in this diagram, and answer the questions which follow.

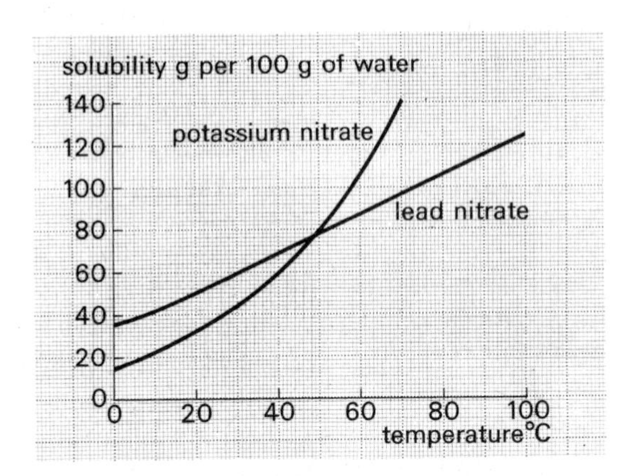


**a** Which substance has the greater solubility at 20 °C?

**b** Which substance has the lower solubility at 70 °C?

**c** What is the solubility of lead nitrate at 50 °C?

**d** At what temperature does potassium nitrate have a solubility of 20 g per 100 g of water?

**e** At what temperature do they have the same solubility?

**f** What is the solubility at this temperature?

**g** Why does the graph not go below 0 °C and above 100° C?

**5** Calculate the solubility of sugar at 40 °C from the following experimental data:

| | |
|---|---|
| Mass of evaporating basin | . . . 23 g |
| Mass of basin + sample of saturated solution | . . . 192 g |
| Mass of basin + solid after evaporation of solution | . . . 142 g |

**6** Why do tanks of tropical fish need to have air bubblers in them?

# 3 Water

## 3.1 Water, water everywhere

There is a great deal of water on the Earth. About four-fifths of its surface is covered with sea; the land is dotted and streaked with lakes and rivers. A visitor from outer space might well think he had come to a very wet planet.

It is just as well that we have all this water, because the human race – and the animals and plants that live with us, totally depend on water. In areas where there is plenty of water, crops and livestock flourish, but in desert areas, little grows – both animals and people are scarce. Britain is lucky in this respect, because we receive just about the right amount. The following statistics are published by the Thames Water Authority, which controls the water supply to about one quarter of the population of England.

## 3.2 How much water?

The Authority estimates that it supplies about 1780 million litres of water through its pipe lines each day and that the total length of the pipes is about 33 600 kilometres. Its sewage and water treatment plants deal with about 2 410 million litres each day (including rain water collected from drains) and their sewage pipes stretch for 38 500 kilometres. That is something to think about.

People use a lot of water in their homes. On average, each one of us uses over 150 litres each day. The figures work out like this:

| | |
|---|---|
| washing and bathing | . . . 57 litres |
| W.C. | . . . 57 ,, |
| laundry | . . . 15 ,, |
| dishwashing and cleaning | . . . 15 ,, |
| gardening | . . . 7 ,, |
| drinking and cooking | . . . 5 ,, |
| car washing | . . . 2 ,, |

Industry uses a great deal of water on our behalf. These are some estimated quantities of water used in different processes:

| | |
|---|---|
| To make a gallon of beer | . . . 68 litres |
| To make a tonne of paper | . . . 90 920 ,, |
| To make a tonne of steel | . . . 45 500 ,, |

## 3.3 The water cycle

The way in which water circulates round the earth can be described by the *water cycle*. Figure 1 is a diagrammatic representation of the water cycle.

The driving force for the water cycle is the sun. All the time, the sun is evaporating water from the sea, causing huge amounts of water vapour to float into the air. The vapour condenses to form millions of minute droplets of water – clouds. Water rains down from them onto the land and the sea. The part that falls onto land soaks into the ground, to return later to the surface as springs. These fill the rivers, which return the water to the sea. But before it reaches the sea, some of it is collected in reservoirs. It is taken in pipes to cities where people, animals and industry, use it and make it dirty. This sewage is treated before it is passed back into rivers, so that it can go back to the sea reasonably clean. All the water that left the sea is thus returned, and the water goes on round and round in the Water Cycle.

This account of the Water Cycle is illustrated in figure 1. Have a really good look at it before you read on.

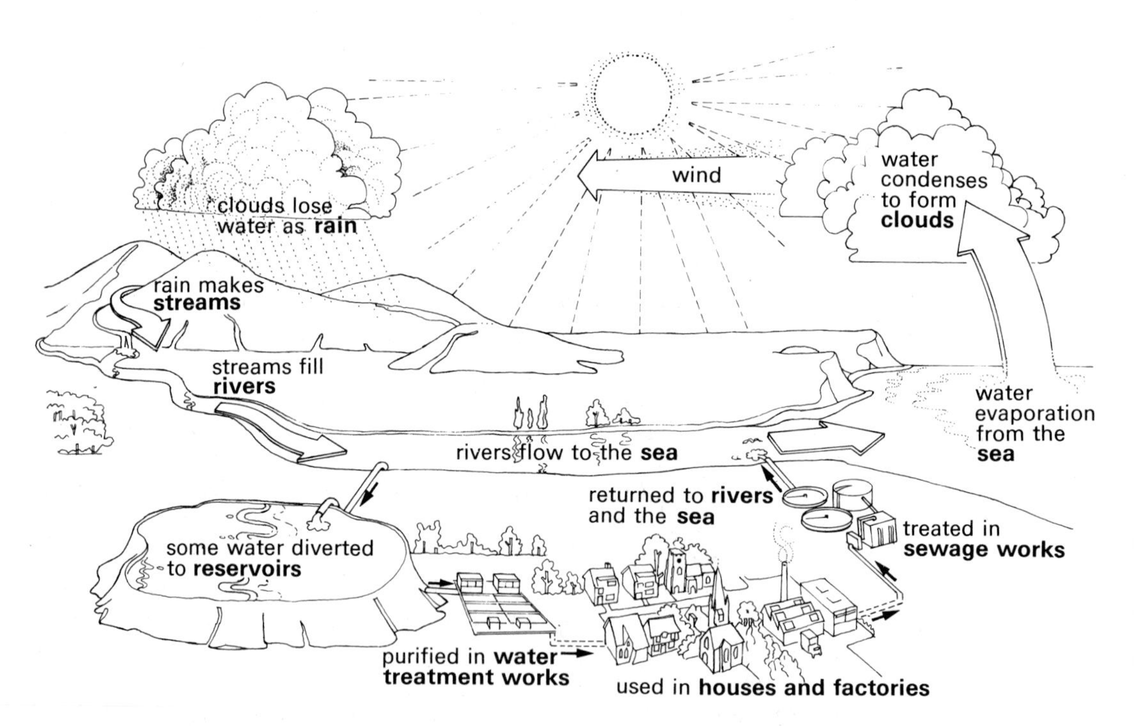


**Figure 1** *The water cycle.*

## 3.4 Sources of impurity

Water is a good solvent. It will dissolve a greater variety of substances than any other liquid. For this reason water contains a lot of dissolved impurities, and most of them must be removed before the water is fit for use.

As streams and rivers flow over the land, they erode and dissolve small amounts of rocks and soil. These substances are generally not harmful to people and animals, but they may cause the water to become resistant to soap, or *hard*.

Dissolved substances can often interfere with industrial processes such as steel making and electroplating, and the water must be purified before use. This can be costly.

A lot of water finds its way into rivers, not from springs and lakes, but by direct drainage from the land. Water which has drained from roofs and roads contain dust and grit. In heavy traffic areas it will also contain oil and lead dust from the exhaust fumes of cars. But this is less of a worry than some of the impurities that drain from farm land.

In 1970, chemical industries in the United Kingdom produced 2 660 000 tonnes of artificial fertilizer and much of this was applied directly to the land, to increase the growth rate and quality of crops. Unfortunately, during heavy rain, the soluble parts of the fertilizer, such as sodium and ammonium nitrates, are dissolved and carried into the rivers. This means a loss for the farmer, and pollution for the rivers. Instead of encouraging useful crops, the nitrogen compounds encourage the growth of small, floating plants called algae which prevent the normal growth of plants and fish in the river. The River Lee, which is a tributary of the Thames, runs through an important market garden area. In 1973 the nitrate level rose above that recommended by the World Health Organization, so it had to be diluted with purer water from the Thames.

Industry can often be responsible for pollution of rivers, so the quantity and type of waste that can be deposited in rivers is usually strictly controlled. Detergent foams are all too evident on rivers in some areas, but solvents and metals such as mercury, lead and copper are not so easy to detect. They are far more dangerous because they are poisonous.

Humans are also responsible for a lot of waste. We produce vast quantities of waste water and solids, and cheerfully forget about them as we pull the plug in the bath and flush the W.C. All this must be treated before it is discharged into rivers or the sea.

So the treatment of water falls into three main types:

**1** Provision of clean water for household and industrial use.

**2** Control of impurities added to rivers and the land.

**3** Treatment of sewage.

## 3.5 Drinking water

Most of the water that comes out of our taps is taken from rivers at places where pollution is low. This usually means above towns and factories, and away from places where dangerous impurities are likely to reach the water.

Figure 2 shows the lay-out of a typical 'water works'.

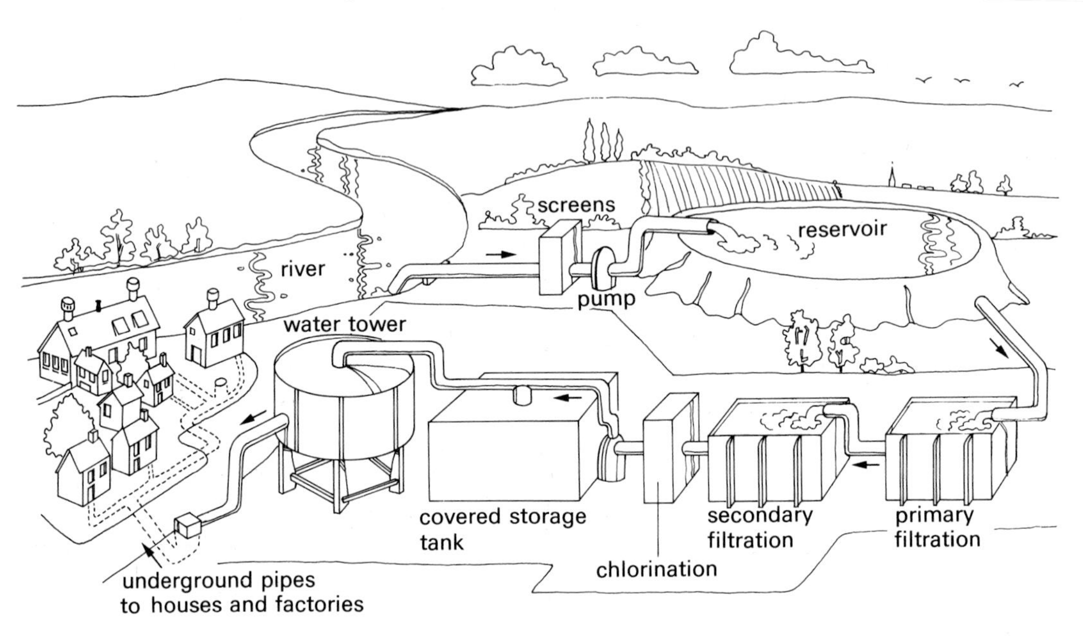


**Figure 2** *A water works.*

The river water is first 'screened'. That is, it is passed through screens which filter out floating debris. It is then pumped up into a reservoir or sometimes high storage tanks. These need to be high up, so the water will flow under the pull of gravity.

The water is then filtered through beds of coarse sand to remove the larger particles of suspended matter. These sand beds have to be washed clean at the end of each day.

After this, a chemical called alum is sometimes added to help any finer particles of dirt settle out of suspension. Finally, the water is passed through more beds of even finer sand to remove the smallest particles of suspended solids. Sometimes, tiny organisms are grown in this sand to feed on any bacteria in the water and so remove them. The sand in these beds is replaced periodically when it gets clogged.

After all this, the water is clean, but may still contain harmful bacteria which cannot be filtered out. To kill these, a small amount of chlorine gas is dissolved in the water. Chlorine kills bacteria – but is added in small amounts so that it cannot be smelled or tasted by the people who are to drink the water.

At last the water is ready for use, so it is pumped into covered storage tanks or reservoirs on hills, or even water towers if the land is flat, so that consumers' taps are supplied by natural gravity pressure.

## 3.6 Sewage

Sewage is a different matter from river water. It contains washing water, waste products from our bodies, and anything else that goes down the sink, drain or W.C. People in the Thames Water Authority area produce 189 million litres of sewage each day! If it was allowed to flow straight into rivers without treatment it could well be a health hazard. Consequently all sewage must pass through a sewage works. So sewage works are an important part of our towns.

Figure 3 shows the layout of a typical sewage works.

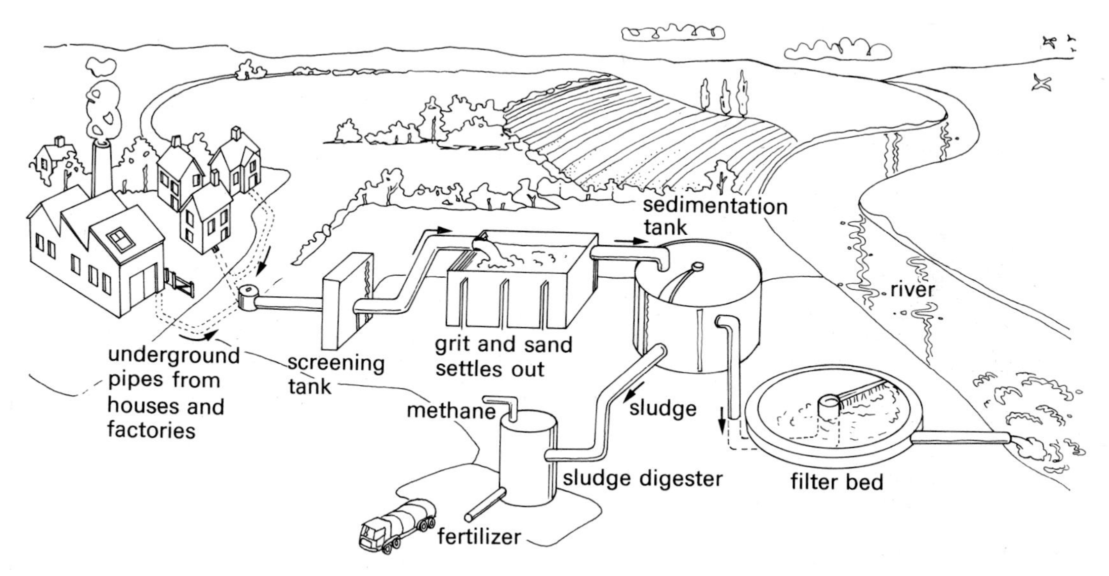


**Figure 3** *A typical sewage works.*

The raw sewage is first passed through screens to remove larger debris such as pieces of wood and rags, and other objects which might damage pumps and other machinery.

Next, grit and sand, which have been carried along from roads and gardens, are allowed to settle out as the sewage slowly flows along specially designed channels.

Solids, or 'crude sludge' are then removed by allowing them to settle to the bottom of sedimentation tanks. This sludge is pumped away to 'digesters' where it is stored in tanks. Slowly, microbes turn the sludge partly into methane gas (which is a fuel, and is used to heat the sewage works and produce electricity), and partly into 'digested sludge' which is sold as a low grade fertilizer, or sometimes taken out to sea and dumped.

The impure water left after the sludge has been removed is sprayed into a bed of clinker or gravel. As it trickles through, microbes act on the bacteria in the water and destroy them. At the same time, the water absorbs oxygen from the air. This process is known as aeration.

Finally, the water is pure enough to be returned to the river. If it were chlorinated, it would be fit for drinking in many cases.

## 3.7 Washing

One important use for water is washing. If we want to clean something, we add soap, and agitate it in water. Soap is a compound called sodium stearate. You can read more about sodium stearate and its structure in chapter 13, but the shape of its molecules is shown in figure 4.

**Figure 4**
*The shape of a soap molecule.*

At one end of the molecule is a group of atoms which dissolve in water. At the other end, is a chain of atoms which do not dissolve in water, but do dissolve in oily compounds.

Soap helps to wash clothes by breaking down the *surface tension* of pure water. Water acts as though it has a thin skin over its surface, and this prevents it from wetting certain surfaces. (Try running cold water over the back of your hand without using any soap, and see how it forms into globules that do not spread out.)

When soap is added, the water soluble end of the molecule dissolves in the water, destroying the surface tension, helping the water to spread out and penetrate the cloth. This loosens the dirt particles when the dirty garment is agitated. Similarly, the oil soluble ends dissolve in any grease or oil, and break it down so that it is removed. (See figure 5.)

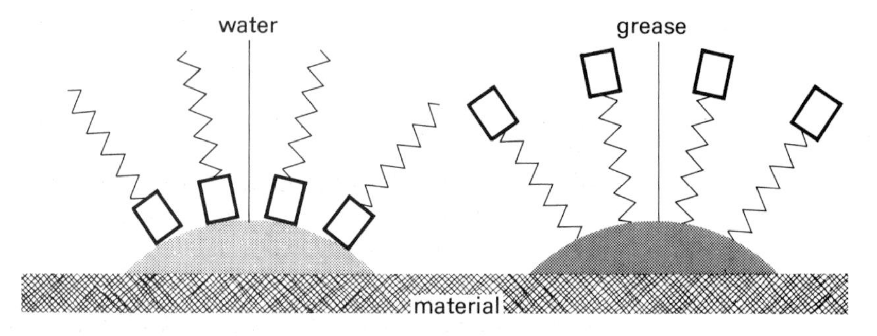


**Figure 5**
*The washing action of soap.*

## 3.8 Hardness of water

When soap dissolves in water, it forms a lather. If the water is hard, then the soap is destroyed, and a scum is formed on top of the water instead. You must have seen the scum floating on top of your bath water. It isn't all dirt! Some of it is destroyed soap.

Hardness of water is caused by calcium and magnesium compounds in the water. They are called calcium hydrogen carbonate, calcium sulphate and magnesium sulphate. Take one example:

$$\underset{\text{(the hardness)}}{\text{calcium sulphate}} + \underset{\text{(soap)}}{\text{sodium stearate}} \longrightarrow \underset{\text{(scum)}}{\text{calcium stearate}} + \text{sodium sulphate.}$$

The soap and the hardness join to form scum. The other compound which is formed dissolves in the water and is no use as soap.
(You will learn more about equations and formulae in chapter 7.)

**Temporary hardness.** This is caused by calcium hydrogen carbonate in the water. When it rains, tiny amounts of carbon dioxide from the air dissolve in the rain and form a very weak acid called carbonic acid:

carbon dioxide + water $\longrightarrow$ carbonic acid

(this can be shown as: $CO_2 + H_2O \longrightarrow H_2CO_3$).

If this acid falls on limestone rocks, it slowly dissolves them:

carbonic acid + limestone $\longrightarrow$ calcium hydrogencarbonate solution
($H_2CO_3 + CaCO_3 \longrightarrow Ca(HCO_3)_2$).

(Remember, more about equations later.)

This is how the hardness gets into the water. The limestone is slowly eroded. In fact, limestone rocks often contain caves and deep pot-holes where the erosion has been going on for millions of years.

Another feature of hardness of water is the formation of stalagmites and stalactites. (See figure 6.)

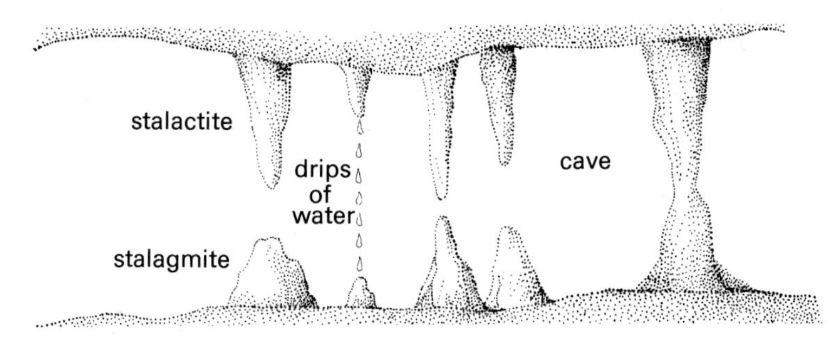


**Figure 6**
*Stalagmites and stalactites.*

As the hard water drips from the roof of a cave, some of the water evaporates to leave a solid deposit of limestone, releasing the carbon dioxide again. This happens very slowly. It takes millions of years for stalactites to grow down from the ceiling. As they do so, stalagmites grow up from the floor:

calcium hydrogencarbonate (hardness) $\longrightarrow$ calcium carbonate (limestone) + carbon dioxide + water.

($Ca(HCO_3)_2 \longrightarrow CaCO_3 + CO_2 + H_2O$).

Hardness of water caused by calcium hydrogencarbonate is called *temporary hardness*, because it can be removed from the water by heating. The same thing happens as when water evaporates in caves. The limestone formed appears as a hard solid on the inside of the vessel in which the water is heated. We say it *precipitates*.

This brings a problem. If temporary hard water is used in boilers or hot water systems, the limestone forms on the inside of the boiler or pipes, reducing the effective size of the pipes. If left untreated, it can stop the flow of water all together, and cause the system to break down.

You can see the limestone (often called scale, or fur) on the inside of kettles in any hard water area.

**Permanent hardness.** This is caused by water dissolving small amounts of calcium and magnesium sulphate from ground rocks as it travels on its way to reservoirs or rivers. It cannot be removed by heating, but it destroys soap just like temporary hardness.

Quite often, water will contain the compounds which cause both permanent and temporary hardness. In areas where the water flows over chalk or limestone, the water is very hard. Where the water flows over insoluble rocks like sandstone or granite, the water is very soft because no chemicals dissolve.

## 3.9 Cures for hardness

There are a number of ways of removing the hardness in water.

**1** Boiling. As you have seen, 'temporary' hard water can be softened by boiling or heating it. This will not cure 'permanent' hardness.

**2** Distillation. Any type of hard water can be purified by distilling it. In this method steam boils off from the water, leaving the dissolved solids behind. This is how 'distilled water' is made, but it is expensive to prepare in large quantities.

**3** Addition of washing soda. This is a compound called sodium carbonate, and its formula is $Na_2CO_3$. When it is dissolved in hard water, the unwanted calcium and magnesium compounds react with the washing soda and precipitate out of solution as insoluble solids, thus leaving the water soft:

$$\text{sodium carbonate} + \text{calcium sulphate} \longrightarrow \text{calcium carbonate} + \text{sodium sulphate.}$$

$$(Na_2CO_3 + CaSO_4 \longrightarrow CaCO_3 + Na_2SO_4).$$

Washing soda was used with soap before detergents were invented.

**4** Use of detergents. These are not destroyed by hardness of water. You can read more about them in chapter 13.

**5** Addition of calcium hydroxide. This is another chemical which may be added to water to cure hardness. Like washing soda it makes the calcium compounds precipitate out of solution, but only calcium hydrogencarbonate, or temporary hardness is removed.

$$\text{calcium hydroxide} + \text{calcium hydrogencarbonate} \longrightarrow \text{calcium carbonate} + \text{water}$$

$$(Ca(OH)_2 + Ca(HCO_3)_2 \longrightarrow 2CaCO_3 + 2H_2O).$$

**6** Calgon. This is the name of a commercial brand of water softener. It contains a compound which takes the calcium and magnesium compounds out of the water and combines with them without leaving any insoluble substances at the bottom of the vessel. Consequently, it is often used as a domestic water softener. The water

*A blocked pipe caused by precipitation.*

flows through a container of calgon, and comes out soft, with the hardening compounds still there but not liable to affect the soap.

**7** Ion exchange resins. These are very efficient methods of softening or completely purifying water, and you can read more about them in the extra time section of chapter 9.

**Summary**

At the end of this chapter, you should be able to:

**1** Describe the route which water takes as it goes round the 'water cycle'.

**2** Describe the quantities of water used by various industries and in processes in the home.

**3** Suggest sources of impurities in water.

**4** Describe how drinking water is prepared from river water.

**5** Describe how sewage is treated.

**6** Describe the washing action of soap and water.

**7** Explain how hardness of water is formed and how it destroys soap.

**8** Distinguish between temporarily and permanently hard water.

**9** Explain how caves, stalagmites and stalactites are formed in limestone areas.

**10** List ways of curing hardness of water.

# Water, water .....where?

*Water is something that we all take for granted – or at least we used to. Over the last ten years or so, people have come to think differently. In 1976 something happened to make them think very differently indeed . . .*

The Summer of 1976 was long and hot, and Great Britain was in the grips of a drought. By August, things were so serious that the Government created emergency powers to control the use of water. It became against the law to:

**1** Water ornamental gardens, lawns, recreation grounds, playing fields, golf and race courses;

**2** Operate mechanical car washes, wash road vehicles for any reason other than safety or hygiene or to clean the exteriors of any buildings;

**3** Operate mechanical or ornamental fountains or cascades, (even including types where water was recycled) or to operate any cistern which flushed automatically during any period when the premises were wholly or substantially unoccupied.

Local Authorities had to decide which measures were needed in their areas, and apply for appropriate orders. The newspapers were full of stories about the drought . . .

**It never rains . . .**

In a country famous for its rain and constant supplies of water, we are in imminent danger of running out. Domestic supplies may have to be cut off before the autumn. In some areas, housewives may have to queue at makeshift standpipes at the street corner for their rations of water.

There is normally no shortage of water in Britain. In an average year, something like 400 000 million litres of the stuff rains down on us. If it could all be used, just over a week's average rainfall could supply the whole country with water for a year.

But things are not as simple as that. Much of the rain falls on the mountains in Wales, the Lake District and the Scottish Highlands. Rainfall is measured by calculating the depth of water that would accumulate if none of it drained away or evaporated. On this basis, Snowdon and Scafell have around 5 metres a year. But London, the East and South-East have a bare 0·6 metres a year. Even rainy Manchester has under one metre a year. In general, the places where people live and work have the least rainfall.

Even so, there is enough rain. In London each person uses on average 300 litres of water a day, which is the average daily rainfall on a plot about 12 metres square: say half the size of a tennis court. But much of it is wasted. In hilly Wales and Cumbria, rocks lie just below the surface, and the rain cannot penetrate them so it runs off in torrents to quick-flooding, fast-flowing rivers rushing wastefully to the sea. All this water is lost, unless it can be trapped in natural or man-made lakes and routed to cities where it is needed.

In normal years enough water is saved, either in reservoirs or in porous rocks beneath

*The river Thames at its lowest level for forty years.*

the surface of the earth. But this winter the rainfall was the lowest on record. In the South and East, which normally have 40 cm of winter rain, dropping to 28 cm in a typical drought year, there was a mere 17 cm this year.

**Liquid assets down the drain**

As the nation is lobbing bricks and any other handy obstructions into its cisterns, the men at the Building Research Station are stepping up a water conservation programme that could mean a radical change in the type of equipment we install in our bathrooms. They are investigating a variety of ways of cutting down the amount of water used by the chief culprit in the home, the toilet cistern, and even considering systems which do not rely on water at all.

Eddie Ball, the senior scientific officer heading the team says, 'We are putting pure drinking water into the cistern and then proceeding to contaminate it as badly as possible. It doesn't make sense.'

Flushing a lavatory accounts for 35 per cent of the water used daily in the average house. But it's not just the present shortage of water that makes this a fact of life which needs rethinking. Before the last two year's low rainfall made the problem acute, the Building Research Station, which is a Department of the Environment establishment, had started on its project largely because of the quickly increasing cost of producing pure water.

Economists don't quite agree about how much water costs. Some insist on including in their final figure the cost involved in disposing of the used water through sewage works. But the current average cost of water supplied to the tap is roughly 10p a cubic metre, although it may vary locally.

'We are looking at a large number of toilets which either use very little water or none at all', says Eddie Ball, 'But most of them use energy instead which is no more economical in the long run.

'There's the Swedish composting system; the matter drops into a tray where it turns into compost which eventually needs emptying out. We have one of these here – we're having a bit of trouble with the flies at the moment.

'So there is more to be said for the electric incinerator or the deep freeze method which seals the waste into plastic bags, or the vacuum type which draws waste into a tank. The disadvantage is that all of them have to be emptied by somebody sooner or later, and the question of disposal has to be tackled all over again.'

But at very least, the Building Research Team has a new improved version of the brick in the cistern. 'Try a plastic bag of water', says Eddie, 'Start with a litre or two in it and conduct your own bit of research adjusting the water level until you discover the maximum amount of water the bag can contain without affecting the efficiency of the flush.'

## Questions

**1** Write a short paragraph to describe how a molecule of water can start off in the sea, and eventually arrive back in the sea after having visited a cow and a baby on its travels.

**2** Take a typical day in your life in which you wash your hair and take a bath. Estimate how much water you would use between getting up and going to bed.

**3** Make a list of your local industries – one of them might be farming. Try to find out what they make, and what their raw materials are. Can you guess or find out what they might be emitting into nearby rivers, streams or sewers? They probably won't tell you if you ask them.

**4** Why does tap water from different areas of the country taste different, even though it has been purified in the same way at the water works?

**5** Hot water will wash salt off a plate, but if the plate has grease on it the water will not have a great deal of effect. If some soap is added however, the grease disappears. Explain why this is so.

**6** Explain how caves and stalactites and stalagmites are formed in limestone areas.

**7** What is the difference between temporary and permanent hardness of water?

**8** A sample of 10 $cm^3$ of distilled water needed 1 $cm^3$ of soap solution to form a lasting lather. The same amount of tap water needed 10 $cm^3$ of soap solution to produce a similar lather. A second sample of the water was boiled. After cooling, it was found that it needed only 5 $cm^3$ of soap solution to form a lather. Even though a third sample of the water was boiled for a long time, it was not possible to reduce the amount of soap needed further. Finally, some washing soda was added to a fourth sample of the water, and after shaking, only 1 $cm^3$ of the soap was needed. Explain this sequence of events.

**9** What would be the best method for softening:

**a** water needed to supply a central heating system,

**b** water needed for washing up, if no detergent is available.

# 4 The structure of the atom

## 4.1 Models of the atom

We saw in chapter 1 that atoms can be thought of as very small spheres which cannot be broken down into anything simpler. That *model* or way of describing the atom is good enough most of the time, but occasionally we need to consider what is inside that tiny sphere.

From the beginning of this century, scientists knew that atoms could be made to give off tiny electrically charged particles which they called *electrons*. They knew that these were negatively charged, and that the remainder of the atom was positively charged. J. J. Thomson, a scientist who did many important experiments trying to work out the structure of the atom, suggested that it looked like a miniature plum pudding with electrons as 'plums' embedded in the rest of the atom.

A famous scientist called Lord Rutherford who was working at Manchester University in 1911, did some experiments and worked out that Thomson's model was not quite right. He showed that the atom was made of two parts: a central *nucleus*, surrounded by electrons revolving round it rather like planets around the sun. Later, in 1932, Sir James Chadwick found that the nucleus contained not only positively charged particles which he called *protons*, but also neutral particles which he called *neutrons*. The model of the atom which we will use is shown below.

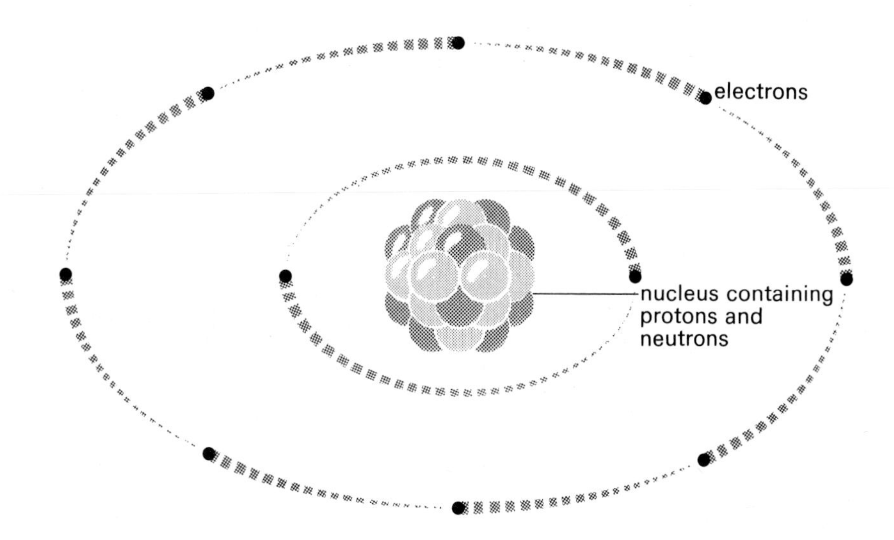

**The facts in brief.** The atom consists of a nucleus containing protons and neutrons, surrounded by tiny electrons which orbit around it. Most of the atom consists of empty space! Imagine an atom magnified until it is the size of a big football stadium. The nucleus is best represented by a marble at the centre of the pitch. The electrons will be like specks of dust at various parts of the stands. Everything in between would be space – nothingness!

If you shrink all that down until the atom is only one thousand millionth of a metre across, you have an atom. So you can see that the diagram on the preceding page shows the nucleus and the electrons very large compared to the size of the actual atom.

Scientists have measured the mass of these particles and their electrical charges. Because they are so very small, the ordinary units of mass and charge which we use in everyday life cannot be used. So they are simply compared with the smallest atom – hydrogen. One unit of mass is chosen to be the mass of a hydrogen atom. These units are called *atomic mass units.* A similar method is used for the electrical charge: one unit of charge is equal to the charge on the nucleus of a hydrogen atom. The results are shown in figure 2.

**Figure 2**
*The masses and charges of the atomic particles.*

| **particle** | **mass** | **electrical charge** |
|---|---|---|
| proton | 1 | +1 |
| neutron | 1 | neutral |
| electron | 1/1840 | −1 |

Think about these figures:

**1** The proton and neutron have the same mass, of one atomic mass unit.

**2** The electron is so much smaller that its mass can be ignored when it is compared with protons and neutrons.

**3** The charges on the proton and electron are equal but opposite.

**4** The neutron has no charge. It is neutral.

## 4.2 Atomic number

To tell us more about the structure of the various elements we need to use two more expressions. The first of these is called *atomic number.* The atomic number of an element tells us how many protons are present in one atom of that element.

An atom is not electrically charged, but the protons and electrons inside it are. So the positive charges of the protons must equal the negative charges of the electrons so that they cancel out. The only way for this to happen is for the number of protons to be the same as the number of electrons.

**In an atom, the number of protons is equal to the number of electrons. This number of protons or electrons is called the Atomic Number.**

For example: the atomic number of aluminium is 13. An atom of aluminium contains 13 protons and 13 electrons.

## 4.3 Mass number

The second expression which we need to use is *mass number*.

The *mass number* of an element tells us how heavy an atom of that element is. Since electrons are very light compared with protons and neutrons, their mass is ignored.

**The mass number of an element is the sum of the number of protons plus the number of neutrons, in one atom of that element.**

For example: the mass number of aluminium is 27. This means that in an atom of aluminium there are 27 protons and neutrons altogether.

These two pieces of information are quite often written with the symbol like this:

$$^{\text{mass number}}_{\text{atomic number}}\text{SYMBOL}$$

For example: $^{27}_{13}Al$

This means that an atom of aluminium has:
Atomic number = 13 which means that there are 13 protons and 13 electrons.
Mass number = 27, so the sum of protons plus the neutrons = 27.
But, the number of protons = 13, so the number of neutrons = 27 − 13 = 14.
An atom of aluminium contains 13 protons, 14 neutrons and 13 electrons.
Figure 3 shows the way in which the elements are built up of protons, neutrons and electrons. For each element:
the number of protons = the number of electrons = atomic number;
the number of neutrons = mass number − the atomic number.
This information may be summarized for all the elements. The figures for the first twenty elements are shown in figure 3 on the next page.

| element | atomic number | number of protons | number of electrons | mass number | number of neutrons (mass number −atomic number) |
|---|---|---|---|---|---|
| $^{1}_{1}H$ | 1 | 1 | 1 | 1 | 0 |
| $^{4}_{2}He$ | 2 | 2 | 2 | 4 | $4 - 2 = 2$ |
| $^{7}_{3}Li$ | 3 | 3 | 3 | 7 | $7 - 3 = 4$ |
| $^{9}_{4}Be$ | 4 | 4 | 4 | 9 | $9 - 4 = 5$ |
| $^{11}_{5}B$ | 5 | 5 | 5 | 11 | $11 - 5 = 6$ |
| $^{12}_{6}C$ | 6 | 6 | 6 | 12 | $12 - 6 = 6$ |
| $^{14}_{7}N$ | 7 | 7 | 7 | 14 | $14 - 7 = 7$ |
| $^{16}_{8}O$ | 8 | 8 | 8 | 16 | $16 - 8 = 8$ |
| $^{19}_{9}F$ | 9 | 9 | 9 | 19 | $19 - 9 = 10$ |
| $^{20}_{10}Ne$ | 10 | 10 | 10 | 20 | $20 - 10 = 10$ |
| $^{23}_{11}Na$ | 11 | 11 | 11 | 23 | $23 - 11 = 12$ |
| $^{24}_{12}Mg$ | 12 | 12 | 12 | 24 | $24 - 12 = 12$ |
| $^{27}_{13}Al$ | 13 | 13 | 13 | 27 | $27 - 13 = 14$ |
| $^{28}_{14}Si$ | 14 | 14 | 14 | 28 | $28 - 14 = 14$ |
| $^{31}_{15}P$ | 15 | 15 | 15 | 31 | $31 - 15 = 16$ |
| $^{32}_{16}S$ | 16 | 16 | 16 | 32 | $32 - 16 = 16$ |
| $^{35}_{17}Cl$ | 17 | 17 | 17 | 35 | $35 - 17 = 18$ |
| $^{40}_{18}Ar$ | 18 | 18 | 18 | 40 | $40 - 18 = 22$ |
| $^{39}_{19}K$ | 19 | 19 | 19 | 39 | $39 - 19 = 20$ |
| $^{40}_{20}Ca$ | 20 | 20 | 20 | 40 | $40 - 20 = 20$ |

**Figure 3** *The structure of atoms of the first twenty elements.*

## 4.4 Isotopes

The number of neutrons in the atom of an element can vary.

For example: an atom of chlorine can contain either 18 or 20 neutrons. The symbols for these different atoms are written as $^{35}_{17}Cl$ and $^{37}_{17}Cl$.

$^{35}_{17}Cl$ contains 17 protons, 17 electrons and 18 neutrons;

$^{37}_{17}Cl$ contains 17 protons, 17 electrons and 20 neutrons.

They are both chlorine atoms because they both have the same number of protons and electrons, but the second one has two more neutrons. They are called *isotopes* of chlorine.

**Isotopes are different forms of the same element. They have the same number of protons and electrons but different numbers of neutrons. They have the same atomic number but different mass numbers.**

If you look in the list of accurate atomic masses at the back of the book, you will see that the atomic mass of chlorine is given as 35·453. This is the *relative atomic mass* of chlorine and it is the average of the mass numbers of the two isotopes of chlorine in the proportion in which they exist in naturally occurring chlorine. Naturally occurring chlorine contains 75% of $^{35}_{17}Cl$, and 25% of $^{37}_{17}Cl$. So you can see that the relative atomic mass is going to be much nearer 35 than 37.

From the table at the back of the book, you can see that the relative atomic mass for many of the elements is not a whole number. These elements consist of a mixture of naturally occurring isotopes.

For elements which are a mixture, we must always use the relative atomic mass. Mass numbers are only used for pure isotopes.

Another example of an element which has isotopes is hydrogen. There are three isotopes of hydrogen:

| $^{1}_{1}H$ | $^{2}_{1}H$ | $^{3}_{1}H$ |
|---|---|---|
| hydrogen | deuterium | tritium. |

You can see from the mass numbers that an atom of hydrogen has no neutrons in its nucleus; deuterium has one neutron, and tritium has two neutrons. Each isotope has one proton and one electron. Naturally occurring hydrogen contains 99·985% of $^{1}_{1}H$ and 0·015% of $^{2}_{1}H$. Tritium is an artificial isotope. It does not occur naturally and has only been made in nuclear reactors. So the relative atomic mass of hydrogen will be the average of the mass numbers of hydrogen and deuterium according to the percentages in which they occur, and the value comes to 1·008. It is close to 1 because hydrogen is by far the most common isotope.

Most of the time, if an element is made of several isotopes, one of them is present in a very large proportion so that when the relative atomic mass is taken to the nearest whole number this gives the mass number of that isotope. But not always. Chlorine is one example, and copper is another.

## 4.5 The arrangement of electrons

The electrons in an atom orbit round the nucleus. Electrons differ from each other in the amount of energy that they have. The electrons which are near the nucleus have a low energy, the ones further away have high energy. Electrons orbit the nucleus in special regions called *shells*. The position of each shell is like the skin of a balloon round a nucleus in the centre of the balloon. Electrons stay very close to each shell. The shells are positioned at fixed distances from the nucleus. (See figure 4).

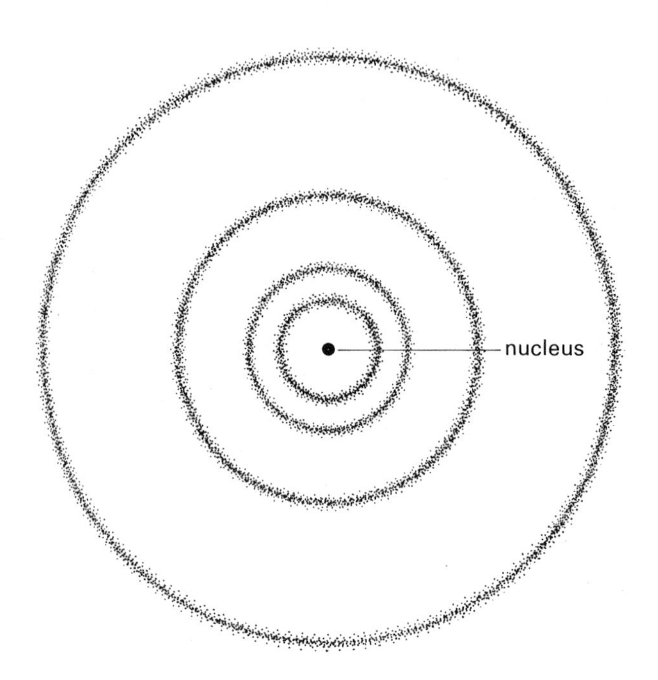


**Figure 4**
*Shells of electrons around the nucleus.*

Each shell can contain only a certain number of electrons. We shall study the first 20 elements. As the number of electrons builds up, you will see that:

the first shell contains a maximum of 2 electrons;
the second shell contains a maximum of 8 electrons;
the third shell contains a maximum of 8 electrons;

The structure of the fourth shell is complicated and shall not be studied here.

Figure 5 shows how the electrons fit into the shells. Figure 6 shows the electron arrangement of a selection of atoms.

| element | atomic number | number of electrons | 1st shell | 2nd | 3rd | 4th |
|---|---|---|---|---|---|---|
| H | 1 | 1 | 1. | | | |
| He | 2 | 2 | 2. | *1st shell full* | | |
| Li | 3 | 3 | 2. | 1. | | |
| Be | 4 | 4 | 2. | 2. | | |
| B | 5 | 5 | 2. | 3. | | |
| C | 6 | 6 | 2. | 4. | | |
| N | 7 | 7 | 2. | 5. | | |
| O | 8 | 8 | 2. | 6. | | |
| F | 9 | 9 | 2. | 7. | | |
| Ne | 10 | 10 | 2. | 8. | *2nd shell full* | |
| Na | 11 | 11 | 2. | 8. | 1. | |
| Mg | 12 | 12 | 2. | 8. | 2. | |
| Al | 13 | 13 | 2. | 8. | 3. | |
| Si | 14 | 14 | 2. | 8. | 4. | |
| P | 15 | 15 | 2. | 8. | 5. | |
| S | 16 | 16 | 2. | 8. | 6. | |
| Cl | 17 | 17 | 2. | 8. | 7. | |
| Ar | 18 | 18 | 2. | 8. | 8. | *3rd shell full* |
| K | 19 | 19 | 2. | 8. | 8. | 1. |
| Ca | 20 | 20 | 2. | 8. | 8. | 2. |

**Figure 5** *The electron arrangements of the first twenty elements.*

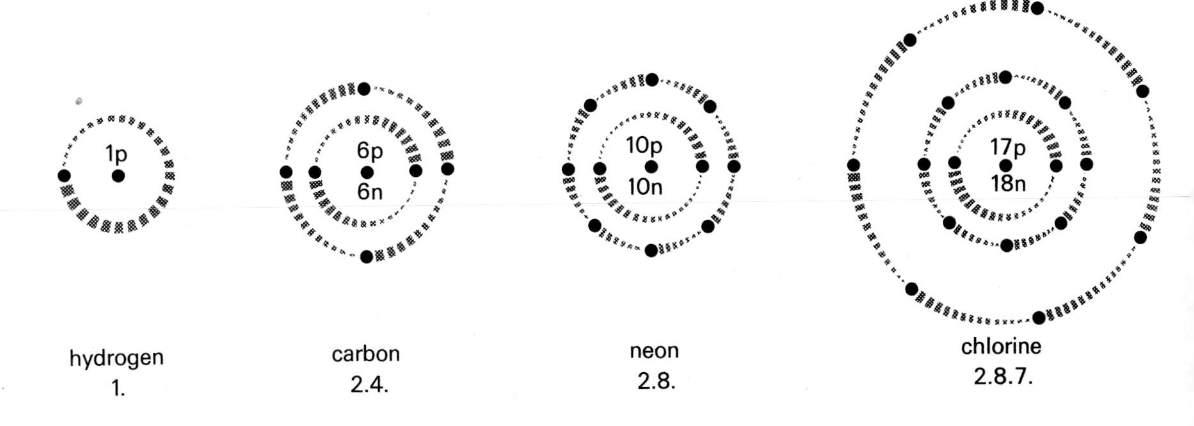


**Figure 6** *The electron arrangements of some atoms.*

4.6

## The Periodic Table

Once the arrangement of the electrons in atoms was understood, a scientist called Niels Bohr finalized a diagram showing most of the information known about the elements. He started with element number 1 – hydrogen, and followed it with number 2 – helium.

H He

He listed the atoms in order of increasing atomic number, putting those with filled shells on the right hand side, and those with only one electron in their outermost shell, on the left. For example, helium has a full shell of electrons, so he put it at the end of the first line. The next element – lithium (Li) had one electron in its outside shell. So he started a new line with it, and followed it with the elements in order of their atomic numbers, until he again had a full outside shell on the right hand side – this time with neon (Ne).

H He

Li Be B C N O F Ne

On the next line, he started with the next element, sodium (Na), which has one electron in its outside shell, and ended the line with argon (Ar), which again has a full outside shell.

H He

Li Be

Na Mg

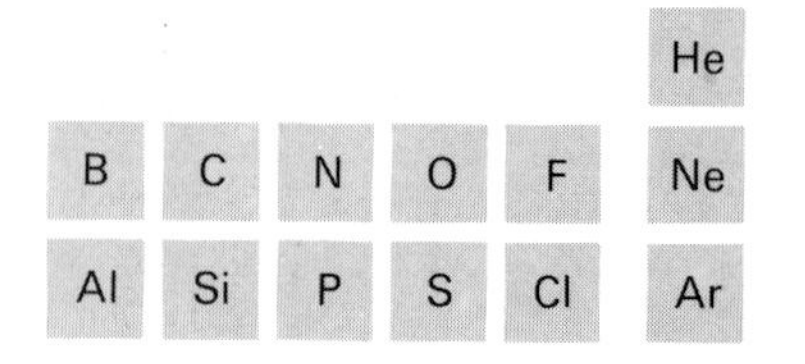


The rest of the elements were built up in a similar way. The full version of this diagram can be seen in figure 7. It is called the *periodic table*. The periodic table can tell chemists a great deal about an element just from its position in the table.

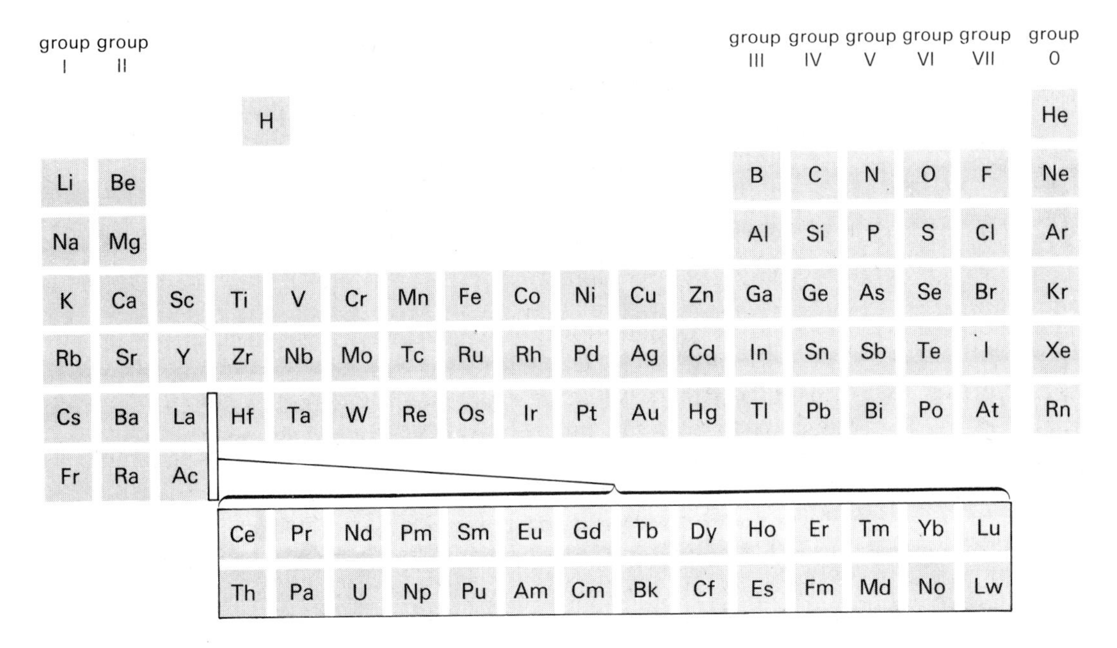


**Figure 7** *The Periodic Table.*

Look at the first vertical column on the left-hand side of the table. A vertical column in the table is called a *Group*. So this is called Group I. Group I contains the elements lithium, sodium, potassium, rubidium and caesium. Francium is radioactive and not naturally occurring. You can see from figure 5 (and the fact that the table is drawn up with this in mind!) that each of these elements has only one electron in its outside shell. Because of this they all have very similar properties and chemical reactions.

Similarly, all the members of Group II of the periodic table (look at figure 5) have two electrons in their outside shells, and consequently, magnesium, calcium, strontium and barium have similar properties and compounds. At the other side of the table, Group VII elements, fluorine, chlorine, bromine and iodine are also very alike. They all have seven outside shell electrons.

The last group of all is called Group 0, to distinguish it from all the others. The elements helium, neon, argon krypton and xenon, all have full outside shells of electrons, and this makes them very unreactive. Because of this, they are called the *inert*, or *noble gases*. Chapter 5 will explain in more detail how the number of outside shell electrons affects the way in which the elements react.

You can now see why this table is called the periodic table. All of the elements which have the same number of electrons in their outside shell occur at regular or periodic intervals. They are in columns in the same Group. The horizontal rows in the table are called *Periods*. All the elements in the same Period have the same number of shells, with the last shell just filling up.

## 4.7 Radioactivity

The isotopes of most elements have atoms which are *stable.* In other words, electrons may be removed from, or added to their outside shells in a chemical reaction, but the number of protons and neutrons in the nucleus does not change. Some isotopes are *unstable.* Their nuclei break apart, forming smaller atoms. Small particles and energy are released. Isotopes which do this are called *radioisotopes*, and the particles and energy which they give out are given the general name, *radioactivity.*

## 4.8 Radioactive particles

There are three different types of radioactive particles. The first type are $\alpha$ (pronounced alpha) particles. These are atoms of helium which have lost their electrons. They are called helium *ions.* $\alpha$-particles are unable to penetrate more than one or two sheets of paper, so they are easily stopped. They cannot go through skin and provided an isotope is not emitting $\alpha$-particles too quickly, it is not dangerous. If it were to be inside the body however, it would be quite a different matter.

The second type of radioactive particles are called $\beta$ (Beta) particles. They are very fast moving electrons which come from inside the nuclei of radioactive atoms. $\beta$-particles can pass through several millimetres of metal foil before they are stopped.

The third type of radioactive particles are called $\gamma$ (gamma) particles. $\gamma$-particles are very much like X-rays, but are much more penetrating and dangerous. $\gamma$-particles can pass through 15 cm of lead before they are stopped. Like $\alpha$- and $\beta$-particles they damage cells when they pass through human tissue and can cause bad burns which take a long time to heal. In extreme cases of exposure to radioactivity, tumours or cancers can be started in bones or vital organs of the body like the lungs.

## 4.9 Uses of radioactivity

Radioactivity is used in many different ways. Some radioisotopes are used in medicine and industry; uranium is used in the production of nuclear power.

**Medicine.** Radioisotopes are very important in the medical world. Small quantities of a radioactive material can be swallowed by a patient so that the radioactivity gets into the blood stream and is carried to parts of the body such as the kidneys or the liver. These otherwise inaccessible parts of the body can then be studied. The radiation coming from the body can be detected by instruments and displayed on a TV screen. Obviously, the amount of radiation used must be small and easily removed afterwards. By contrast, large doses of $\gamma$-radiation can be used both to destroy cancer tumours in the body, and to sterilise surgical instruments before they are used. The small hypodermic needles that doctors use are sterilised in this way after they have been packed and sealed in plastic.

*This is a 'gamma camera' picture of a human brain. It is a side view, with the patient's face looking to the left. The black area is the brain, and the two white spots are brain tumours.*

**Industry.** Radioisotopes also have many industrial uses. They are often used for monitoring the flow of materials in a production line.

In the paper mill shown as a diagram in figure 8, the thickness of the paper depends on how far apart the rollers are set. If the paper for some reason becomes either too thin or too thick, the detector receives more or less α-particles, and so it adjusts the rollers automatically. The thickness of the paper on which this book was printed was controlled in this way.

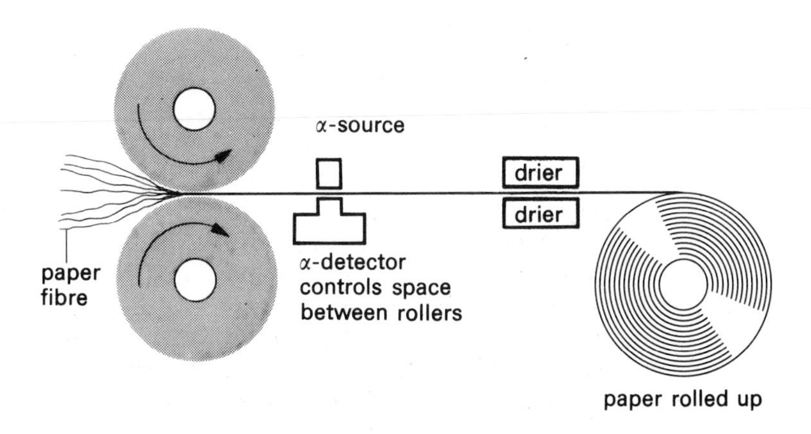


**Figure 8**
*Use of an α-source to control an industrial process.*

**Atomic Energy.** Probably the first uses of radioactivity that most people think of are the atom bomb, and the nuclear reactor.

Before the second world war, scientists found that an atom of uranium could be made to decay into two smaller atoms when it was bombarded by a neutron:

$$^{235}_{92}U + n \longrightarrow {}^{90}_{36}Kr + {}^{142}_{56}Ba + 3n + \text{heat}.$$

Each uranium atom releases 3 neutrons when it splits. These in turn split 3 more uranium atoms and these 3 split a total of 9 more. This process is called *nuclear fission* (See figure 9.) A *chain reaction* builds up in a fraction of a second, and if a large piece of uranium is used, an enormous amount of heat energy is released within seconds. This is what happens when an atom bomb explodes.

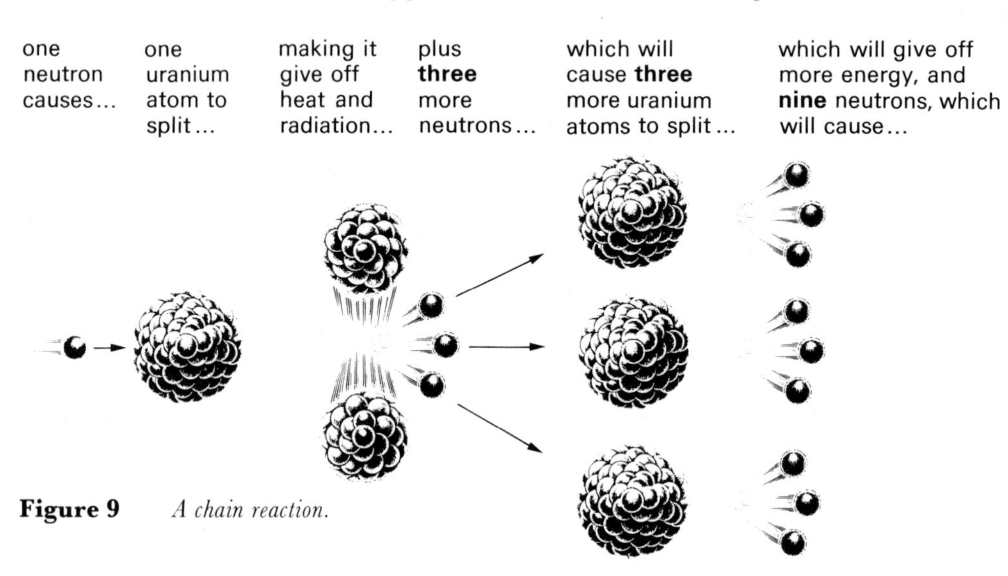


**Figure 9** *A chain reaction.*

In a nuclear reactor the chain reaction is not allowed to reach explosive point. This is done using a *moderator* which slows down neutrons. Graphite is a moderator, and this surrounds the uranium oxide fuel which is contained in special tubes.

In addition, *control rods* made of boron are lowered into or raised out of the centre of the reactor (called the *core*) to absorb excess neutrons. In this way, the temperature of the core is controlled – lowering the rods absorbs more neutrons, so the reaction slows down, and raising them makes it go faster.

The heat produced by the nuclear reaction in the core is carried away by a continuous stream of carbon dioxide. The extremely hot gas is then used to make steam in giant boilers. The steam drives turbines, which rotate generators. These feed electricity into the National Grid.

There must be no radiation leak from the core of the reactor, so it is surrounded with steel and concrete to a thickness of several metres. Machines are used to load and unload the fuel tubes. The used fuel is highly radioactive, so it is left in giant underground water tanks for a while. In the water, the rods glow with a strange, blue light (shown on the front cover). After a while they are put into permanent storage. Figure 10 shows the layout of a typical nuclear power station.

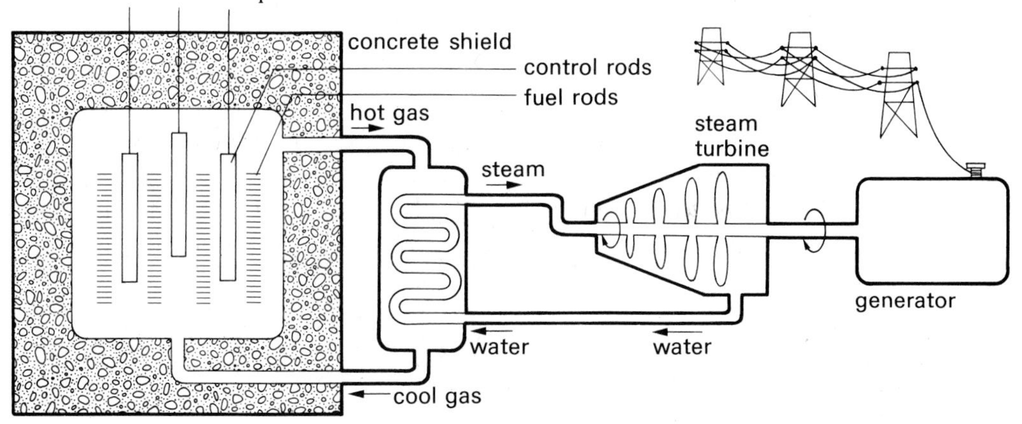


**Figure 10** *A nuclear power station.*

**Summary**

At the end of this chapter you should be able to:

**1** Name the three particles found in the atom, and give their relative masses and electrical charges.

**2** Describe how these particles are arranged in the atom.

**3** Explain what is meant by the terms atomic number and mass number.

**4** Calculate the mass number and atomic number of any element from details about the number and type of particles in its atom, and vice-versa.

**5** Explain the arrangement of electrons in shells around the nucleus of an atom.

**6** Work out the number of electrons in the shells of all of the first twenty elements.

**7** Explain the meaning of the term isotope, and give an example of an element which has isotopes.

**8** Distinguish between the terms mass number and relative atomic mass.

**9** Sketch the approximate shape of the periodic table and put the first 20 elements into their correct places.

**10** Explain what is meant by Groups and Periods in the table.

**11** Describe the three types of radioactivity, and compare their penetrating power.

**12** Describe the uses of radioisotopes in medicine and industry.

**13** Describe in simple terms how the fission of a uranium atom can lead to the production of electrical energy by means of a nuclear reactor.

# Nuclear power–promise or threat?

*On the sixth of August 1945, the atomic age began in earnest. An atomic bomb was dropped on Hiroshima in Japan, which, with its heat and radiation blast, killed over a hundred thousand people. A second bomb at Nagasaki a few days later ended the Second World War. Since then, the potential use of nuclear energy for both good and evil has presented the whole world with one of the greatest problems it has ever had to face. The two articles that follow present two opposite extremes. The first defends the development of peaceful nuclear energy; the second is from the start of a 'diary' of the tragedy at Hiroshima. The characters involved in both articles are fictitious, but the facts they describe are all true.*

**Nuclear power: the promise**
In a recent speech, the famous scientist Dr. John Price argued that it will be absolutely necessary for Britain to develop many more nuclear power stations before the year two thousand.

**Reactors are safe**
'There is too much talking without enough use of facts', claimed Dr. Price. 'I am tired of seeing headlines such as "Nuclear Reactors may explode", or "Nuclear Reactors may poison our food supplies". There are two things that must be said. First, that nuclear reactors are just as safe as any other energy source, and second, that without them we shall never be able to provide the energy that we need for the twenty-first century – when the school children of today are 40 years old.'

**Death rate low**
'Let me quickly put to you a few of the facts about the safety of modern reactors. The chance of an accidental breakdown causing a leak of radiation is minute – the danger to the population from this hazard is much lower than the possibility of being struck by lightning.

'If you think that atomic energy is dangerous, you should look at the figures for "safe" fuels such as coal and oil. Each day, each modern, coal fired power station belches out 20 tons of ash, 500 tons of sulphur dioxide, 50,000 tons of carbon dioxide – and 4 tons of carbon monoxide, a few sniffs of which can be lethal.

'Even the actual workers in atomic plants are safe – I have got the figures to prove it. Over the period from 1962–74 the death rate for atomic plant workers was 25% *less* than the national average. The rate for coal face workers has been as high as 3500% *greater* than the national average.'

**Nuclear energy is a must**
'The supplies of oil and gas are very limited, so it is becoming more and more expensive (some might say, stupid) to use them for making electricity in power stations. The cost of making electricity from coal and oil is going up and up, but the cost of making it from nuclear fuel is staying low.

*Dounreay atomic power station, Scotland.*

'There is enough uranium to produce 10 times the present total world output of electricity for at least 300 years. There is no practical alternative to nuclear energy as a successor for oil and gas. None of the other possibilities such as solar or fusion power or tidal energy, are anywhere near ready for exploitation; even if they were, they are too small in total potential to make a major contribution.'

**All change, please?**

'Anyone who argues that we can do without nuclear energy is arguing, with or without realizing it, that the whole structure of our industrial society is going to change. I for one do not want to have to live off a few meagre half-eaten plants that I've grown in my garden. I am quite sure that opponents of nuclear power are not aware of the full consequences of what they are saying.'

**Nuclear Waste – a tiny problem**

'The world will be hugely overpopulated by the year 2000. Unless we use nuclear energy, our children will not have the energy they will need to drive the industry upon which they will depend.

'The problems caused by a few thousand tons of very carefully contained radioactive waste are tiny compared with the alternative of insufficient energy which the world needs for its survival in its present form.'

**Nuclear power: the threat**

At the station, on my way to work, I met one of the women I work with.

'Mr. Shizuma,' she said, brushing back a wisp of hair with her fingers, 'I shouldn't really ask you in a place like this, but we need your seal on those papers you prepared for us the other day so – '

At a point three metres to the left of the waiting train, I saw a ball of blindingly intense light, and simultaneously I was plunged into total, unseeing darkness. The next instant, the black veil in which I seemed to be enveloped was pierced by cries and screams of pain. I was flung onto the tracks on the opposite side from the platform. Another body landed heavily on top of me. Other people were piled up on either side of me. A cry of pain and rage escaped me, to be echoed in my ear by a similar cry, in a heavy local accent, from a man whose head was jammed against mine.

Then, those who could, began to run about in sheer panic. Scarcely aware of what I was doing, I wrapped my arms around a pillar, and clung to it with all my might. . . .

Eventually, quiet descended around me. Slowly and fearfully, I tried opening my eyes. Everything within my field of vision seemed to be obscured with a light brown haze, and a white, chalky powder was falling from the sky. Not a soul was in sight on the platform. Despite the uproar a moment ago, now not

*A typical scene at Hiroshima.*

even a station official was to be seen in the building. I must have stayed clinging to the pillar with my eyes shut for considerably longer than I had realized.

I ran my hand over my face. The left hand came away wet and sticky. I looked, and found the left palm had something bluish-purple like little shreds of damp paper on it. I stroked my cheek again, and again some sticky substance came off on my hand.

It was extremely odd – I had no recollection of hitting my face on anything. It must be ash or dust or something else that rubbed off like tiny rolls of dead skin. It was like the shreds from a rubber eraser, but more slippery to the touch.

Something recalled to my mind a phrase from a propaganda leaflet that an enemy plane had dropped early the previous month. Words to the effect of: 'We'll be along sometime soon with a little present for the people of Hiroshima. . . .'

Some hours later, I witnessed a sight that I shall not forget. A steady flow of people were leaving Hiroshima, going to the hills. Need I say how varied in appearance and condition were the hundreds, the thousands of people I saw on the way? I feel compelled – unnecessary though it may be – to set down here some of my memories of them, just as they come back to my mind today:

The countless people who had blackish dried blood clinging to them where it had flowed from their faces onto their shoulders and down their backs, or over their chests and down their bellies. Some were still bleeding, but they seemed to have no energy to do anything about it.

The people staggering along in whatever direction the crowd carried them, their arms dangling purposelessly by their sides.

The people who walked with their eyes shut, swaying to and fro as they were pushed by the crowd.

The woman leading a child by the hand who realized that the child was not hers, shook her hand free with a cry, and ran off. And the child – a boy of six or seven – running, crying plaintively, after her.

The father leading his child by the hand who lost hold of him in the crush. He pushed through the crowd calling the child's name over and again, till finally he was struck brutally and repeatedly by someone he had thrust out of the way.

But this was only the beginning of the tragedy. The fatal 'radiation sickness' had not yet even begun . . .

## Questions

**1** From the information in the chapter and at the back of the book, find out the atomic numbers and mass numbers of each of the following elements: fluorine, calcium, lithium, sulphur and neon.

Make a sketch of their atoms, showing the numbers of protons and neutrons in their nuclei, and the arrangement of their electrons in shells.

**2** An element, X, has the mass number of 14 and its atom contains 7 neutrons. Calculate the number of protons and electrons that it will contain, and then find out which element X really is. What will the arrangement of its electrons be in shells?

**3** Two atoms both have 12 neutrons. The first has 11 protons and 11 electrons and the second has 12 protons and 12 electrons. Are they isotopes? Explain your answer.

**4** Carbon's structure may be written as ${}^{12}_{6}C$. It has an isotope which has the mass number of 13. Write the symbol for this isotope and describe its structure.

**5** Aluminium has three electrons in its outside shell. Which Group of the periodic table will it be in? It has three shells of electrons: Which Period of the periodic table will it be in?

**6** Argon is one of the noble gases. It has a full outside shell of electrons. Which Group of the periodic table will it be in?

**7** Make an outline sketch of the periodic table and put the following elements into their correct places.

| *element* | *electron arrangement* |
|---|---|
| **a** | 2. 2. |
| **b** | 2. 8. 4. |
| **c** | 2. 8. 7. |
| **d** | 2. |
| **e** | 2. 8. 8. 1. |

**8** List the three types of radioactivity in order of their penetrating power. If you had to do some experiments with $\gamma$-radiation, describe the sort of precautions you would take.

**9** Some radioisotopes are quite safe enough to be put into the water supply in small quantities. Can you think of a way in which the Water Board might find a leak in a pipe-line without digging up tonnes of earth to locate it?

# 5 How atoms combine

## 5.1 How do atoms combine?

We saw in the last chapter that each shell around the nucleus of an atom contains a certain number of electrons, up to a maximum.

| | |
|---|---|
| The first shell can contain up to | 2 electrons; |
| the second shell can contain up to | 8 electrons; |
| the third shell can contain up to | 8 electrons. |

After that, things become complicated.

The elements which have an outside shell containing a maximum number of electrons are very unreactive. We put them in Group 0 of the periodic table and call them the inert, or noble gases. They do not react because they have no reason to do so. Their outside shell of electrons is full. All the other elements are reactive because they have incomplete outside shells. In reacting, they are aiming to have full outside shells.

They can do this in one of two ways. Look at figure 1. Find the Group I metals, lithium, sodium and potassium. They each have only one electron in their outside shell. When they react, they do so in order to lose this electron. The shell disappears because it is only a place where electrons are. The next shell in becomes the outside one, and it is full.

All the Group I metals react in this way, and that is why they all have similar chemical properties and compounds.

Group II metals react in a similar way. Look at figure 1 again. Each Group II metal has two electrons in its outside shell. It loses those two electrons in order to make its outermost shell full.

On the other hand, Group VII elements have seven electrons in their outside shells, so they want to take in one electron to make them up to eight – a full shell. All the Group VII elements react in this way, and that is why they all have similar chemical reactions and compounds. Compounds made by elements gaining or losing electrons are given a special name. They are called *electrovalent* compounds.

| element | atomic number | number of electrons | 1st shell | 2nd | 3rd | 4th |
|---|---|---|---|---|---|---|
| H | 1 | 1 | 1. | | | |
| He | 2 | 2 | 2. | *1st shell full* | | |
| Li | 3 | 3 | 2. | 1. | | |
| Be | 4 | 4 | 2. | 2. | | |
| B | 5 | 5 | 2. | 3. | | |
| C | 6 | 6 | 2. | 4. | | |
| N | 7 | 7 | 2. | 5. | | |
| O | 8 | 8 | 2. | 6. | | |
| F | 9 | 9 | 2. | 7. | | |
| Ne | 10 | 10 | 2. | 8. | *2nd shell full* | |
| Na | 11 | 11 | 2. | 8. | 1. | |
| Mg | 12 | 12 | 2. | 8. | 2. | |
| Al | 13 | 13 | 2. | 8. | 3. | |
| Si | 14 | 14 | 2. | 8. | 4. | |
| P | 15 | 15 | 2. | 8. | 5. | |
| S | 16 | 16 | 2. | 8. | 6. | |
| Cl | 17 | 17 | 2. | 8. | 7. | |
| Ar | 18 | 18 | 2. | 8. | 8. | *3rd shell full* |
| K | 19 | 19 | 2. | 8. | 8. | 1. |
| Ca | 20 | 20 | 2. | 8. | 8. | 2. |

**Figure 1**
*The electron arrangements of the first twenty elements.*

Not all elements lose or gain electrons when they react. Group IV elements have four outside shell electrons. They cannot lose those four, neither can they gain four more in order to fill their outside shell. Instead, they react by sharing electrons with other elements. Compounds made by elements sharing electrons are said to be *covalent* compounds. Most of the groups towards the middle of the table make covalent compounds.

The next two sections deal separately with electrovalent and covalent compounds.

## 5.2 Electrovalency

The best way to understand this type of bonding is to look at some examples. The next four paragraphs explain the structures of some typical electrovalent compounds.

**The structure of sodium chloride.** When sodium is heated in chlorine gas, the compound sodium chloride is formed and heat is given off. A solid and a gas form a hot, white crystalline solid:

sodium + chlorine $\longrightarrow$ sodium chloride (salt).

We know from the last section that both sodium and chlorine (Groups I and VII), react by losing and gaining electrons respectively. Look at the arrangement of the electrons:

| Na | + | Cl |
|---|---|---|
| 2. 8. 1. | | 2. 8. 7. |

In both cases, the two inner shells are full, so they are not involved in the bonding. Only the outside shell electrons are important, and they can be shown more easily on a diagram:

Na•     ×Cl× (seven × electrons)

• and × represent electrons from the outside shells of the elements.

When the two elements react, the sodium atom loses its one outside shell electron, and the chlorine atom takes it in. In this way, they both get a full outside shell.

Na• + ×Cl× $\longrightarrow$ Na + •×Cl×

Both atoms have become unreactive now, because they both have full outside shells of electrons, but they are no longer neutral atoms. Because it has lost its outer electron, the sodium atom is no longer electrically balanced. It has one electron too few. It has become a *sodium ion.* In the same way, the chlorine atom is no longer balanced. It has one electron too many. It has become a *chloride ion.*

Ions are atoms which have either lost or gained electrons and so have become positively or negatively charged. The number of charges depends upon the number of electrons they have lost or gained. Back to sodium chloride:

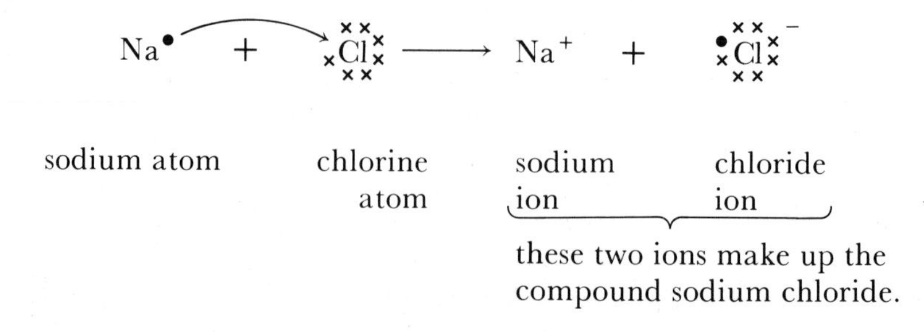

The sodium has gained one positive charge to become an ion, and the chloride has gained one negative charge. The charges on the sodium ion and the chloride ion are equal and opposite. They balance each other out, and the formula of sodium chloride is NaCl. You will learn more about how to write formulae in the next chapter.

**The structure of lithium fluoride.** Lithium metal will burn in fluorine gas to produce a white crystalline substance called lithium fluoride:

lithium + fluorine ⟶ lithium fluoride
2. 1. 2. 7.

$$\mathrm{Li}\bullet + \overset{\times\times}{\underset{\times\times}{{}_{\times}\mathrm{F}_{\times}^{\times}}} \longrightarrow \mathrm{Li}^{+} + \left[\overset{\times\times}{\underset{\times\times}{{}_{\times}^{\bullet}\mathrm{F}_{\times}^{\times}}}\right]^{-}$$

The same thing happens. Lithium and fluoride ions are formed. Lithium fluoride is made of lithium ions and fluoride ions, and the formula of lithium fluoride is LiF.

**The structure of magnesium oxide.** Magnesium burns in oxygen with a bright white flame to form a white ash called magnesium oxide:

magnesium + oxygen ⟶ magnesium oxide.
2. 8. 2. 2. 6.

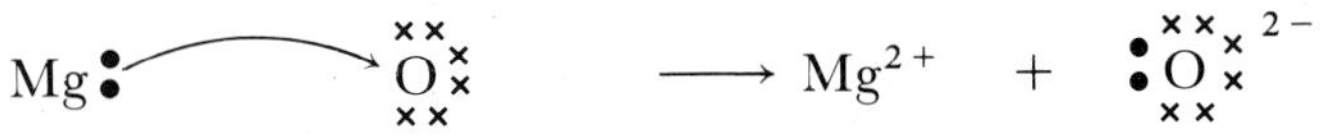

$$\mathrm{Mg}: + \overset{\times\times}{\underset{\times\times}{\mathrm{O}_{\times}^{\times}}} \longrightarrow \mathrm{Mg}^{2+} + \left[\overset{\times\times}{\underset{\times\times}{{}^{\bullet}_{\bullet}\mathrm{O}_{\times}^{\times}}}\right]^{2-}$$

The magnesium atom needs to lose two electrons to empty its shell, and the oxygen atom needs to gain two electrons to fill its shell, so two electrons are transferred. The magnesium atom has now lost two negative electrons and become a magnesium ion with two positive charges, and the oxygen atom has gained two negative electrons and become an oxide ion with two negative charges.

Magnesium oxide is made of magnesium ions and oxide ions, and its formula is MgO.

**The structure of potassium sulphide.** Potassium and sulphur react together violently to make the compound potassium sulphide:

potassium + sulphur ⟶ potassium sulphide.
2. 8. 8. 1. 2. 8. 6.

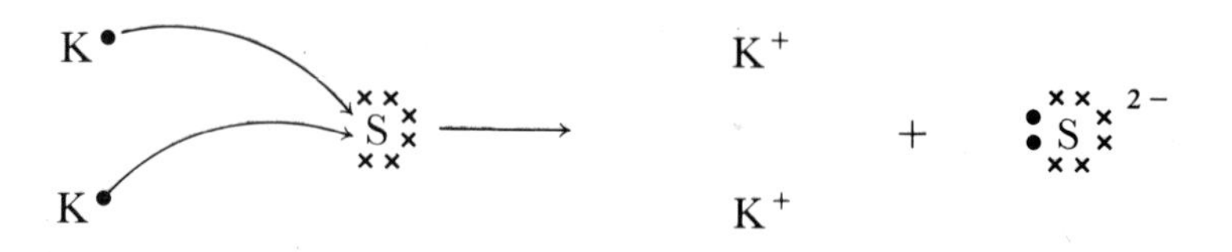


Each potassium atom has only one outside shell electron to lose. The sulphur atom needs to gain two electrons to fill its outside shell, so two potassium atoms lose one electron each. Two potassium ions and one sulphide ion are formed. They join to make potassium sulphide, and its formula is $K_2S$.

Sodium chloride, lithium fluoride, magnesium oxide, and potassium sulphide, are all substances containing ions. They are called *electrovalent compounds*. (There are more examples of this sort for you to try in the questions at the end of the chapter.)

Figure 2 shows which of the first twenty elements form ions and the charges that their ions have.
This diagram tells us a lot. You can see that except for the inert gases, the elements that form ions are mainly on the left hand side or the right hand side of the periodic table. In other words, they have only a few electrons to lose or gain in order to fill their outside shells. The elements in the middle have not got enough energy to lose or gain large numbers of electrons, and so do not form ions.

**Figure 2** *The ions of the first twenty elements.*

## 5.3 Covalency

You can see from figure 2 that some elements do not form ions. When they react, they have to combine with other elements by making *covalent* compounds. Carbon, nitrogen and phosphorus are good examples of elements which form covalent compounds.

**The structure of methane.** Carbon atoms and hydrogen atoms are joined together in a compound called methane:

$$C \quad + \quad 4H \longrightarrow CH_4.$$

2. 4.    1.

(This reaction does not actually take place; methane is made by another method.)

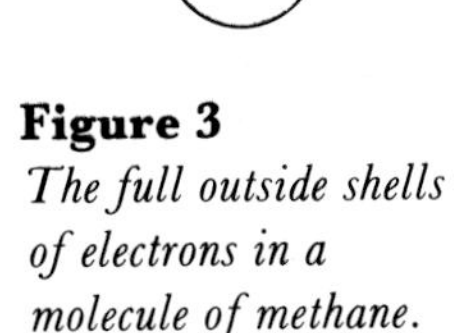


**Figure 3**
*The full outside shells of electrons in a molecule of methane.*

```
 •                       H
                         •×
•C•  +  4H×  ⟶   H ×• C •× H      a methane molecule.
 •                       ×•
                         H
```

The carbon atom and four hydrogen atoms share electrons in order to fill their outside shells. The carbon atom now has eight electrons, and each of the hydrogen atoms has two electrons in its outside shell – that is all each of them needs for a full shell. The compound has been formed by sharing shell space. (See figure 3.)

**The structure of ammonia.** Ammonia is a gas which is made from nitrogen and hydrogen:

nitrogen + hydrogen $\longrightarrow$ ammonia

$$N \quad + \quad 3H \longrightarrow NH_3$$

2. 5.    1.

```
 ••                 ••
•N•  +  3H×  ⟶  H •× N •× H
 •                  •×
                    H
```

One nitrogen atom shares electrons with three hydrogen atoms to form one molecule of ammonia.

**The structure of phosphorus(III) chloride.** Phosphorus and chlorine react to form a colourless liquid called phosphorus(III) chloride:

phosphorus + chlorine $\longrightarrow$ phosphorus(III) chloride.

$$P \quad + \quad 3Cl \longrightarrow PCl_3$$

2. 8. 5.    2. 8. 7.

```
 ••           ××              ××   ••   ××
•P•  +   3 ×Cl××   ⟶      ××Cl ×• P •× Cl××
 •            ××              ××   ×•   ××
                                 ××Cl××
                                   ××
```

A molecule of phosphorus(III) chloride is formed and the phosphorus and the three chlorine atoms get full outside shells by sharing electrons.

(There are more examples like these in the questions at the end of the chapter.)

## 5.4 Differences between electrovalent and covalent compounds

**Electrovalent compounds are made of ions.** For this reason, they are sometimes called *ionic compounds.*

Covalent compounds are made of molecules and do not contain any ions.

**Electrovalent compounds have a regular structure.** This arrangement makes a crystal *lattice.* Figure 4 shows the ways in which a sodium chloride crystal lattice is made up.

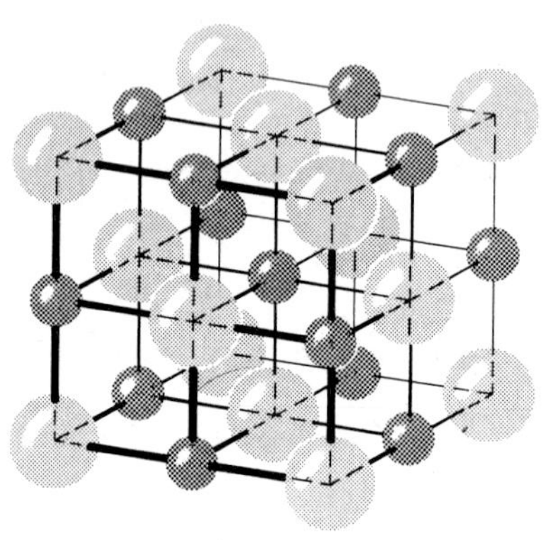

**Figure 4** *The sodium chloride crystal lattice.*

The lattice is held together very tightly because the positively charged sodium ions are attracted to the negatively charged chloride ions. The shape of the lattice is a cube because with the sodium and chloride ions pulling on one another, they fit together best this way. A single crystal of sodium chloride contains millions of ions.

In a crystal lattice, the attraction between the ions is very strong. This force of attraction is called an *ionic bond.* Solid covalent substances also form crystals, but the molecules cannot be held together by ionic bonds, because there are no ions present. Instead, they are held together by much weaker forces called *Van der Waal's* forces. In covalent liquids, the molecules are even more weakly attracted, and in covalent gases the forces are almost non-existent.

**Melting points and boiling points.** Electrovalent substances have much higher melting points and boiling points than covalent substances. This is because of their structure. In order to melt an electrovalent substance, it must be given heat energy to pull the ions in the crystal lattice apart from one another. The ionic bonds are so strong that a lot of energy is needed, so this makes a high temperature necessary. Even when the substance is molten, the charged ions are still attracted to each other. A lot more energy is needed to make the ions move about enough to leave the liquid and form a gas. So the boiling point is very high.

Covalent substances have very small forces between the molecules, and so the amount of energy needed to separate them is small. Consequently their melting and boiling points are generally lower than those of electrovalent substances.

Figure 5 compares the melting and boiling points of some electrovalent and covalent compounds.

| **electrovalent** | **covalent** | **melting point** °C | **boiling point** °C |
|---|---|---|---|
| aluminium oxide ($Al_2O_3$) | | 2040 | 2980 |
| calcium chloride ($CaCl_2$) | | 782 | 1600 |
| sodium chloride (NaCl) | | 808 | 1465 |
| | ammonia ($NH_3$) | –78 | –33 |
| | phosphorus(III) chloride ($PCl_3$) | –64 | 76 |
| | methane ($CH_4$) | –183 | –161 |

**Figure 5** *Melting and boiling points of some electrovalent and covalent substances.*

**Covalent substances generally have a smell.** Electrovalent substances generally do not. If you think of smells as being molecules that get up your nose, it is clear that covalent substances can put molecules into the vapour state more easily than electrovalent substances.

**Most electrovalent compounds will dissolve in water.** They will not dissolve in covalent solvents such as petrol, tetrachloromethane, propanone and ethanol.

Most covalent compounds will *not* dissolve in water, but will dissolve in covalent solvents.

The secret lies in the water molecule. Although water is a covalent molecule, it does have small positive and negative charges on it. (See figure 6.)

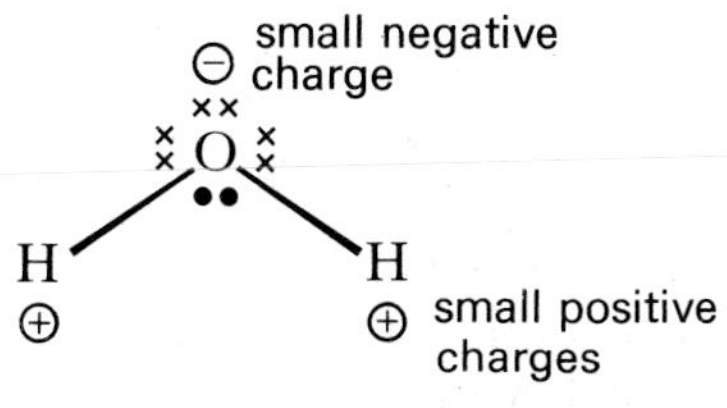


**Figure 6**
*The water molecule.*

These charges enable water molecules to attract individual positive and negative ions and dissolve the electrovalent substance.

Covalent substances have no charges on them that the water molecules can be attracted to, so they do not dissolve. They will sometimes dissolve if the molecules can simply spread out.

**Electrovalent compounds will conduct electricity.** Electricity will flow through liquids if there are ions to carry the current. The ions are charged and will carry the electricity. A molten electrovalent substance contains ions which are free to move about. If two electrodes are placed in the liquid, and are connected to a battery and a bulb, the bulb will light up as the current flows between the electrodes. Figure 7 shows the apparatus that would be used to show this using sodium chloride.

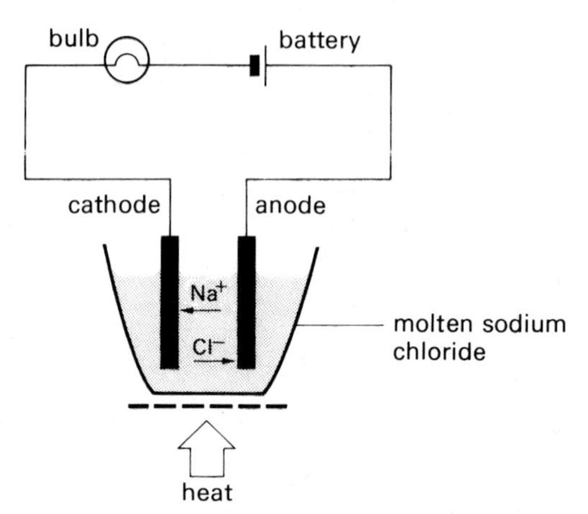


**Figure 7**
*Electricity will flow through molten sodium chloride.*

The positive electrode is called the *anode*, and the negative electrode is called the *cathode*. The positive sodium ions are attracted towards the cathode, and the negative chloride ions are attracted towards the anode. The ions carry the electric current and the bulb lights up. Solid sodium chloride will not conduct an electric current because its ions are held tightly in the crystal lattice, and they can only vibrate. Covalent substances cannot carry an electric current either, because they contain no ions.

## Summary

At the end of this chapter you should be able to:

**1** Explain why the noble gases do not react.

**2** Explain why the other elements do react.

**3** Describe, by means of diagrams, how two elements react to form a compound containing electrovalent bonds.

**4** Describe, by means of diagrams, how two elements form a compound containing covalent bonds.

**5** Work out the ions formed by elements, according to their positions in the periodic table.

**6** Describe the shape of a sodium chloride crystal lattice and explain how it is constructed from its ions.

**7** Describe and explain the differences between electrovalent and covalent compounds.

# X-rays

*Roentgen and his apparatus.*

*Many of you will have had an X-ray at one time or another. It may have been to see if your lungs were all right, or it may have been to see if you had broken a bone. In either case, you will have found that the process of being X-rayed is quick and painless. The following article tells you more about these 'mysterious rays'.*

What about the name, 'X-rays'? X stands for 'unknown'. This name was chosen because even the man who discovered them could not explain what they were. Scientists had been making X-rays for some time before they realized they were doing so. The first man to actually notice them was a German physicist called Wilhelm Röentgen. In November 1895, he was doing a routine experiment with a glass tube containing gas at a very low pressure. When he applied a high voltage to this tube, the gas glowed brightly. This was well known. However, when Röentgen covered the tube with a black cloth so that he could no longer see the glowing gas, some crystals that were on the table near the apparatus continued to glow by themselves. He realized that some sort of invisible ray must have been coming out of the tube, through the cloth, and into the crystals.

**Skeleton appears**

Röentgen tested these rays to see what they would go through. 'I placed my hand before a screen put in the path of the rays. I could see the outline of the bones framed by the flesh. Afterwards, I took a photographic plate, and putting it in the place of the screen, made a photograph of the hands with the bones clearly outlined.'

He called these strange rays *X-rays*. It was soon realized how useful they were for looking inside people. By 1900, doctors were using them to locate bullets in wounded soldiers; shortly after this, dentists in America examined their patients' teeth using them.

Some people were scared of the new rays, and in London, tailors advertised X-ray proof clothes. They weren't any good, because tests have shown that only thick pieces of lead will stop X-rays.

**X-rays to the rescue**

Much use of X-rays is made in modern medicine, for examining bones, vital organs such as the heart, and searching for cancer. X-rays do not go through bones or metal. If they are shone through you onto a photographic film, the bones or any metal objects inside you show up as shadows on the film. If you need to have your stomach X-rayed, you must first drink a suspension of barium sulphate. This substance does not allow X-rays to pass through it and so a shadow of your stomach appears on the film.

X-rays are very important in industry. Oxygen cylinders are X-rayed before they are refilled to see if they have any tiny cracks in them, and they can reveal cracks in metal aeroplane parts. Scientists also use them to find out the internal structures of crystals. When X-rays are shone through a crystalline sub-

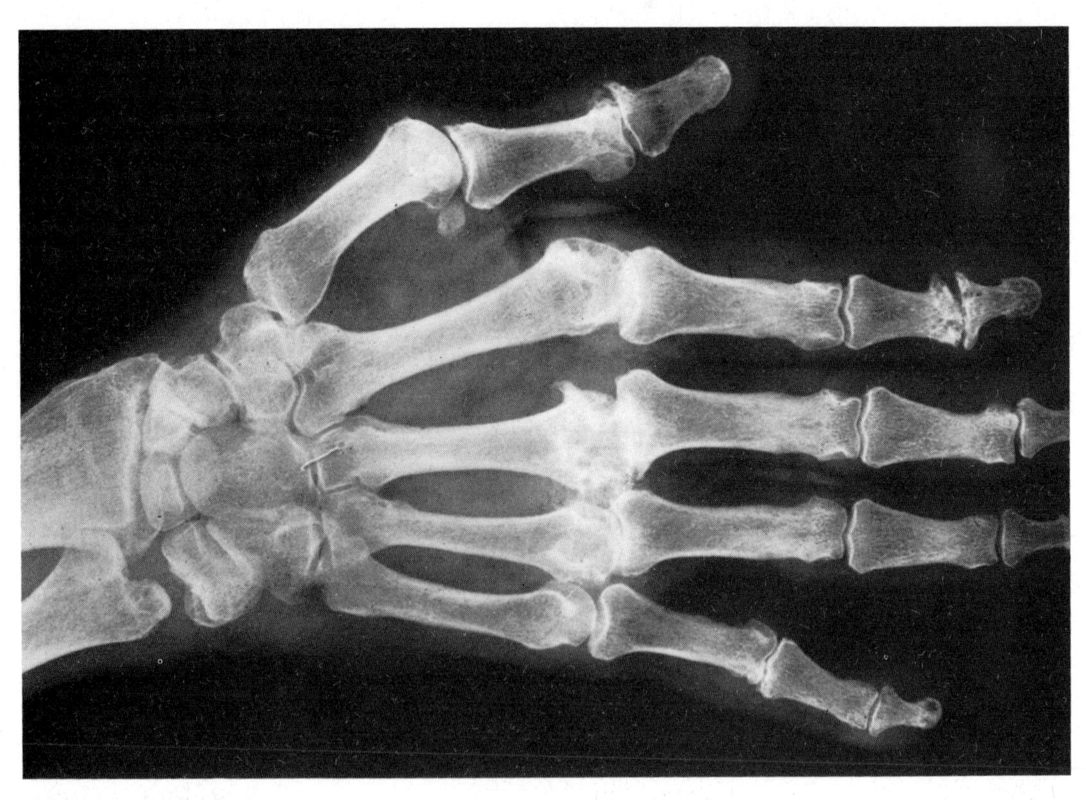

*An X-ray of a hand.*

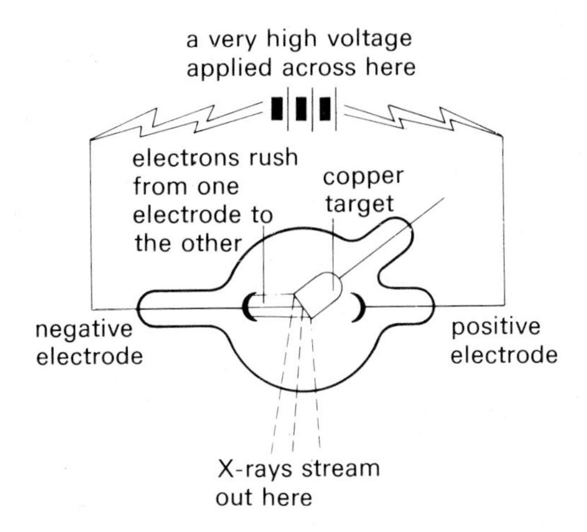


stance, they are 'reflected' by the ions inside. As they emerge, they are photographed and analysed, and this can tell a scientist which substances are present, and how the ions are arranged inside the crystal.

Many substances can be analysed in this way. Forensic scientists may use this technique of 'X-ray crystallography' to identify otherwise useless clues found at the scene of a crime.

### X-rays can be dangerous

X-rays can destroy living tissue. Although it is safe if you have an X-ray now and then, X-ray machine operators have to take care that they are not exposed to the rays too often. They wear lead-lined clothes when they work X-ray machines, and have film badges pinned to their clothes all the time. The badges contain small pieces of photographic film. At the end of each week, the piece of film for each X-ray operator is developed and examined to see how much radiation they have received. If they have received too much, then they must stop working with X-rays for a time.

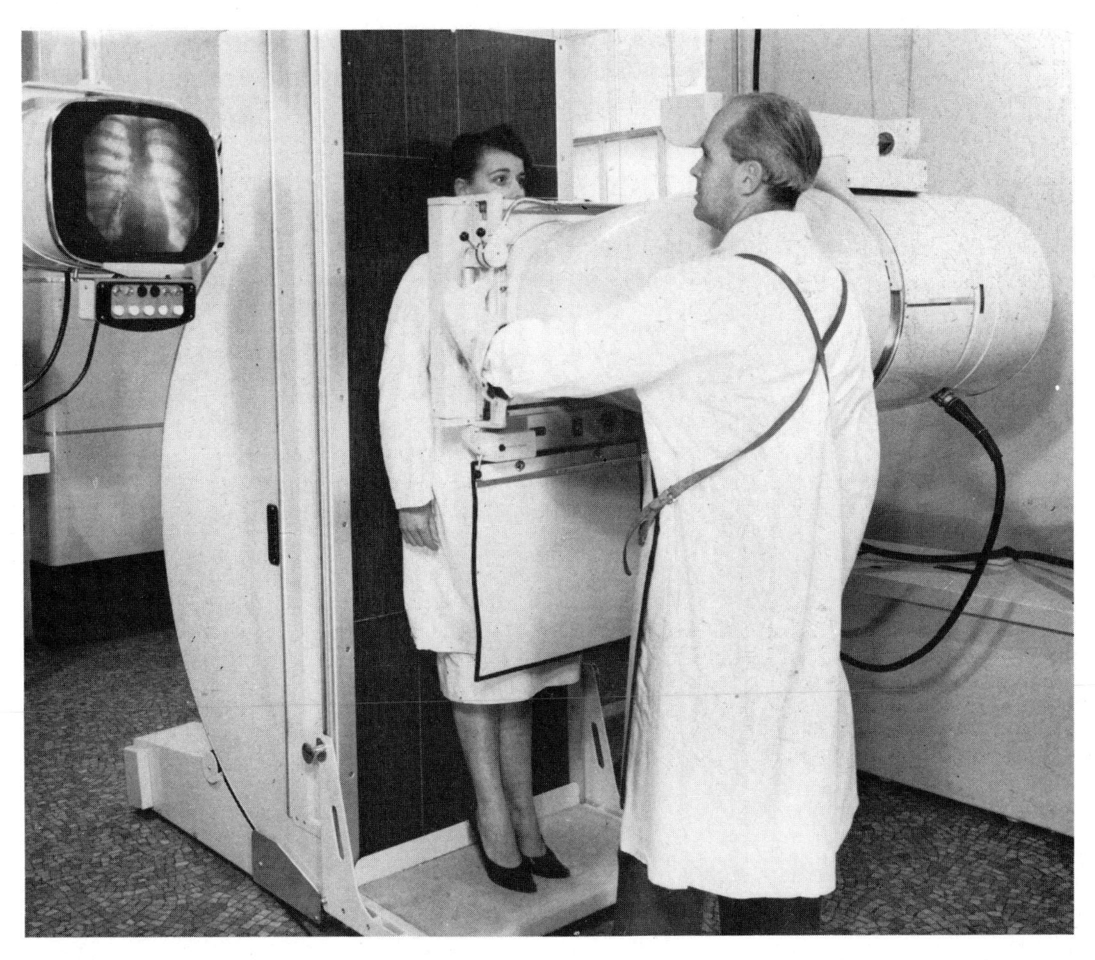

*Modern X-ray equipment.*

## Questions

**1** Use the periodic table in chapter 4 to help you with this question.
**a** Find the element caesium.
**b** What group is it in?
**c** How many outside shell electrons does it have?
**d** When it reacts, how many electrons will it lose or gain in order to get a full outside shell?
**e** Write down the ion it will form, complete with its charge.
**f** Now do the same for the elements bromine, aluminium, magnesium and selenium.

**2** Show, by means of diagrams, how the following electrovalent compounds are formed, and work out their formulae:
sodium oxide, calcium chloride, and aluminium fluoride.

**3** Tetrachloromethane ($CCl_4$), chlorine ($Cl_2$), hydrogen ($H_2$), and water ($H_2O$), are all covalent substances, and contain molecules. Draw diagrams to show how their atoms share electrons to get full outside shells.

**4** Use the periodic table to help you put the following compounds into two lists – one of electrovalent compounds, and the other of covalent compounds.
Potassium oxide, barium sulphide, phosphorus oxide, nitrogen oxide and carbon dioxide.

**5** Substance A has a melting point of 2000 °C. When it is melted, and two electrodes which have been connected to a battery and bulb are put into the liquid, the bulb lights up.
Substance B is a liquid with a strong smell. When it is added to water, it does not dissolve in it, but floats on the surface.

**a** Which substance is likely to have the higher boiling point?
**b** Which substance is likely to exist as a crystal lattice of ions?
**c** If two electrodes, connected to a battery and bulb are put into liquid B, what will happen? Why?
**d** Which substance is electrovalent?
**e** Suggest a possible solvent for substance A.
**f** Suggest a possible solvent for substance B.
**g** Which one contains molecules?

# 6 Formulae and equations

## 6.1 Formulae

You saw in the last chapter that the metal elements (on the left-hand side of the periodic table) and some non-metal elements (from the right-hand side) can form ions. Ionic compounds are formed when these ions fit together in a crystal lattice. Figure 1 is a list of most of the ions that you are likely to meet. Some of them are not single elements, but groups of elements combined to make an ion.
You will use many of these ions as you progress through the book, so you will find it very helpful if you learn their names, formulae and charges.

| **positively charged ions** | | **negatively charged ions** | |
|---|---|---|---|
| $Li^+$ | lithium ion | $OH^-$ | hydroxide ion |
| $Na^+$ | sodium ion | $NO_3^-$ | nitrate ion |
| $K^+$ | potassium ion | $HCO_3^-$ | hydrogen carbonate ion |
| $NH_4^+$ | ammonium ion | $Cl^-$ | chloride ion |
| $H^+$ | hydrogen ion | $Br^-$ | bromide ion |
| $Ag^+$ | silver ion | $I^-$ | iodide ion |
| $Cu^+$ | copper(I) ion | | |
| $Mg^{2+}$ | magnesium ion | $SO_4^{2-}$ | sulphate ion |
| $Ca^{2+}$ | calcium ion | $CO_3^{2-}$ | carbonate ion |
| $Ba^{2+}$ | barium ion | $SO_3^{2-}$ | sulphite ion |
| $Cu^{2+}$ | copper(II) ion | $S^{2-}$ | sulphide ion |
| $Zn^{2+}$ | zinc ion | $O^{2-}$ | oxide ion |
| $Pb^{2+}$ | lead ion | | |
| $Hg^{2+}$ | mercury ion | | |
| $Fe^{2+}$ | iron(II) ion | | |
| $Fe^{3+}$ | iron(III) ion | $PO_4^{3-}$ | phosphate ion |
| $Al^{3+}$ | aluminium ion | | |

**Figure 1**
*Some common ions.*

By putting these ions together, according to certain rules, you can work out the formula of almost any ionic substance you are likely to meet. (Don't worry about covalent substances. You do not need to know many of these, and you can learn them as you go along.) There is a routine for writing a formula. It is best shown using some examples.

**The formula of copper(II) chloride.**

| | | |
|---|---|---|
| **1** Write down the name of the substance. | Copper(II) chloride. | |
| **2** Write down the ions used in the substance. (Don't forget their charges.) | $Cu^{2+}$ | $Cl^-$. |
| **3** Add positive ions or negative ions until the number of positive charges is the same as the number of negative charges. | 2 +charges: | So we need $2Cl^-$'s altogether to give 2 −charges. |
| **4** Write down the ions again, and how many were needed for a balance. | one $Cu^{2+}$ | two $Cl^-$. |
| **5** Write down the formula without the charges in the proper way (There is no need to write in the charges because the ions now balance.) | $Cu\,Cl_2$. | |

(When a number is written on the line it refers to the ion immediately in front of it. The 2 in the $CuCl_2$ means $2Cl^-$ ions only.)

Now look at these other examples.

**The formula of lithium sulphate.**

**1** Lithium sulphate.

**2** $Li^+$ $SO_4^{2-}$.

**3** The 2− charge on the sulphate ion needs two lithium ions to balance it.

**4** Two $Li^+$ and one $SO_4^{2-}$ are needed for a balance.

**5** $Li_2SO_4$. The $_2$ means two lithium ions only. The $_4$ is part of the sulphate ion.

**The formula of ammonium carbonate.**

**1** Ammonium carbonate.

**2** $NH_4^+$ $CO_3^{2-}$.

3 Two ammonium ions are needed to balance the charge on the single carbonate ion.

4 Two $NH_4^+$ ions and one $CO_3^{2-}$ ion are needed for a balance.

5 $(NH_4)_2CO_3$. The brackets are put round the ammonium ion to show that two whole ammonium ions are used. If the brackets were not used we would have $NH_{42}$, and that would be nonsense. Brackets are used when the formula needs two or more ions which are made up from more than one element.

**The formula of zinc nitrate.**

1 Zinc nitrate.

2 $Zn^{2+}$ $NO_3^-$.

3 The double positive charge on the zinc ion needs two nitrate ions to balance it.

4 One $Zn^{2+}$ and two $NO_3^-$ ions are needed for a balance.

5 $Zn(NO_3)_2$. Again, brackets are needed to show that there are two whole nitrate ions.

**The formula of aluminium oxide.**

1 Aluminium oxide.

2 $Al^{3+}$ $O^{2-}$.

3 Now, the simplest way this can be balanced is to have two aluminium ions and three oxide ions. Then there will be six positive charges to balance six negative charges.

4 Two $Al^{3+}$ ions and three $O^{2-}$ ions are needed.

5 $Al_2O_3$. There is no need for brackets around the Al, because the two letters represent only one element.

**The formula of iron(III) sulphate.**

1 Iron(III) sulphate.

2 $Fe^{3+}$ $SO_4^{2-}$

3 Take two iron(III) ions and three sulphate ions to make the positive and negative charges total the same amount.

4 Two $Fe^{3+}$ ions and three $SO_4^{2-}$ ions are needed.

5 $Fe_2(SO_4)_3$. The brackets are needed because the sulphate ion is made up of more than one element.

Now try more examples from the questions at the end of the chapter.

## 6.2 Equations

An equation is a chemical sentence. It tells you what is going on in a chemical reaction. It tells you both what is reacting and what is being formed, and also about the quantity of the substances involved.

$$\text{REACTANTS} \xrightarrow[\text{(e.g. temperature)}]{\text{conditions}} \text{PRODUCTS}$$

Again, there is a routine for writing equations.

**Equation for iron-sulphur reaction.**

**1** Write down the reaction in words.

Iron reacts with sulphur to form iron(II) sulphide.

Iron + sulphur $\longrightarrow$ iron(II) sulphide.

**2** Change the words into the correct formulae.

$Fe + S \longrightarrow FeS$.

(check that these are the correct symbols and formulae)

**3** Balance the numbers of atoms of all elements on each side.

| *Left hand side:* | *Right hand side:* |
|---|---|
| one atom Fe, | one atom Fe, |
| one atom S. | one atom S. |

The equation is balanced.

**4** Write down the equation with the numbers of atoms balanced.

$Fe + S \longrightarrow FeS$.

**5** After each reactant and product, write in its physical state.

$Fe(s) + S(s) \longrightarrow FeS(s)$.

(s) in this case stands for solid. If one of the substances was a liquid, you would put (l) after it, and if it was a gas, you would use (g). If the substance was dissolved in water, you would use (aq), standing for aqueous.

**Equation for copper oxide-hydrogen reaction.**

**1** Copper(II) oxide reacts with hydrogen to give copper and steam.

Copper(II) oxide + hydrogen $\longrightarrow$ copper + steam.

**2** $CuO + H_2 \longrightarrow Cu + H_2O$.

Check that these are the correct formulae. Hydrogen has to be $H_2$ because a molecule of hydrogen gas contains two atoms.

**3** Are there the same number of atoms of each element on each side of the equation? There are, so the equation is balanced.

**4** $CuO + H_2 \longrightarrow Cu + H_2O$.

**5** $CuO(s) + H_2(g) \longrightarrow Cu(s) + H_2O(g)$.

**Equation for zinc-hydrochloric acid reaction.**

**1** Zinc dissolves in dilute hydrochloric acid to form zinc chloride and hydrogen.

Zinc + hydrochloric acid (dilute) ⟶ zinc chloride + hydrogen.

**2** $Zn + HCl \longrightarrow ZnCl_2 + H_2$.
The symbols and formulae are correct.

**3** But the number of atoms of each element on each side of the equation is different. There are two chlorine atoms and two hydrogen atoms on the right hand side, but only one of each on the left hand side. The numbers must be equal, so double the number of chlorine atoms and hydrogen atoms on the left hand side. You can't do this by writing $HCl_2$, because this would change the formula of the compound. Instead you must put a 2 in front of the formula: 2HCl. This only changes the number of whole molecules, and that's allowed. The equation now balances.

**4** $Zn + 2HCl \longrightarrow ZnCl_2 + H_2$.

**5** $Zn(s) + 2HCl(aq) \longrightarrow ZnCl_2(aq) + H_2(g)$.
Remember, when you are balancing equations you can only put numbers in front of formulae.

**Equation for aluminium chloride-sodium hydroxide reaction.**

**1** Aluminium chloride reacts with sodium hydroxide to form aluminium hydroxide and sodium chloride.

Aluminium chloride + sodium hydroxide ⟶ aluminium hydroxide + sodium chloride.

**2** $AlCl_3 + NaOH \longrightarrow Al(OH)_3 + NaCl$.
The symbols and formulae are correct.

**3** There are three (OH)'s in the aluminium hydroxide, so there must have been three on the left hand side, too. This means that there must be three Na's, for the formula of sodium hydroxide to be correct. These three Na's react with the three Cl's from the aluminium chloride, so 3 NaCl will be formed, to give:

**4** $AlCl_3 + 3NaOH \longrightarrow Al(OH)_3 + 3NaCl$.

**5** $AlCl_3(s) + 3NaOH(aq) \longrightarrow Al(OH)_3(s) + 3NaCl(aq)$.

Now try the examples in the questions at the end of the chapter.

## 6.3 Relative molecular mass

You saw in chapter 5 that elements can have a relative atomic mass. Similarly a compound can have a relative molecular mass. Just as with relative atomic mass, relative molecular mass is a number, without units, which compares the mass of one molecule of a compound with the mass of one molecule of other compounds. You can calculate it for any compound by adding up the relative atomic masses of the elements in the compound. There is a routine for this too.

### Relative molecular mass of water.

| | | |
|---|---|---|
| **1** Write down the name of the compound. | Water. | |
| **2** Turn the name into a formula. | $H_2O$. | |
| **3** Write down how many atoms of each element are present in one molecule. | 2 atoms of hydrogen; | 1 atom of oxygen |
| **4** Look up the relative atomic masses of each of the elements in the compound. | H = 1 | O = 16 |
| **5** Use the information in 3 and 4 to find the total mass of each element present. Add them all together: | | |

for the H: $2 \times 1 = 2$
for the O: $1 \times 16 = 16+$
rel.mol.mass $= \underline{18}$

So the relative molecular mass of water is 18.

### Relative molecular mass of iron(II) sulphide.

**1** Iron(II) sulphide.

**2** FeS.

**3** One atom of iron; one atom of sulphur.

**4** Fe = 56 S = 32

**5** Fe: $1 \times 56 = 56$
S: $1 \times 32 = 32+$
So FeS $= \underline{88}$

**Relative molecular mass of copper(II) oxide.**

**1** Copper(II) oxide.

**2** CuO.

**3** One atom of copper; one atom of oxygen.

**4** Cu = 64 O = 16

**5** Cu: $1 \times 64 = 64$
O: $1 \times 16 = 16+$
So CuO $= \underline{80}$

**Relative molecular mass of ammonium chloride.**

**1** Ammonium chloride.

**2** $NH_4Cl$.

**3** One atom of nitrogen, four atoms of hydrogen and one atom of chlorine.

**4** N = 14
H = 1
Cl = 35·5

**5** N: $1 \times 14 = 14$
H: $4 \times 1 = 4$
Cl: $1 \times 35{\cdot}5 = 35{\cdot}5+$
So $NH_4Cl = \underline{53{\cdot}5}$

**Relative molecular mass of aluminium sulphate.**

**1** Aluminium sulphate.

**2** $Al_2(SO_4)_3$.

**3** Two atoms of aluminium. The $_3$ refers to everything inside the bracket. So there are 3 sulphur atoms. In the case of oxygen, it is already $O_4$, so the $_3$ outside the brackets multiplies 4 by 3 to give 12. There are 12 oxygen atoms.

**4** Al = 27
S = 32
O = 16

**5** Al: $2 \times 27 = 54$
S: $3 \times 32 = 96$
O: $12 \times 16 = 192+$
So $Al_2(SO_4)_3 = \underline{342}$

## 6.4 The mole

You already know that the relative atomic mass of the element carbon is 12. If in the laboratory you weigh out twelve *grams* of carbon, then you will have weighed out an amount called *one mole* of carbon. In the same way, if you weigh out sixty-four grams of copper and 56 grams of iron filings, you have weighed out one mole of copper, and one mole of iron.

Here are some more examples:

1 mole of sodium = 23 g. 1 mole of hydrogen gas ($H_2$) = 2 g.

1 mole of calcium = 40 g. 1 mole of chlorine gas ($Cl_2$) = 71 g.

But although different elements have moles of different masses, one mole of *any* substance contains the same number of particles. A scientist called Avogadro was the first person to suggest this. The actual number of particles present in one mole of any substance is 602 000 000 000 000 000 000 000 or $6{\cdot}02 \times 10^{23}$.

The same thing applies to compounds. If you look back to the example on calculating relative molecular mass, you will see the relative molecular mass of water is 18. 18 grams of water make one mole of water. In 18 grams of water there are $6{\cdot}02 \times 10^{23}$ water molecules. Here are some examples of how to calculate how many grams are needed for one mole of different compounds:

**1** The relative molecular mass of calcium oxide (CaO) is

Ca: 40
O: 16+
56

**So 1 mole of calcium oxide is 56 g.**

**2** The relative molecular mass of zinc nitrate ($Zn(NO_3)_2$) is

Zn: 65
N: $2 \times 14$ = 28
O: $6 \times 16$ = 96+
189

**So 1 mole of zinc nitrate is 189 g.**

We can also turn moles into grams, and see what mass of chemicals actually takes part in a chemical reaction. Here are two examples:

**The reaction of copper(II) oxide with hydrogen.**

$$CuO(s) + H_2(g) \longrightarrow Cu(s) + H_2O(g)$$

means 1 mole of CuO + 1 mole of $H_2$ ⟶ 1 mole of Cu + 1 mole of $H_2O$

means $(64+16)g + (2 \times 1)g \longrightarrow 64\,g + (2 \times 1 + 16)g$

$80\,g$ of CuO + $2\,g$ of $H_2$ ⟶ $64\,g$ of Cu + $18\,g$ of $H_2O$.

**The reaction of calcium carbonate with hydrochloric acid.**

$$CaCO_3(s) + 2HCl(aq) \longrightarrow CaCl_2(aq) + CO_2(g) + H_2O(l)$$

means:

$$\begin{matrix}\text{1 mole} \\ \text{of } CaCO_3\end{matrix} + \begin{matrix}\text{2 moles} \\ \text{of HCl}\end{matrix} \longrightarrow \begin{matrix}\text{1 mole} \\ \text{of } CaCl_2\end{matrix} + \begin{matrix}\text{1 mole} \\ \text{of } CO_2\end{matrix} + \begin{matrix}\text{1 mole} \\ \text{of } H_2O\end{matrix}$$

means:

$$(40+12+48)\text{g} + 2(1+35{\cdot}5)\text{g} \longrightarrow (40+71)\text{g} + (12+32)\text{g} + (2+16)\text{g}$$

$$\begin{matrix}\text{100 g} \\ \text{of } CaCO_3\end{matrix} + \begin{matrix}\text{73 g} \\ \text{of HCl}\end{matrix} \longrightarrow \begin{matrix}\text{111 g} \\ \text{of } CaCl_2\end{matrix} + \begin{matrix}\text{44 g} \\ \text{of } CO_2\end{matrix} + \begin{matrix}\text{18 g} \\ \text{of } H_2O.\end{matrix}$$

We can now use this to work out problems.

**Problem 1**

How much iron(II) sulphide can be made by heating 64 g of sulphur with excess iron filings?
Use this routine:

**1** Write the balanced reaction:

$$Fe(s) + S(s) \longrightarrow FeS(s).$$

**2** Write down the moles involved:

$$\begin{matrix}\text{1 mole} \\ \text{of Fe}\end{matrix} + \begin{matrix}\text{1 mole} \\ \text{of S}\end{matrix} \longrightarrow \begin{matrix}\text{1 mole} \\ \text{of FeS.}\end{matrix}$$

**3** Pick out the substances that concern the question:

The question concerns only the sulphur and the iron(II) sulphide.

$$\begin{matrix}\text{1 mole} \\ \text{of S}\end{matrix} \longrightarrow \begin{matrix}\text{1 mole} \\ \text{of FeS.}\end{matrix}$$

(there is as much Fe as the reaction wants)

**4** Convert the moles to grams:

1 mole of S $\longrightarrow$ 1 mole of FeS.
32 grams S $\longrightarrow$ (56 + 32) grams FeS.
32 grams $\longrightarrow$ 88 grams.

**5** Introduce the given facts:

if 32 grams of S $\longrightarrow$ 88 grams of FeS,
then 1 gram of S $\longrightarrow$ $\frac{88}{32}$ grams of FeS.

So 64 grams of S $\longrightarrow$ $\frac{88 \times 64}{32}$ grams of FeS = 176 grams.

**Answer:** 64 grams of sulphur will give 176 grams of iron(II) sulphide.

That problem involved simple proportion. If you did not follow it you had better read the next section. If you had no difficulty, miss it out.

**Elephants and Mars bars** – or How to succeed in mathematics without really trying.
5 elephants have 10 Mars bars. How many Mars bars does each elephant have? The answer is 2. Easy, but how did you do it? You probably divided the 10 by 5 in your head.

Now, let's make it more difficult.
Suppose you have 5 elephants and only 9 Mars bars (and a knife). How many do they each have now? Do the same as before. Divide the 9 by the 5 and you get $\frac{9}{5}$.

Now try this one.
10 elephants have 5 Mars bars. How many do they each get? Do the same. Share the Mars bars between the elephants and you get $\frac{5}{10} = \frac{1}{2}$. What happens if you have 9 elephants and 5 Mars bars?

What is the answer now? Did you get $\frac{5}{9}$?

**Problem 2**

**How much copper(II) oxide is needed to react with dilute sulphuric to produce 40 g of copper(II) sulphate?**

**1** $CuO(s) + H_2SO_4(aq) \longrightarrow CuSO_4(aq) + H_2O(l)$.

**2** 1 mole of CuO + 1 mole of $H_2SO_4$ $\longrightarrow$ 1 mole of $CuSO_4$ + 1 mole of $H_2O$.

**3** The question concerns only the copper oxide and the copper sulphate.

1 mole of CuO $\longrightarrow$ 1 mole of $CuSO_4$.

**4** $(64+16)$ g $\longrightarrow$ $(64+32+64)$ g
80 g of CuO $\longrightarrow$ 160 g of $CuSO_4$. (80 Mars bars. 160 elephants. How much for one elephant?)

**5** $\frac{80}{160}$ g of CuO $\longrightarrow$ 1 g of $CuSO_4$.

$\frac{80}{160} \times 40$ g of CuO $\longrightarrow$ 40 g of $CuSO_4$.

So: 20 g of CuO $\longrightarrow$ 40 g of $CuSO_4$.

**Answer:** 20 g of copper(II) oxide are needed.

### Problem 3

**How much magnesium is needed to make 55 g of magnesium oxide?**

1 $2Mg(s) + O_2(g) \longrightarrow 2MgO(s)$.

2 2 moles of Mg + 1 mole of $O_2$ $\longrightarrow$ 2 moles of MgO.

3 The question concerns only the magnesium and the magnesium oxide.

2 moles of Mg $\longrightarrow$ 2 moles of MgO.

1 mole of Mg $\longrightarrow$ 1 mole of MgO.

4 24 g of Mg $\longrightarrow$ 40 g of MgO.

5 $\frac{24}{40}$ g of Mg $\longrightarrow$ 1 g of MgO.

So: $\frac{24 \times 55}{40}$ g of Mg $\longrightarrow$ 55 g of MgO

33 g of Mg $\longrightarrow$ 55 g of MgO.

**Answer:** 33 g of magnesium are needed.

### Problem 4

**What mass of ammonia will be formed when 12 g of hydrogen is reacted with nitrogen?**

1 $N_2(g) + 3H_2(g) \longrightarrow 2NH_3(g)$.

2 1 mole of $N_2$ + 3 moles of $H_2$ $\longrightarrow$ 2 moles of $NH_3$.

3 The question concerns only the hydrogen and the ammonia:

3 moles of $H_2$ $\longrightarrow$ 2 moles of $NH_3$.

4 6 g $\longrightarrow$ 34 g.

5 1 g $\longrightarrow$ $\frac{34}{6}$ g.

So: 12 g $\longrightarrow$ $\frac{34 \times 12}{6}$ g.

12 g $\longrightarrow$ 68g.

**Answer:** 68 g of ammonia are produced.

Now try the questions at the end of the chapter.

## 6.5 Empirical formula

The formula of a compound tells you how many moles of each element would combine together to make one mole of the compound. For example, $H_2O$ means two moles of hydrogen atoms combines with one mole of oxygen atoms to make 1 mole of water. $NH_3$ means one mole of nitrogen atoms combines with 3 moles of hydrogen atoms to make 1 mole of ammonia. But you can also use this information the other way round. If you find out in an experiment that 1 mole of nitrogen atoms combines with 3 moles of hydrogen atoms, then the formula of ammonia must be $NH_3$. Experimental results can be used to work out the empirical formula of a compound:

### 1. Empirical formula of calcium oxide.

40 g of calcium is found to react with 16 g of oxygen to form calcium oxide.

This means that 1 mole of calcium has reacted with 1 mole of oxygen, because 1 mole of calcium is 40 g and 1 mole of oxygen is 16 g. We can write down this information as $Ca_{1\ mole}O_{1\ mole}$: in other words, the formula must be $Ca_1O_1$ or CaO. The *empirical formula* of calcium oxide is CaO. 'Empirical' formula just means 'simplest whole number' formula found by doing an experiment.

### 2. Empirical formula of iron sulphide.

28 g of iron combined with 16 g of sulphur to make a compound. What was the empirical formula of the compound?

28 g Fe $= \frac{28}{56}$ mole $= \frac{1}{2}$ mole because there are 56 g of Fe in 1 mole.

16 g S $= \frac{16}{32}$ mole $= \frac{1}{2}$ mole because there are 32 g of S in 1 mole.

$\frac{1}{2}$ mole Fe combined with $\frac{1}{2}$ mole S. So 1 mole Fe combines with 1 mole S. The empirical formula is $Fe_{1\ mole}S_{1\ mole}$ or FeS.

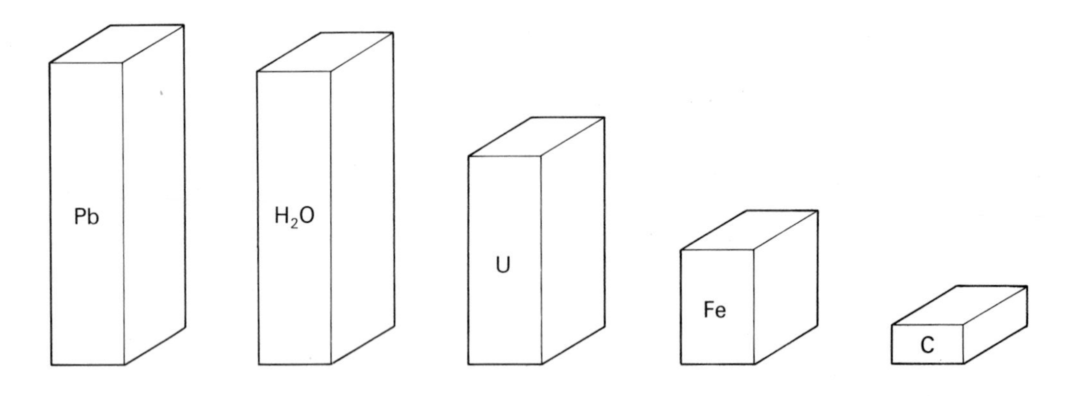


**Figure 2** *The relative sizes of one mole of different substances.*

**3. Empirical formula of lead oxide.**

1242 g of lead combined to make an oxide whose mass was 1370 g. What was the empirical formula of the oxide?

Mass of lead = 1242 g = $\frac{1242}{207}$ moles = 6 moles.

Mass of oxygen combined with the lead = 1370 g − 1242 g = 128 g of oxygen

128 g of oxygen = $\frac{128}{16}$ moles = 8 moles.

Therefore the empirical formula is $Pb_6O_8$, or in simplest whole numbers, $Pb_3O_4$.

Now try the question at the end of the chapter.

**Summary**

At the end of this chapter, you should be able to:

**1** Write the formula for most common ionic compounds.

**2** Write a balanced equation for a chemical reaction.

**3** Work out the relative molecular mass of a compound.

**4** Explain the meaning of the term mole, and write down the number of grams in a mole of any element or compound.

**5** Calculate the amount of substances that combine in a reaction by using equations.

**6** Share out Mars bars between elephants.

**7** Work out empirical formulae from data about reacting masses of elements.

# The man who made diamonds

*This chapter has been about formulae and equations: how to change one set of chemicals into another. Scientists now have a clear idea of which changes are possible and which are not. This hasn't always been the case . . .*

For thousands of years alchemists set about the impossible task of trying to turn lead into gold. In 1880, the Scottish inventor, James Valentine Hannay, set about a task which he knew was possible – to turn carbon, the main chemical in soot, into diamonds.

Hannay was not the first man to try to make his own diamonds. One experimenter who claimed success in 1828 was found to have made nothing more than charcoal, mixed with hard crystalline slag. Another enthusiast thought that he could make diamonds by keeping a solution of phosphorus in carbon disulphide under a layer of water. But nothing happened. Another scientist who experimented with electrodes produced a crystalline dust which could scratch rubies; it proved to be graphite.

Various experimenters tried all sorts of different carbon-based mixtures for their experiments. One experiment included a helping of treacle; Hannay's experiment was no exception. When he announced that he had become the first man in the world to create diamonds, he had to admit that the starting mixture had consisted of paraffin, lithium and bone oil: a stew that has been compared with 'Eye of newt and toe of frog'.

**Explosive announcement**

Hannay's equipment for turning this mixture into diamond was a powerful furnace into which he put extremely strong iron capsules containing the mixture. It was in the year 1881, after a series of disconcerting explosions, that Hannay announced he had made diamonds.

He used 50-cm long tubes of wrought iron for his experiments. After filling them with the mixture he was trying, he cooled one end of each tube in ice while he made the other end white hot, so that he could hammer it closed. He then heated the tubes to temperatures up to 900 °C for 14 hours. After cooling, he bored them open to examine the contents. Nine out of ten times the tubes exploded, and in only three out of 80 experiments did anything resembling a diamond appear. When he opened the tube on these occasions, a great volume of gas was given off. After pouring off the liquid from the tube, Hannay chiselled away the slimy black residue. When he broke this up in a mortar, several tiny crystalline and mostly transparent particles remained at the bottom of the tube. These he dissolved out using acid. They were his diamonds . . . colourless, with the unmistakable lustre and appearance of natural diamonds.

**Diamonds tested**

Hannay did not expect the world to take his experiments at face value. He sent 12 specimens to the Professor of mineralogy at Oxford University, who was the leading authority on diamonds in Britain. Hannay's crystalline particles, wrote the Professor, looked exactly like diamond. With them he had scored deep grooves in the polished surface of a sapphire. When set alight on platinum foil, one of them had glowed and gradually disappeared exactly as the mineral diamond would do.

Not everyone was impressed however. The Royal Society declined to publish a paper that Hannay had written about his experiments, and in 1902, the Encyclopaedia Britannica said that he had not produced diamonds, but carborundum, a very hard compound of sand and coke.

Hannay died in 1931, at the age of 76, and his work was virtually forgotten until 1943. During that year, somebody at the British Museum discovered that they still had the Hannay diamonds, mounted on cotton wool in a matchbox. The crystals were X-rayed and the verdict was that they were, beyond all doubt, diamonds. In 1950, Reginald Jones, a professor at Glasgow University repeated Hannay's experiments, with much more up-to-date equipment, but out of 20 separate runs, he succeeded in producing only one speck of 'promising material', and this one was doubtful, because real diamond dust was later found in a cabinet in the room where the experiment was carried out.

**Diamonds are fakes, naturally**

Finally, in the summer of 1975, the sparks of the Hannay controversy began flying again. Dr. Alan Collins, a lecturer at London University, announced that he had subjected nine of Hannay's original diamonds to a very special testing process. They were bombarded with a steam of very fast moving electrons, and the light that emerged was studied. This method could distinguish between real and synthetic diamonds by revealing their internal structures. The Hannay diamonds proved to be natural and not manufactured, so the British Museum had been hoaxed. It looked as though Hannay had sent real diamonds to the Museum, instead of his own crystalline particles.

Dr. Collins, however was reluctant to believe that Hannay had cheated. He preferred to think that the man had either been careless, and allowed real diamonds that he was studying to get into the experiment, or that some of Hannay's workers, tired of the explosions and failures in the laboratory, planted the real diamonds in an attempt to get the enthusiastic scientist to stop his investigations.

Diamonds *are* made nowadays, but much higher temperatures and pressures are used.

*The General Electric Company's diamond making machine.*

## Questions

A table of relative atomic masses will be found at the back of the book.

**1** Work out the formulae of the following compounds:

lithium hydroxide
iron(III) chloride
calcium iodide
ammonium phosphate
silver nitrate
lead(II) nitrate
iron(II) chloride
iron(II) hydroxide
sodium sulphite
sodium sulphate
magnesium nitrate
sodium hydrogen carbonate
aluminium oxide
calcium hydroxide
zinc hydrogen sulphate
mercury(II) sulphate
sodium hydroxide
ammonium hydroxide
iron(II) sulphide
calcium phosphate

**2** Write balanced equations for the following chemical reactions.

**a** sodium + oxygen ⟶ sodium oxide.

**b** magnesium + sulphuric acid ($H_2SO_4$) ⟶ hydrogen + magnesium sulphate.

**c** hydrogen + oxygen ⟶ steam.

**d** sodium hydroxide + copper(II) sulphate ⟶ sodium sulphate + copper(II) hydroxide.

**e** iron(III) chloride + ammonium hydroxide ⟶ iron(III) hydroxide + ammonium chloride.

**f** zinc + hydrochloric acid (HCl) ⟶ zinc chloride + hydrogen.

**g** calcium carbonate + sulphuric acid ⟶ calcium sulphate + carbon dioxide + water.

**h** copper(II) oxide + sulphuric acid ⟶ copper(II) sulphate + water.

**i** sodium hydroxide + sulphuric acid ⟶ sodium sulphate + water.

**j** iron + chlorine ⟶ iron(III) chloride.

**k** barium chloride + sodium sulphate ⟶ barium sulphate + sodium chloride.

**l** zinc + copper(II) sulphate ⟶ copper + zinc sulphate.

**m** hydrogen + lead(II) oxide ⟶ lead + steam.

**n** calcium + oxygen ⟶ calcium oxide.

**o** magnesium + steam ⟶ hydrogen + magnesium oxide.

**p** copper + oxygen ⟶ copper(II) oxide.

**q** carbon + oxygen ⟶ carbon dioxide.

**r** hydrogen + chlorine ⟶ hydrogen chloride.

**s** copper(II) carbonate + hydrochloric acid ⟶ copper(II) chloride + carbon dioxide + water.

**t** sodium hydroxide + phosphoric acid ($H_3PO_4$) ⟶ sodium phosphate + water.

**3** Work out the relative molecular masses of the following compounds.

sodium sulphate
aluminium chloride
potassium iodide
iron(III) hydroxide
lithium phosphate
calcium nitrate
ammonium sulphate
copper(II) nitrate
calcium hydrogencarbonate
sodium hydrogensulphate

**4** Work out the number of grams contained in:

**a** one mole of: sodium, aluminium, potassium, oxygen gas, sulphur.

**b** two moles of: zinc, iron, magnesium, manganese, silicon.

**c** 0·5 mole of carbon.

**d** 3 moles of boron.

**e** 0·125 mole of sulphur.

**f** 1 mole of hydrogen gas.

**g** 0·1 mole of phosphorus.

**h** 5 moles of zinc.

**i** 10 moles of lithium.

**j** 6 moles of nitrogen gas.

**5** Calculate:

**a** the number of moles in 32 g of sulphur, 56 g of iron, 127 g of iodine.

**b** The number of moles in 8 g of helium, 108 g of aluminium, 414 g of lead, 40 g of bromine.

**6** Calculate the mass of:

**a** one mole of water.

**b** 2 moles of carbon dioxide.

**c** 0·5 mole of calcium carbonate.

**d** 5 moles of sodium sulphate.

**e** 1·5 mole of aluminium oxide.

**7** How much calcium carbonate will have to react with dilute hydrochloric acid to produce 22 g of carbon dioxide?

**8** 446 g of lead(II) oxide is heated in a stream of hydrogen until it has all changed to lead. The steam that forms is condensed. What will its mass be?

**9** 56 g of iron was found to react exactly with 106·5 g of chlorine. What was the empirical formula of the iron chloride that was formed?

**10** When 97·5 g of zinc was heated in oxygen, 121·5 g of zinc oxide was formed. Calculate the empirical formula of the oxide.

# 7 Chemical reactions

## 7.1 Recognising chemical reactions

Chemical reactions are happening all the time, all around us – and even inside us. Cars burn petrol, natural gas is burnt in cookers to turn eggs hard, plants grow – and inside us, food is digested.

Chemical reactions happen when atoms and molecules re-arrange themselves to form new compounds. Magnesium combines with oxygen to burn with a brilliant white flame, leaving a white ash; green plants (like grass and trees) take in carbon dioxide and water combining them to make – more plant, and oxygen.

All chemical reactions have certain characteristics in common. Using some of the reactions that happen in the laboratory, we can see each of the characteristics separately.

**Chemical reactions cause a change of appearance.** The products often have a completely different appearance to that of the reacting substances. For example:

| ammonium dichromate | $\xrightarrow{\text{heat}}$ | chromium(III) oxide | + | nitrogen | + | steam |
|---|---|---|---|---|---|---|
| $(NH_4)_2Cr_2O_7(s)$ | $\longrightarrow$ | $Cr_2O_3(s)$ | + | $N_2(g)$ | + | $4H_2O(g)$. |
| orange crystals | | green powder | | colourless gas | | colourless gas |

**Chemical reactions are difficult to reverse.** Sometimes, they cannot be reversed at all. It is often impossible to get back the things you started with. For example:

$$\text{magnesium} + \text{oxygen} \longrightarrow \text{magnesium oxide.}$$

$$2Mg(s) + O_2(g) \longrightarrow \underset{\text{white ash}}{2MgO(s)}.$$

It is extremely difficult to turn this white ash back to a silver-coloured metal and a colourless gas without using many complicated chemical reactions.

**Chemical reactions involve energy changes.** This may be heat taken in or given out during the reaction, or it may be sound, light, mechanical, or even electrical energy. For example: when ammonium carbonate reacts with ethanoic acid, the reaction gets very cold. This is because heat is being taken in by the molecules from their surroundings in order for the reaction to take place:

$$\text{ammonium carbonate} + \text{ethanoic acid} \xrightarrow[\text{in}]{\text{heat}} \text{ammonium ethanoate} + \text{carbon dioxide} + \text{water.}$$

Reactions which take in heat are called *endothermic* reactions. Most reactions, however, give out heat. For example: when hydrochloric acid is added to sodium hydroxide, the reaction gets very hot. This is because the molecules give out energy as they rearrange:

$$\text{hydrochloric acid (dilute)} + \text{sodium hydroxide} \longrightarrow \text{sodium chloride} + \text{water} + \text{heat.}$$

$$HCl(aq) + NaOH(aq) \longrightarrow NaCl(aq) + H_2O(l)$$

Reactions of this sort are called *exothermic* reactions.
Fuels give out both heat energy and light energy when they react. For example:

$$\text{methane} + \text{oxygen} \longrightarrow \text{carbon dioxide} + \text{steam} \quad + \text{heat and light.}$$

$$CH_4(g) + 2O_2(g) \longrightarrow CO_2(g) \quad + \quad 2H_2O(g)$$

Reactions like this one which produce gases can do mechanical work if the pressure of the gas is used to push against something. When marble chips react with an acid, carbon dioxide gas is produced. This can be used to push back the plunger of a syringe. (See figure 1.)

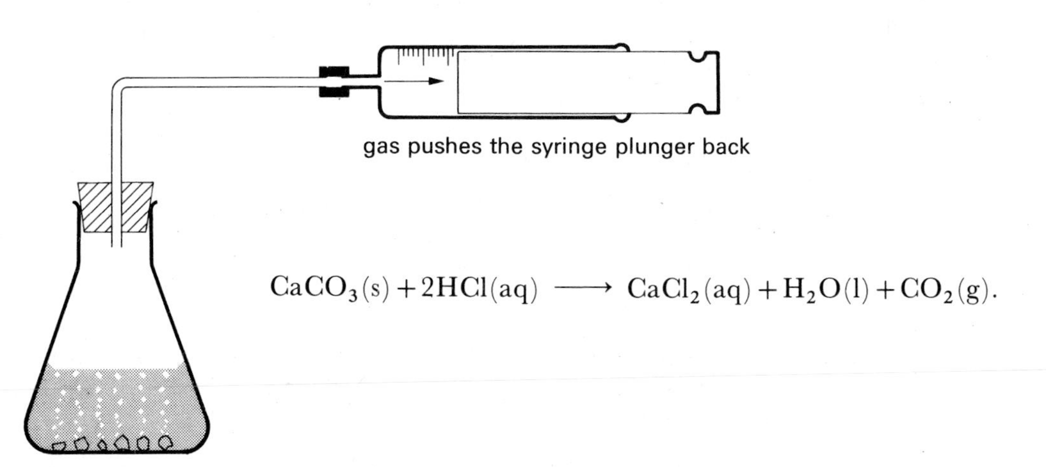


$$CaCO_3(s) + 2HCl(aq) \longrightarrow CaCl_2(aq) + H_2O(l) + CO_2(g).$$

**Figure 1**
*A gas can do work.*

Explosive fuels react to produce a blast which contains both mechanical, and sound energy.

Finally, one very important type of energy that can come from a chemical reaction is electricity. The battery that you put into a torch or radio is nothing more than a chemical reaction in a case. You can find a lot more about chemical electricity in chapter 17.

## 7.2 Physical changes

Things that happen in test tubes are not always chemical reactions. Changes may happen that take in or give out heat, and some may produce mechanical energy, but they do not involve the formation of a new chemical substance. We call these *physical changes.*

The most common physical changes are melting and freezing, boiling and condensing. For example, if you heat a block of ice, it first melts to form water, and then boils to form steam. These changes involve heat, but no new substance is formed, because ice, water and steam are all the same chemical, in a different form. If you allow the steam to cool, it will change back into water, and if you cool this, it will change back into ice. (See figure 2.)

**Figure 2**
*Physical changes.*

HEAT IN $\longrightarrow$

$$\text{ice(s)} \underset{\text{freezing}}{\overset{\text{melting}}{\rightleftharpoons}} \text{water(l)} \underset{\text{condensing}}{\overset{\text{boiling}}{\rightleftharpoons}} \text{steam(g)}$$

$\longleftarrow$ HEAT OUT

When things dissolve, this is a physical change. Unlike a chemical change, the process is easily reversed by evaporation.

## 7.3 Different types of chemical reaction

For convenience, we often give certain types of chemical reactions special names.

**1. Decomposition reactions.** These are reactions in which a compound breaks down into simpler compounds, or even into elements.

For example: when lead nitrate crystals are heated they decompose noisily, giving off brown nitrogen(IV) oxide gas and oxygen gas, leaving behind yellow lead oxide.

$$2Pb(NO_3)_2(s) \xrightarrow{\text{heat}} 2PbO(s) + 4NO_2(g) + O_2(g).$$

**2. Combination or synthesis reactions.** These reactions are the opposite of decomposition reactions. Elements combine to form a compound.
For example: ammonia gas may be made from nitrogen and hydrogen in an industrial process:

$$\text{nitrogen} + \text{hydrogen} \longrightarrow \text{ammonia}.$$
$$N_2(g) + 3H_2(g) \longrightarrow 2NH_3(g).$$

We say that the ammonia has been *synthesised.*

**3. Displacement reactions.** These reactions happen when something is pushed out.
For example: when chlorine gas is bubbled through potassium iodide solution, the colourless liquid first turns red and then black as solid iodine is displaced from the potassium iodide:

chlorine + potassium iodide ⟶ potassium chloride + iodine.

$$Cl_2(g) + 2KI(aq) \longrightarrow 2KCl(aq) + I_2(s).$$

**4. Oxidation and reduction reactions.** When an element or a compound takes in oxygen during a chemical reaction, we say that it has been *oxidised*. On the other hand if a compound loses oxygen during a reaction, we say that it has been *reduced*.

When something is oxidised it gains oxygen.
When something is reduced it loses oxygen.

Look at the following example:

copper oxide + hydrogen $\xrightarrow{\text{heat}}$ copper + steam.

$$CuO(s) + H_2(g) \longrightarrow Cu(s) + H_2O(g).$$

The copper oxide loses oxygen when it reacts, so it has been reduced.

The hydrogen gains oxygen when it turns into steam, so it has been oxidised.

Both *red*uction and *ox*idation have taken place in the same reaction and so it is called a *redox* reaction.

Look at the example once more.

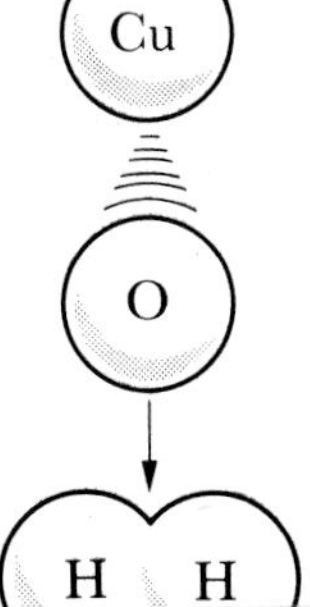


The copper oxide is reduced because the hydrogen takes its oxygen away. The hydrogen is called the reducing agent.

Another way of looking at the same thing is to say that the copper oxide is giving its oxygen to the hydrogen, so it is called the oxidising agent.

In a redox reaction, two things happen at once. The oxidising agent gets reduced and the reducing agent gets oxidised.

The idea of reduction and oxidation can be taken a little further. An element or compound is said to be oxidised if it loses electrons, and it is said to be reduced if it gains electrons, as shown on the next page.

## 7.4 Redox reactions as electron transfer

For example: a solution of iron(II) chloride may be oxidised to iron(III) chloride if chlorine is bubbled through it:

$$2FeCl_2(aq) + Cl_2(g) \longrightarrow 2FeCl_3(aq).$$

Iron(II) chloride contains $Fe^{2+}$ ions.
Iron(III) chloride contains $Fe^{3+}$ ions.
When Iron(II) chloride is oxidised into Iron(III) chloride this is what happens:

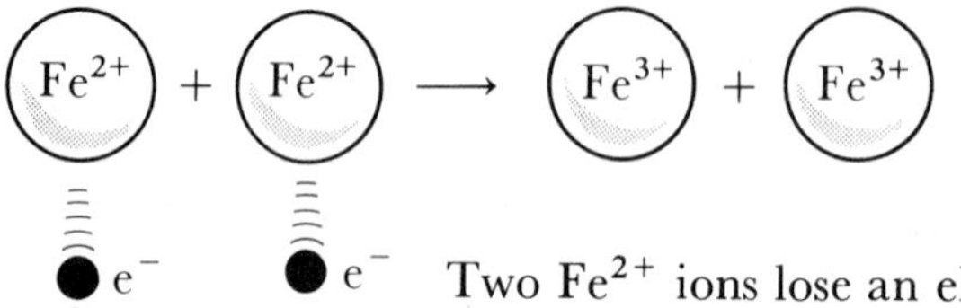

$$2Fe^{2+} \longrightarrow 2Fe^{3+} + 2e^- \quad \text{(oxidation).}$$

Two $Fe^{2+}$ ions lose an electron each. Since they have lost electrons, they are oxidised. They form $Fe^{3+}$ ions.

One chlorine molecule uses the two electrons to form two $Cl^-$ ions. It gains electrons, so it is reduced.

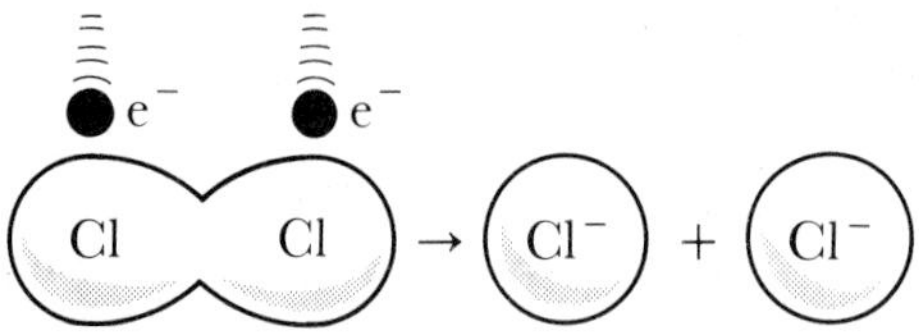

$$Cl_2 + 2e^- \longrightarrow 2Cl^- \quad \text{(reduction).}$$

Putting the two halves together:

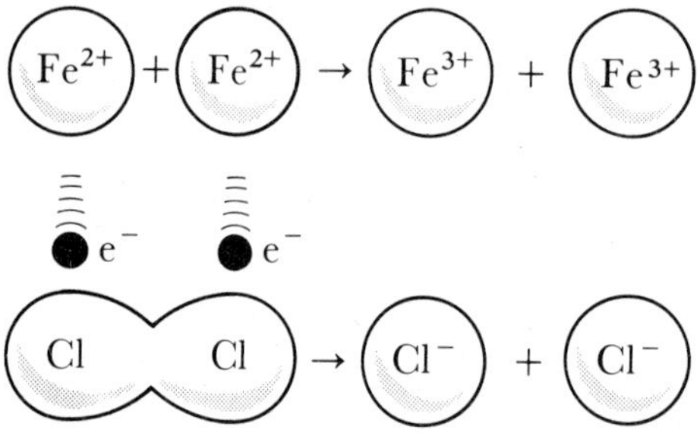

$$2Fe^{2+} \longrightarrow 2Fe^{3+} + 2e^- \quad \text{(oxidation).}$$

$$Cl_2 + 2e^- \longrightarrow 2Cl^- \quad \text{(reduction).}$$

$$Cl_2 + 2Fe^{2+} \longrightarrow 2Fe^{3+} + 2Cl^- \quad \text{(redox).}$$

The $Fe^{2+}$ ions give electrons to the $Cl_2$ molecules so they are called the *reducing agent.* In doing this they are oxidised.

The chlorine molecules take electrons away from the $Fe^{2+}$ ions, so they are called the *oxidising agent.* In accepting them, they are reduced.

The oxidising agent gets reduced.
The reducing agent gets oxidised.

There is much more about oxidation and reduction seen as swopping of electrons in later chapters.

## 7.5 Changing the rate of chemical reactions

Different sorts of chemical reaction take place at different speeds or *rates*. The gunpowder in a firework burns very rapidly, but a piece of iron left in the open air rusts slowly. There are four main ways in which you can increase the rate of a chemical reaction. For each method, you could do the opposite to decrease it.

**1. Increase the concentration.** If zinc is placed in dilute hydrochloric acid, it reacts to produce bubbles of hydrogen, and zinc chloride:

zinc + hydrochloric acid ⟶ zinc chloride + hydrogen.

$$Zn(s) + 2HCl(aq) \longrightarrow ZnCl_2(aq) + H_2(g).$$

If the reaction is repeated with concentrated acid, the hydrogen is evolved much more quickly, making the liquid fizz. This is because the rate of a reaction depends upon how frequently the molecules of the reacting substances collide. The concentrated acid has more molecules for a given volume than the more dilute acid. Because there are more molecules about, the frequency of collisions is greater, and the reaction happens faster.

**2. Increase the amount of reacting surface.** If you have a reaction involving a solid, the reaction can only happen at its surface. Make more surface available, and the reaction will happen faster.
For example: calcium carbonate in the form of marble chips reacts with dilute hydrochloric acid to produce carbon dioxide:

calcium carbonate + hydrochloric acid ⟶ calcium chloride + water + carbon dioxide.

$$CaCO_3(s) + 2HCl(aq) \longrightarrow CaCl_2(aq) + H_2O(l) + CO_2(g).$$

Suppose you start with one 10 g lump of marble. Instead of using it as it is, you break it up into ten 1 g lumps. You have the same amount of marble, but much more surface area for the acid to act upon. Consequently, the reaction is faster. (See figure 3.) The reaction would be even faster if the marble were powdered.

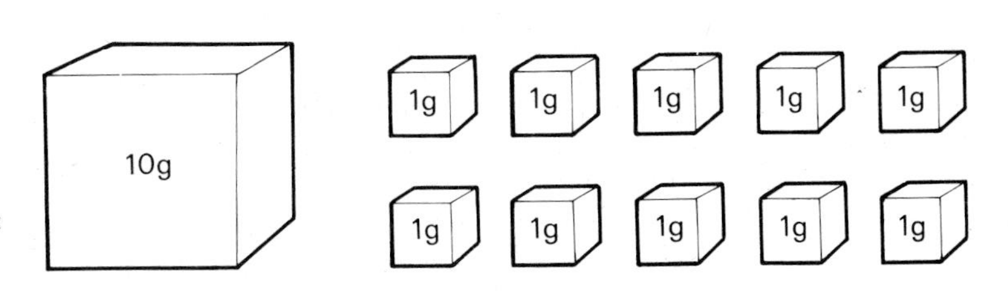


**Figure 3**
*Ten small lumps have more surface area than one large lump.*

**3. Increase the temperature.** A chemical reaction takes place when the molecules of the reacting substances collide with each other. This gives them the energy necessary to react. Heat is a form of energy, so if a reaction is given heat, the molecules will move faster and they will collide with each other more frequently. This means that the rate of the chemical reaction will increase.
For example: magnesium ribbon has hardly any reaction at all with cold water. One or two bubbles may be produced as the reaction tries to take place:

magnesium + water ⟶ magnesium hydroxide + hydrogen.

$$Mg(s) + 2H_2O(l) \longrightarrow Mg(OH)_2(s) + H_2(g).$$

However, if the water is heated to boiling point before the magnesium is added, it fizzes quite vigorously as the hydrogen is evolved.

**4. Add a catalyst.** Catalysts are chemicals which speed up the rate of chemical reactions, but do not get used up themselves. For example: potassium chlorate is a white solid. When it is heated strongly, it first melts, and then decomposes to give off oxygen:

potassium chlorate ⟶ oxygen + potassium chloride.

$$2KClO_3(s) \longrightarrow 3O_2(g) + 2KCl(s).$$

Manganese(IV) oxide is a black powder and acts as a catalyst for this reaction. If a small amount is mixed in with the potassium chlorate before it is heated, then the oxygen is evolved much more quickly, even before the potassium chlorate melts. The reaction takes place more quickly and at a lower temperature because of the catalyst. The manganese(IV) oxide will not have been used up during the reaction and may be used again.

## 7.6 Two laws about chemicals and their reactions

Up to now, you have probably been assuming two things about the way things react. It is now necessary to state them clearly, in the form of two laws.

**The Law of Conservation of Matter:**

**Matter can neither be created nor destroyed in the course of a chemical reaction.**

What this means is that in any chemical reaction, the total mass of all the products must be the same as the total mass of all the starting materials.
For example:

sodium hydroxide + iron(III) chloride ⟶ iron(III) hydroxide + sodium chloride.

This may be written as:

$$3NaOH(aq) + FeCl_3(aq) \longrightarrow \underset{\text{yellow/brown solid}}{Fe(OH)_3(s)} + 3NaCl(aq).$$

A reaction has obviously taken place because two liquids have produced a yellow/brown solid, but really all that has happened is that ions have rearranged themselves in a new way.

If you were to find the total mass of sodium hydroxide and iron(III) chloride before the reaction and the total mass of sodium chloride and iron(III) hydroxide after the reaction, the two totals would be the same.

Reactions involving gases can cause trouble. For example:

$$\text{candle wax(s)} + \text{oxygen(g)} \longrightarrow \text{carbon dioxide(g)} + \text{steam(g)} + \text{soot(s)}.$$

The candle gets smaller as it burns – its mass decreases. As it does so, the carbon dioxide, steam and soot are made. If they were all collected and weighed, the total mass would be the same as the amount of candle and oxygen used. But weighing all those gases would be very difficult!

In fact, the Law as stated above is not quite accurate. It should be the Law of Conservation of Energy, because matter and energy can be inter-converted (see chapter 4). In chemical reactions, minute amounts of matter are converted into energy, some of which appears as heat. However, we could never detect these changes on ordinary laboratory balances.

The second law is about compounds. It concerns the proportions in which elements combine to make compounds.

### The Law of Constant Composition (The Law of Definite Proportions):

**All pure samples of the same chemical compound contain the same elements in the same proportions by mass.**

This means that if you make several samples of the same compound in several different ways, and then analyse them, they will all contain the same proportions of different elements, in other words, the same ratio of moles of each element. All pure samples of the same compound will have the same composition, no matter where in the world they are made.

For example: the formula of copper(II) oxide is CuO. This formula means that in the crystal lattice of copper(II) oxide, there is always one mole of copper ions for every one mole of oxide ions.

1 mole $Cu^{2+}$ combines with 1 mole $O^{2-}$, or in mass units: 64 g Cu combines with 16 g O.

By making copper(II) oxide in three different ways, the law may be verified. This is shown on the next page.

**1** Copper may be heated in oxygen:

$$2Cu(s) + O_2(g) \xrightarrow{\text{heat}} 2CuO(s).$$

**2** Copper(II) nitrate may be decomposed:

$$2Cu(NO_3)_2(s) \xrightarrow{\text{heat}} 2CuO(s) + 4NO_2(g) + O_2(g).$$

**3** Copper(II) carbonate may be decomposed:

$$CuCO_3(s) \xrightarrow{\text{heat}} CuO(s) + CO_2(g).$$

Suppose copper(II) oxide is made by each of these methods and a sample of each is saved. The samples may be analysed by reducing the copper oxide with hydrogen. (See figure 4.)

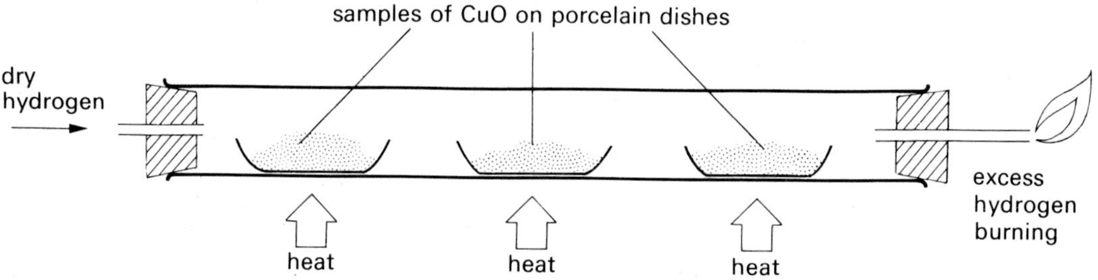


**Figure 4**
*Reducing copper(II) oxide with hydrogen.*

Copper(II) oxide + hydrogen $\longrightarrow$ copper + steam.

$$CuO(s) + H_2(g) \longrightarrow Cu(s) + H_2O(g).$$

Weighed samples of copper(II) oxide are heated in a stream of hydrogen. As the hydrogen reduces the copper(II) oxide, it changes it from black to pink. The steam is carried out in the excess hydrogen, and the samples are allowed to cool down while the hydrogen is still flowing, to prevent re-oxidation.

When they have cooled down, the samples of copper can be weighed. Suppose these were the results.

| | **sample 1** | **sample 2** | **sample 3** |
|---|---|---|---|
| Mass of dish | 5·0 g | 5·2 g | 4·9 g |
| Mass of dish + copper oxide | 13·0 g | 17·2 g | 20·9 g |
| Mass of dish + copper | 11·4 g | 14·8 g | 17·7 g |
| ∴ Mass of copper oxide | 8·0 g | 12·0 g | 16·0 g |
| ∴ Mass of copper | 6·4 g | 9·6 g | 12·8 g |
| ∴ Mass of oxygen combined with copper | 1·6 g | 2·4 g | 3·2 g |

The calculation from these results is shown on the next page.

**Sample 1.** 6·4 g of copper were combined with 1·6 g oxygen.
In moles:

$\frac{6\cdot4}{64}$ moles of copper were combined with $\frac{1\cdot6}{16}$ moles of oxygen.

So the formula for this sample of copper oxide is:

$Cu_{\frac{6\cdot4}{64}}O_{\frac{1\cdot6}{16}}$ or $Cu_{0\cdot1}O_{0\cdot1}$ which gives an empirical formula of CuO.

**Sample 2.** 9·6 g of copper were combined with 2·4 g of oxygen.
In moles:

$\frac{9\cdot6}{64}$ moles of copper were combined with $\frac{2\cdot4}{16}$ moles of oxygen.

So the formula for this sample of copper oxide is:

$Cu_{\frac{9\cdot6}{64}}O_{\frac{2\cdot4}{16}}$ or $Cu_{0\cdot15}O_{0\cdot15}$ which gives an empirical formula of CuO.

**Sample 3.** 12·8 g of copper were combined with 3·2 g of oxygen.
In moles:

$\frac{12\cdot8}{64}$ moles of copper were combined with $\frac{3\cdot2}{16}$ moles of oxygen.

So the formula for this sample of copper oxide is:

$Cu_{\frac{12\cdot8}{64}}O_{\frac{3\cdot2}{16}}$ or $Cu_{0\cdot2}O_{0\cdot2}$ which gives an empirical formula of CuO.

**All three samples have the same empirical formula so the Law of Constant Composition has been demonstrated to hold true.**

**Summary**

At the end of this chapter you should be able to:

**1** Distinguish between a chemical reaction and a physical change.

**2** Describe the main characteristics of chemical reactions.

**3** Explain what is meant by the terms exothermic and endothermic.

**4** Recognise examples of decomposition, synthesis, displacement, and redox reactions.

**5** Describe four ways in which the rate of a chemical reaction may be changed.

**6** Explain what a catalyst does.

**7** State and explain the Law of Conservation of Matter.

**8** State and explain the Law of Constant Composition, and demonstrate its truth from experimental results.

# Seveso–an accident

*This chapter has dealt with 'chemical reactions'. The article that follows tells you about the consequences of one particular chemical reaction getting out of control . . .*

It is Saturday night on the 10th July 1976. The scene is a chemical factory, in the small town of Seveso, in Northern Italy. The machinery is left running, but there is only a very small staff looking after the plant. No-one notices that one of the chemical reactions goes out of control. Tremendous heat builds up inside one of the containers, so an automatic safety valve blows a small cloud of gas into the air. The reaction quietens down.

**A Deadly Cloud**

Two chemicals are present in the cloud. One of them is harmless – it is, in fact, T.C.P. which is used as an antiseptic; but the other compound is highly poisonous and is used in insecticides. The chemist in charge of the plant phones the local police to tell them about the accident. Neither he nor they yet realise that the 2 kilograms of gas which have been released (about the same weight as two bags of sugar) are going to cause horrific problems.

**Cat out of the bag**

Five days later, some animals are found dead in the surrounding fields. One farmer finds a dead cat, and when he picks it up its tail falls off. He buries it, but when he has to dig it up later for a post mortem, he finds that the whole body has dissolved away. The first problems have begun to show themselves. The next day, several children are taken into local hospitals because they have developed strange skin rashes. A week later, 34 people are in hospital showing signs of severe chemical burns and poisoning. Health authorities at the hospitals complain that the staff of the factory will not tell them exactly what chemicals were released. Local authorities play the matter down, but the effects of the poisons are becoming more and more clear. 500 people are moved out of their contaminated homes, and taken to hotels in Milan. Needless to say, they are not very happy about this.

**Health hazard for children and parents**

Two new worries appear. Children appear to be more prone than adults to the effects of the chemical, but doctors suggest that the poison may cause problems for the next generation. They explain that the chemicals may become incorporated in the sperms and eggs of the local men and women, hence causing deformed babies. They are told that it would be best not to have children for up to three years. A woman member of the opposition party in the government calls for an emergency law to allow pregnant women from the disaster area to have abortions because of the possibility of their babies being born deformed. This causes tremendous political and religious problems because abortions are illegal in Italian law, and are deemed to be murder by the Italian church. Things are going from bad to worse.

**Volunteers and £26 million are used**

On the 9th August, 32 workers from the factory volunteer to clean out all parts of the works which have been affected by the explosion. The concentration of the poison here is about 500 times greater than that in the surrounding countryside. They wear protective clothing and gas masks and follow a strict safety code. The next day, the government allocates £26 million for clearing up the Seveso disaster area. It will be used to clear land, build new houses, and to compensate farmers for the crops and animals that have had to be destroyed. Three days after this, three women have abortions in a Milan clinic. These are the first legal abortions in Italy.

**People return to contaminated homes**

Half way through October, 500 people who have been lodging in Milan return to their homes. They come in lorries and cars and burst through the barbed wire fence which the army have built around the area. The mayor and local officials persuade them to return after just one night in their homes. Although the area has been fenced off, nothing has been done to make the area safe for them to live in and to stay there would mean that they slowly poisoned themselves. But the fact that nothing has been done makes the local people very cross indeed.

**Seveso is forgotten?**

By March 1977, very little has been done. 50 babies have been born to mothers who were pregnant at the time of the disaster, and 5 of them are badly deformed. There are 650 more babies on the way; more and more mothers are asking for abortions. The local authorities are still reluctant to permit these. 417 school children are still suffering from the poison.

But at last, there is some hope. Plans for the cleaning up of the Seveso area are approved. They involve removing the bad topsoil and burning it, fencing in less contaminated areas, scrubbing and vacuuming walls, and covering the part of the river that flows through the infected area.

All this, because someone did not take enough care over controlling a chemical reaction. The Italian government are tightening up their safety regulations, to make sure that it doesn't happen again . . .

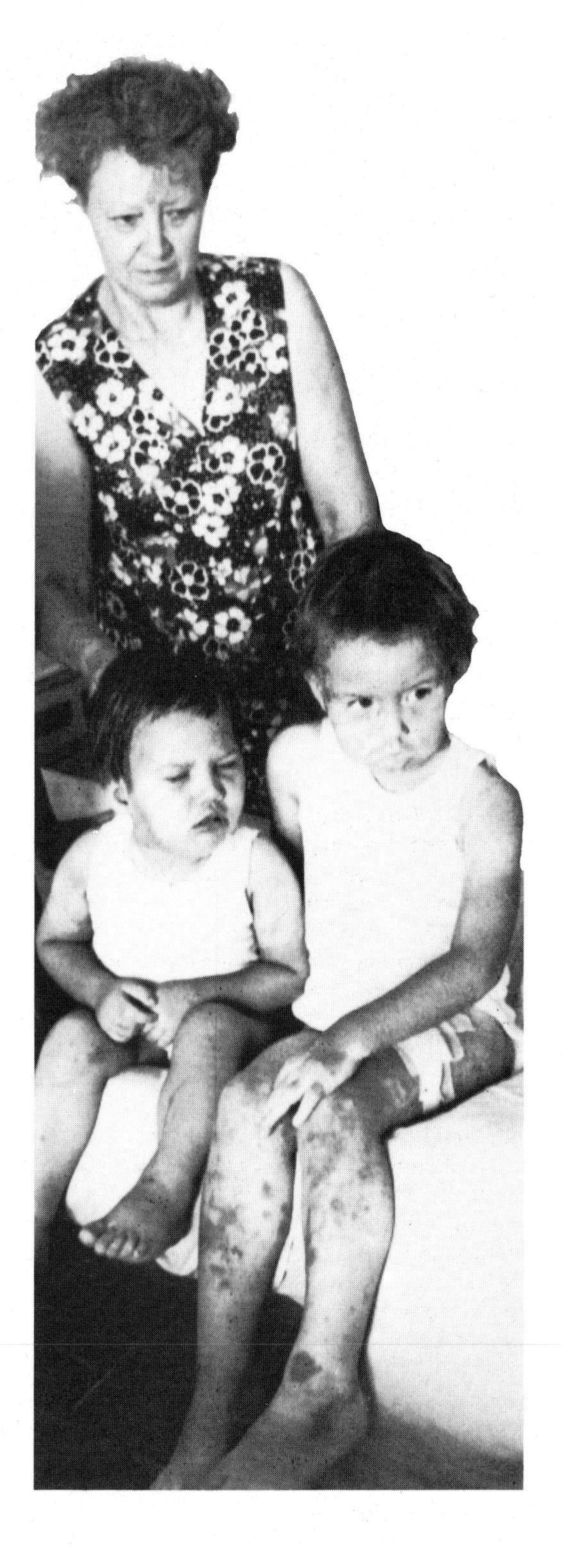

*Children suffered most as a result of the Seveso accident.*

## Questions

**1** Copper is a shiny pink coloured metal. When it is heated in air it turns black because it becomes coated with copper(II) oxide. Give three reasons why you consider this to be a chemical reaction. Write the equation for the reaction.

Chemical reactions are often difficult to reverse, but in this case, the copper can be reclaimed. Read the chapter carefully to find out what chemical you would use. Draw the apparatus you would use for the reaction. What type of reaction would it be?

**2** Look at the following chemical reactions and say whether you consider them to be decomposition, synthesis, displacement or redox reactions. Some of them might be two types at once.

**a** $2Fe(s) + 3Cl_2(g) \longrightarrow 2FeCl_3(s)$.
**b** $CuCO_3(s) \longrightarrow CuO(s) + CO_2(g)$.
**c** $Zn(s) + Pb(NO_3)_2(aq) \longrightarrow Pb(s) + Zn(NO_3)_2(aq)$.
**d** $Mg(s) + SnO(s) \longrightarrow Sn(s) + MgO(s)$.
**e** $FeCl_3(aq) + H_2S(g) \longrightarrow FeCl_2(aq) + S(s) + 2HCl(aq)$.

**3** Put the following list of changes into two groups: one of chemical reactions, and one of physical changes. In each case, explain the reasons for your choice.

**a** Baking a cake,
**b** burning a firework,
**c** painting a piece of wood,
**d** adding sugar to tea,
**e** heating a piece of iron to 2000 °C and letting it cool again,
**f** leaving a piece of iron out in the rain,
**g** lighting a candle and leaving it,
**h** driving a car down the road.

**4** In 7.4 you will find the reaction of hydrochloric acid with marble chips. This reaction was performed twice, using two samples of hydrochloric acid. One of them was dilute, and the other was concentrated. As the gas was evolved in both experiments, it was collected in a syringe so that its volume in $cm^3$ could be noted at regular intervals. On the same piece of paper, with time on the horizontal axis, use the following results to plot the two graphs of volume of gas against time. Compare them to find out which sample of acid was the concentrated one.

**Sample A**

| volume ($cm^3$) | 8·5 | 26·5 | 52·5 | 88·5 |
|---|---|---|---|---|
| time (secs) | 1 | 3 | 6 | 10 |

**Sample B**

| volume ($cm^3$) | 9·5 | 23·0 | 36·0 | 54·0 |
|---|---|---|---|---|
| time (secs) | 2 | 5 | 8 | 12 |

**5** List all the chemical reactions which you can think of that are speeded up by increasing their temperature.

**6** 'When potassium chlorate is heated, oxygen is evolved much more easily if manganese(IV) oxide is used as a catalyst.'

What is a catalyst?

Describe the experiment you would perform to show that the rate at which oxygen is evolved is speeded up by using the catalyst. Draw the apparatus you would use, and say what measurements you would make.

A catalyst is not supposed to be used up during a reaction. If potassium chlorate and potassium chloride are soluble in water, but manganese(IV) oxide is not, how could you show that this is true?

**7** When methane is burned in oxygen, carbon dioxide and steam are formed. This is the reaction:

$$CH_4(g) + 2O_2(g) \longrightarrow CO_2(g) + 2H_2O(g).$$

In an experiment some methane was burned in air, and the gases that were produced were sucked through two U tubes containing calcium chloride and soda lime. Calcium chloride absorbs water, and soda lime absorbs carbon dioxide. The following diagram shows the apparatus.

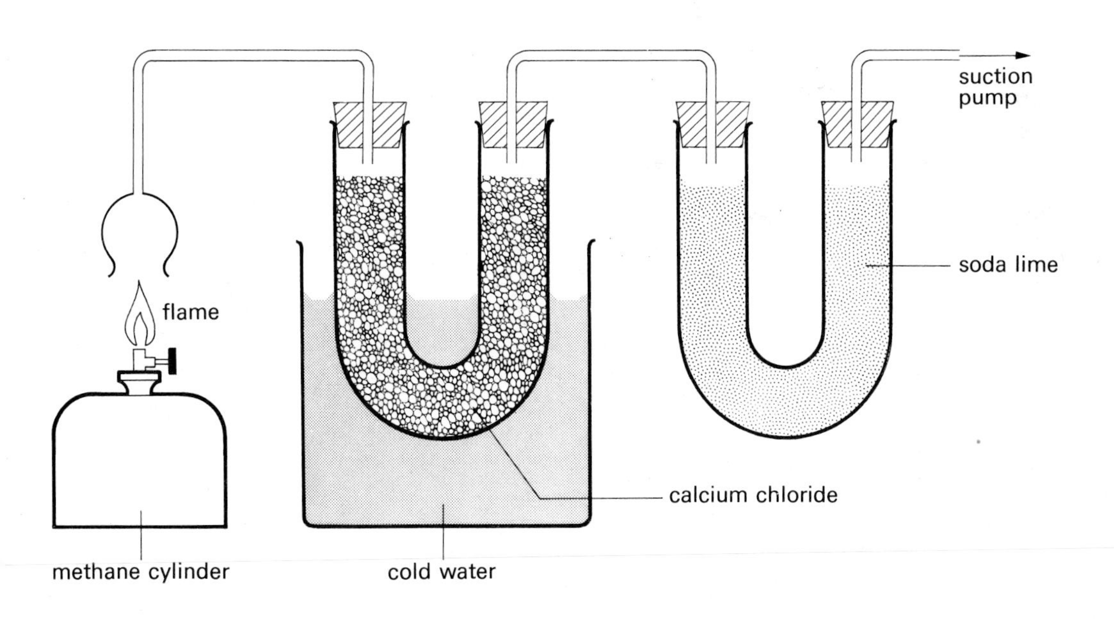


The following measurements were made:

| | |
|---|---|
| mass of methane cylinder at start of experiment | . . . 418·0 g |
| mass of methane cylinder after experiment | . . . 402·0 g |
| mass of calcium chloride U tube before experiment | . . . 31·0 g |
| mass of calcium chloride U tube after experiment | . . . 67·0 g |
| mass of soda lime U tube before experiment | . . . 48·5 g |
| mass of soda lime U tube after experiment | . . . 92·5 g |

**a** What mass of methane was used up?
**b** What mass of water was produced?
**c** What mass of carbon dioxide was produced?
**d** Write down the Law of Conservation of Mass.
**e** What mass of oxygen must have been used if the Law is correct?

**8** Write down the Law of Constant Composition.
Two samples of lead(II) oxide were made by different methods. The first sample started with a mass of 223 g. After it had been reduced to pure lead by passing hydrogen over it, the mass had gone down to 207 g.
The second sample was found to contain 414 g of lead combined with 32 g of oxygen.
Do the results uphold the Law?

# 8 Air

## 8.1 Respiration

We all need air. If we are deprived of it for more than a few minutes we become unconscious. This leads on to brain damage – which may mean loss of speech, or paralysis, or even death. Our bodies need air, or more correctly, oxygen.

As we breathe in, air passes into the lungs through two large tubes called bronchi. These branch into many other small tubes called bronchioles. At the end of the bronchioles there are millions of tiny hollow bags called air sacs which are surrounded by blood vessels. (See figure 1.) The blood absorbs oxygen from the air sacs and this oxygen is then pumped by the heart to the organs of the body and the millions of cells that need it in order to function. In these cells, the oxygen reacts with compounds taken from the food we eat. These reactions produce energy, and carbon dioxide, which is carried by the blood back to the lungs, passed into the air sacs, and exhaled.

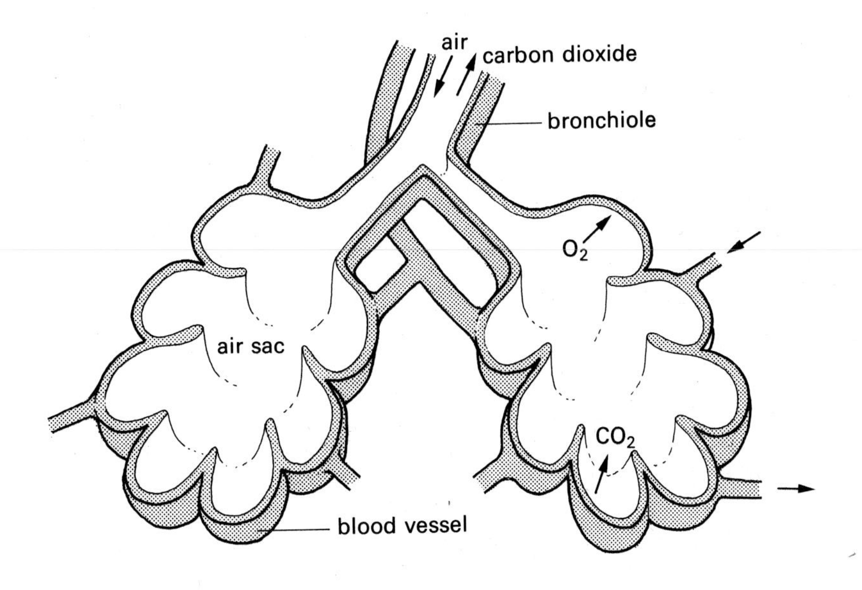


**Figure 1**
*Air sacs in the lung.*

All green plants absorb carbon dioxide through their leaves, and water through their roots. They then need sunlight as a source of energy, and the green pigment called *chlorophyll* as a catalyst. With raw materials, energy, and catalyst, they produce carbohydrates which are the main building blocks that they need. At the same time, they release oxygen, which they don't want.

$$CO_2(g) + H_2O(l) \xrightarrow[\text{chlorophyll}]{\text{light}} \text{'}CH_2O\text{'} + O_2(g).$$

'$CH_2O$' is the building block for many different starches and sugars (carbohydrates). For example:

$$6CH_2O \longrightarrow C_6H_{12}O_6.$$

glucose (a typical carbohydrate)

The process of turning carbon dioxide and water into carbohydrates and oxygen is called *photosynthesis*.

## 8.2 The composition of the air

Air is a mixture of several different gases. The composition of 'pure' air changes very little despite the fact that we, along with fires, and car engines, use up oxygen and replace it with carbon dioxide. The balance is kept because green plants take in carbon dioxide, and give out oxygen.

Figure 2 shows the amount of various gases in unpolluted air.

| **component of the air** | **% by volume** |
|---|---|
| nitrogen | 78·08 |
| oxygen | 20·95 |
| carbon dioxide | 0·03 |
| water vapour | depends on where you are |
| the noble gases: | |
| argon | 0·93 |
| neon | 0·002 |
| helium, krypton and xenon: | less than that |
| other gases such as | |
| methane, ozone and hydrogen: | even less than that. |

**Figure 2** *The composition of the air.*

## 8.3 The percentage of oxygen in the air

This experiment removes the oxygen from the air and makes it possible to measure how much gas remains.

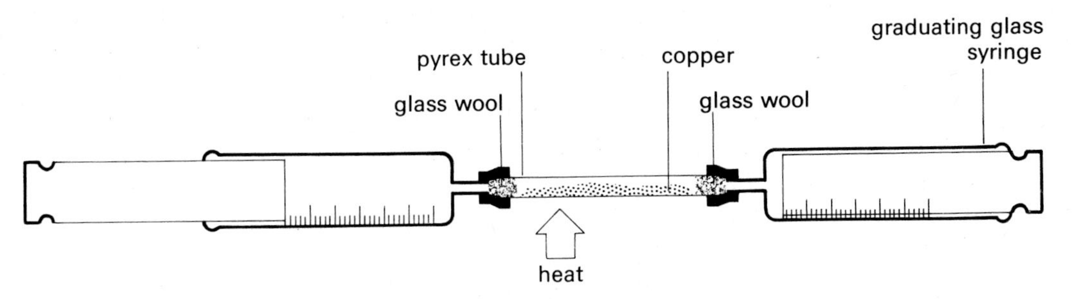


**Figure 3** *Finding the percentage of oxygen in the air.*

**1** The pyrex tube is packed with copper wire. This is a special porous type of copper which has an uneven surface and therefore a large surface area. There is a small, loosely fitting glass plug in each end of the tube. These prevent small pieces of copper from getting into the syringes.

**2** One syringe is full of air (plunger out) and the other is empty (plunger in). The pyrex tube of copper is heated at one end, and the air is pushed from one syringe to the other. As it passes through the pyrex tube, the oxygen in the air reacts with the pink copper, changing it to black copper(II) oxide:

$$\underset{\text{pink}}{2Cu(s) + O_2(g)} \longrightarrow \underset{\text{black}}{2CuO(s)}.$$

**3** As the copper goes black, the bunsen is moved to a fresh portion of pink copper. Eventually, no matter how much the copper is heated and how much the air is pushed from syringe to syringe, no more copper oxide is formed. This means that there is no more oxygen left in the sample of air.

**4** The apparatus is left to cool down and the volume of air left in the full syringe is noted.

A typical result for the experiment might be:

volume of air at the beginning $= 100\ cm^3$

volume of air at the end $= 79\ cm^3$

$\therefore$ volume of air used up (i.e. the oxygen) $= 21\ cm^3$

therefore the percentage by volume of oxygen in the sample of air is

$$\frac{\text{volume of } O_2}{\text{volume of air}} \times 100 = \frac{21}{100} \times 100\% = 21\%.$$

## 8.4 Separating the mixture

Air is a mixture, so we should be able to separate it using one of the methods described in chapter 2. Fractional distillation is used, after the air has been liquefied.

There are three main stages in the process:

**1** The carbon dioxide and water vapour have to be removed first. This can be done by cooling the air in a refrigeration plant, because although carbon dioxide and water vapour are both gases under normal conditions, they can be quite easily frozen and removed. If this is not done, they solidify whilst the air is being cooled, and block up the pipes.

**2** The remaining air is then compressed to about 150 times atmospheric pressure. As this is done it gets very hot (feel your bicycle pump as you pump up your tyres). It is then allowed to cool off. After the air has lost its heat, it is allowed to expand again. As it does so, it gets very cold indeed (feel the air rushing out as you let your bicycle tyre down).

**3** This is continued until the temperature is as low as −200 °C (73 K), by which time all the gases except neon and helium are liquefied. The mixture of liquids can then be separated by fractional distillation.

**Figure 4** *The products of the fractional distillation of air.*

| gas | degrees centigrade °C | Kelvins K |
|---|---|---|
| xenon | −108 | 185 |
| krypton | −153 | 120 |
| oxygen | −183 | 90 |
| argon | −186 | 87 |
| nitrogen | −196 | 77 |
| neon | −246 | 27 |
| helium | −269 | 4 |

## 8.5 Uses of the separated gases

**Oxygen.** This gas has many uses. Oxy-acetylene burners produce very high temperature flames for cutting and welding metal. When steel is being made, oxygen is used to remove the impurities from the molten iron. Pure oxygen stored in cylinders, is used to support breathing – to help patients in hospitals, and for high altitude or underwater work. Space rockets have to carry liquid oxygen in order to burn their fuels in space.

**Nitrogen.** This gas is used in liquid form for freezing 'frozen' food. Bulls' semen is frozen in liquid nitrogen at cattle artificial insemination centres. Nitrogen is also a very important ingredient in the manufacture of ammonia.

**The noble gases.** These make up only about 1% of the air by volume. Argon is the most plentiful of these gases and is used as an inert atmosphere to surround aluminium, titanium and certain types of steel when they are welded. This means that the argon stops the metals from burning, or forming outside coatings of oxide, which would make it difficult for them to be joined successfully. Light bulbs also contain argon to prevent the filaments from burning out.

Neon is extensively used for making advertizing signs. Tubes filled with the gas glow red when a high voltage is applied across the gas.

Helium is used as an alternative to hydrogen in meteorological balloons because it is a light gas, which, unlike hydrogen, does not burn. It is also used to dilute the oxygen which deep sea divers use because both pure oxygen and nitrogen are dangerous to breathe at the high pressures which exist under water.

Krypton and xenon are put into some electrical valves and T.V. tubes and in high powered lamps in lighthouses and miners' lamps.

Each day, air distillation plants separate about 100 tonnes of air. It is comforting to know that our atmosphere weighs at least 150 million million tonnes, so there is little chance that we will run out of air to breathe.

## 8.6 Pollution

It is well known that our air is polluted, and that the things responsible for this pollution are part of our everyday lives. We live in an industrial world, and factories produce waste products such as smoke, poisonous gases, steam and heat. Although the amount of smoke from chimneys is now controlled by regulations (in many parts of Britain, it is against the law to make smoke from a domestic fire), the air around industrial towns is far from clean.

Worse than that, chemical plants producing essential products such as sulphuric acid add tonnes of sulphur dioxide to the air each year. This gas escapes into the atmosphere only to be washed out by rain, often hundreds of miles away. It then damages plants, and corrodes buildings. Mixed with fog it can be lethal for people with breathing difficulties.

Petrol and diesel engines are great polluters. Their exhaust fumes contain lead vapour, oxides of nitrogen, and carbon monoxide, small amounts of which are poisonous. Other compounds have also been detected which are possibly carcinogenic (cancer forming). Just watch a lorry or bus next time it climbs a steep hill and think about the pollution it creates.

Even carbon dioxide might be dangerous. Some scientists have calculated that if we carry on adding more and more carbon dioxide to the atmosphere, green plants will not be able to use it all. As it builds up, it will have a noticeable effect on the temperature of the air, because carbon dioxide makes the atmosphere act like a giant 'greenhouse'. Scientists estimate that by the year 2000, the average temperature of the Earth's atmosphere may have risen by 0·5 °C. This may not seem very much, but a rise of 1 °C could cause the Polar ice caps to start melting, which would cause flooding in many parts of the World.

## 8.7 Combustion

When things burn in air, they combine with oxygen and give out hot gases. These are usually seen as flames.

For example: When carbon is burned, carbon dioxide is produced:

$$C(s) + O_2(g) \longrightarrow CO_2(g).$$

If methane is burned in plenty of air, carbon dioxide and steam are given off:

$$CH_4(g) + 2O_2(g) \longrightarrow CO_2(g) + 2H_2O(g).$$

Most substances that burn contain carbon and hydrogen, so steam and carbon dioxide are the two main gases produced. However, if the air supply is restricted, complete combustion may not occur and carbon monoxide may be made. If the burning substance contains nitrogen, sulphur, or chlorine, then hydrogen cyanide, sulphur dioxide, and hydrogen chloride are likely to be produced. If you are caught in a fire where plastics or modern synthetic materials are burning, you are more likely to die of poisoning than burning.

## 8.8 Fire!

For burning, three things are needed:

FUEL HEAT OXYGEN

If any of these are removed, the burning stops. So remember, if you find a fire, there are three things to try to do:

**1** Cut off the fuel. This might mean turning off the gas or electricity, or covering oil with sand or soil. Take care, especially with electricity. If in doubt, do not touch.

**2** Reduce the temperature by covering the flames with water or foam.

**3** Cut off the air supply with a blanket, sand, or foam.

There are many different sorts of fire.

If petrol or oil is burning, never try to put it out using water. This only spreads the flames because oil and petrol float on water. Instead, use foam, carbon dioxide, or even sand or soil.

If the fire has been caused by electricity, never use water or foam. These conduct electricity at high voltages. Instead, use carbon dioxide or dry sand.

If someone is on fire, try to roll them in a blanket or carpet, or a special fireproof blanket if one is available.

Fires in laboratories are sometimes caused by chemicals – or more often, by careless people using them. Always use chemicals with care. Highly inflammable substances should never be kept in hot places, or near unprotected flames – in the laboratory, this means near bunsen burners. If they do catch fire, try to cut off their air supply by covering them with a cloth, or a mat. If you have to use an extinguisher, make sure it is a carbon dioxide one.

If the fire is too much for you, abandon it. Shut all windows and doors to cut off the air supply; get everyone out and dial 999.

**Summary**

At the end of this chapter, you should be able to:

**1** Describe the process of breathing.

**2** Explain what happens during 'photosynthesis'.

**3** List the main gases in the air and the rough percentages in which they occur.

**4** Describe an experiment to measure the percentage of oxygen in the air.

**5** Explain how air can be separated into its component gases by fractional distillation.

**6** Describe some of the chemicals which pollute our atmosphere and say where they come from.

**7** Describe the most likely products of burning.

**8** Draw up a list of fire rules for your laboratory.

# The killer fog of '52

*Between the fifth and ninth of December, 1952, a fog descended on London which was so full of chemical pollution that thousands of people were poisoned by it. About four thousand more deaths happened than was usual for the time of year. There was another 'killer fog' in 1956, but since then, the Clean Air Act has made London one of the cleanest cities in the world. The following articles are all taken from newspaper reports made at the time.*

The great fog blanket which paralysed London over the week-end was expected last night to persist for another 24 or 36 hours.

Weather men, hoping for a wind to clear the murk, said there was little sign of even a slight breeze.

The result in most London homes today is likely to be that morning deliveries of bread, milk, mail and newspapers will be late. At one large dairy an official said yesterday, 'All our milk has been delivered to the bottling depot, but we are 12 hours behind with the actual bottling. Deliveries of house coal may also be held up. Thousands of householders who used up their stocks over the chilly week-end may get no more until tomorrow at the earliest.

In London, road and rail traffic were paralysed, aircraft were grounded and cinemas were half empty. Heroes of the capital were the Ambulance men, who dealt with more than 200 accident calls, and 210 general calls. Time taken for each averaged two hours instead of the normal 25 minutes. Late last night ambulance headquarters said they had run out of flares.

At least two London babies were born as their mothers were on the way to hospital, with the ambulance men acting as midwives.

On the Thames, where all shipping was at a standstill, Port of London police patrolled the Docks wearing life jackets because so many people had walked into the water.

**London's Night of Terror in Record Fog**

A terrifying new wave of crime hit London last night as the gunmen and cosh boys cashed in on the 'Worst ever' fog which blacked-out the Capital.

A policeman fought an armed raider . . . a teenaged couple held up a newsagent with a gun . . . a cinema manager brutally coshed . . . a shopkeeper and his wife were attacked . . . a post office was raided.

The Policeman, who had been in the force only six months, was keeping a watch on parked cars outside a West End Hotel when he saw a man behaving suspiciously. He tackled the man and there was a violent struggle . . .

A girl bandit described as 'pretty and refined', took part in an armed hold-up at a newsagents in Queen's Road Surrey. She stood guard as a youth of about 18 pointed a a pistol at Mrs Gladys Tubbs. 'The gunman said 'Stick 'em up', said Mrs Tubbs late last night. 'I was absolutely terrified and I screamed at the top of my voice and the youth fled with another man without taking anything.'

Two bandits attacked a tobacconists last night and robbed him of the £400 he was carrying in a brief case.

**A 'Bunny Nose' is Mrs. H.'s answer**

One woman who is keeping in step with doctors on the fog problem is dressmaker Mrs. Pamela Hickey, daughter of an inventor. She has designed a 'Bunny nose' mask to combat the 'killer' fog. Inside the 'Bunny nose' of the mask, which she has patented, is a special filter. And this, says Mrs. Hickey, traps the smoke-laden fog particles before they can damage the chest and lungs.

She said yesterday, 'The mask is washable and it can be made in any colour. I have made it in all sizes, for babies, for adults with big noses and people with little noses. It should sell in the shops for about 5/- (25p).

Mrs. Hickey has experimented with the mask by holding her head over burning sulphur and by staying out all night in an open car on Salisbury Plain during the thick fog. The mask, which she claims is foolproof, 'worked perfectly'.

**Breath of Whisky Puts Beef into Them**

Best way to breathe in the fog – through a mask of whisky fumes. That's the way cattle were breathing at Earl's Court yesterday, and they were thriving on it.

Fog, seeping into the hall, had affected some of the leading entries in the Smithfield Show. The 22-month-old heifer, owned by Mr. Frank Parkinson, of West Tisted, Hants., was so distressed that she had to be slaughtered.

So several Scottish breeders got together and consulted 50-year-old herdsman Willie Stewart, of Huntly, Aberdeenshire. Together, they found a remedy.

'I've been among cattle all my life,' Willie said. 'We used to treat pneumonia cases by making a respirator out of a sack, soaking it with eucalyptus, and putting it on the nose of the sick beast.

'There was not enough eucalyptus available this time, but I had heard of whisky being used. So we soaked a sack in whisky and one heifer came round quite nicely.

'Other folk were quick to ask about our cure and that's why a dozen or so animals here are wearing whisky-soaked respirators today.'

## Questions

**1** Explain what is meant by the term 'Photosynthesis'. What do you think might happen if a new big city was built, but no trees or parks were included in it, and the houses had no gardens?

**2** You want to measure the percentage by volume of oxygen in the air, but you have not got the apparatus used in the experiment shown in figure 3.

You have got a candle, a bowl and a large sweet jar. Describe the experiment you would perform. Remember that things use up oxygen when they burn.

**3** A sample of air has a volume of 130 $cm^3$. It is liquefied and distilled. Which gas will boil off first?

Altogether, there is 104 $cm^3$ of nitrogen in the sample. What is the percentage by volume of nitrogen in the air sample?

**4** One of the dangerous components of car exhaust fumes is carbon monoxide. Use the index to find the main section on carbon monoxide in this book.
Find out what its chemical reactions are.

Now suppose that you have to design a new exhaust system for a car which prevents carbon monoxide from being expelled into the atmosphere. Write down a few ideas about the method you might use.

**5** What gases might be formed when the following substances burn in a fierce fire:

**a** propanone $CH_3.CO.CH_3$
**b** poly vinylchloride $(CHCl.CH_3)_n$.
**c** Nylon $(-NH.(CH_2)_6.NH.CO.(CH_2)_4.CO-)_n$.

**6** Look at the fire extinguishers near your laboratory or class room. What type are they? Do you know how to work them? (Don't!) What kind of fires can they be used for and what kind of fires must they not be used for? Ask teachers of other subjects if they know where their nearest extinguisher is, and if they know how to use it.

# 9 Acids, bases and salts

## 9.1 Acids

Most people are familiar with acids and some of the things that acids do, such as dissolving away metals and causing 'burns' if it contacts the skin. Figure 1 shows some naturally occurring acids and the places where they are found.

| acid | place where found |
|---|---|
| acetic acid (ethanoic acid) | vinegar |
| formic acid (methanoic acid) | ants' stings and stinging nettles |
| citric acid | citrus fruits such as lemons |
| oxalic acid | rhubarb |
| hydrochloric acid | in your own stomach |
| tartaric acid | grape juice |
| lactic acid | sour milk |

**Figure 1**
*Some naturally occurring acids.*

You may have been surprised to see hydrochloric acid in that list. It is used by the body to digest food. It is of course very dilute. Even the 'dilute' acid in the laboratory is much more concentrated.

In the laboratory, the main acids used are:

| acid | chemical formula | |
|---|---|---|
| hydrochloric acid | $HCl$ | these are often used |
| nitric acid | $HNO_3$ | |
| sulphuric acid | $H_2SO_4$ | |
| phosphoric acid | $H_3PO_4$ | |
| carbonic acid | $H_2CO_3$ | these are sometimes used |
| nitrous acid | $HNO_2$ | |
| sulphurous acid | $H_2SO_3$ | |

**Figure 2**
*Acids used in the laboratory.*

## 9.2 Bases

You may not be so familiar with *bases*. Bases are the chemical opposi hydroxides of metals. Some of the laboratory are shown in figure 3.

| base | chemical formul | |
|---|---|---|
| copper(II) oxide | $CuO$ | ses are *uble* in water |
| iron(II) oxide | $FeO$ | |
| zinc oxide | $ZnO$ | |
| sodium hydroxide | $NaOH$ | these bases are *soluble* in water |
| potassium hydroxide | $KOH$ | |
| calcium hydroxide | $Ca(OH)_2$ | |

**Figure 3**
*Common bases.*

The bases which dissolve in water are given a special name. They are called *alkalis*.

Sodium hydroxide, potassium hydroxide, and calcium hydroxide are all alkalis. One other alkali which is a little unusual is ammonium hydroxide, $NH_4OH$. Here, the ammonium ion $NH_4^+$ acts in place of a metal ion. But it is still an alkali.

## 9.3 Indicators

The presence of acids or alkalis may be shown using *indicators*. An indicator changes from one colour when mixed with an acid, to another colour when mixed with an alkali. Figure 4 shows some indicators that you might use.

| indicator | colour in acid | colour in alkali |
|---|---|---|
| methyl orange | orange | yellow |
| phenolphthalein | colourless | pink |
| bromothymol blue | yellow | blue |
| litmus | red | blue |

**Figure 4**
*Some common indicators.*

The best known of these indicators is litmus. It is red when added to an acid, and blue when added to an alkali.

n
gth

**Concentration.** Acids and alkalis are usually dissolved in water. The concentration tells you how much water has been added. Add more water, and the solution becomes less concentrated. A solution containing 10 g of nitric acid per cubic decimetre ($dm^3$) is more concentrated than one which contains only 1 g per $dm^3$.

**Strength.** There is, however, another term which may be used about acids and alkalis. This is the *strength* of an acid or an alkali. A strong acid is one which completely or nearly completely breaks up into its ions. Figure 5 tells you which are strong acids and which are weak.

| acid | strength of acid |
|---|---|
| hydrochloric acid<br>nitric acid<br>sulphuric acid<br>phosphoric acid | are almost completely ionised. They are strong acids. |
| carbonic acid<br>sulphurous acid<br>acetic acid (ethanoic acid) | are not completely ionized. They are weak acids. |

**Figure 5**
*Strong and weak acids.*

The same applies to alkalis. Some of them completely break up into ions. They are strong alkalis. Others are not so ready to break up. They are weak alkalis. Figure 6 tells you which are which.

| alkali | strength of alkali |
|---|---|
| sodium hydroxide<br>potassium hydroxide | are strong alkalis |
| calcium hydroxide<br>ammonium hydroxide | are weak alkalis |

**Figure 6**
*Strong and weak alkalis.*

We are able to distinguish between strong and weak acids and alkalis by using a special mixture of indicators called *Universal Indicator*. Universal Indicator has a different colour for each strength of acid and alkali. Figure 7 shows the usual colours.

red pink beige yellow green light blue dark blue

STRONG ACID | WEAK ACID | NEUTRAL | WEAK ALKALI | STRONG ALKALI

**Figure 7**
*The colours of Universal Indicator.*

Each different strength of acid or alkali is given a different pH number. This symbol comes from a German word which means 'strength of acid'. Each colour of the Universal Indicator corresponds to a different pH number. (See figure 8.)

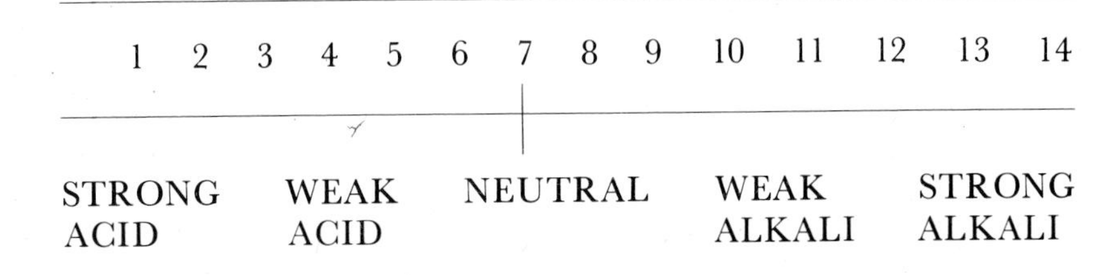


**Figure 8**
*pH numbers.*

Figure 9 shows some common solutions and their pH numbers.

| solution | pH number |
|---|---|
| dilute hydrochloric acid | 1 |
| dilute sulphuric acid | 1 |
| dilute nitric acid | 1 |
| lemon juice (citric acid) | 2 |
| health salts (tartaric acid) | 3 |
| vinegar (acetic acid) | 4 |
| pure water | 7 |
| sodium chloride solution | 7 |
| baking powder (sodium hydrogen carbonate) | 9 |
| ammonium hydroxide | 11 |
| dilute sodium hydroxide solution | 14 |
| dilute potassium hydroxide solution | 14 |

**Figure 9**
*Some solutions and their pH numbers.*

Indicators come in different forms. They can be used as liquids, or in the case of litmus, they can be soaked onto paper and dried. Different brands of Universal indicators have different colours – but they all have the same pH numbers, so these are always used when the strength of an acid or alkali is quoted.

## 9.5 Properties of acids

**1. Acids have a sour taste.** Think of vinegar and lemon juice. DO NOT taste any chemical substance in the laboratory.

**2. Acids turn litmus paper and solution red.**

**3. All acids contain hydrogen ions.** These are the only positive ions present in the acid. For example:

hydrochloric acid $HCl(aq) \longrightarrow H^+(aq) + Cl^-(aq)$.

sulphuric acid $H_2SO_4(aq) \longrightarrow 2H^+(aq) + SO_4{}^{2-}(aq)$.

nitric acid $HNO_3(aq) \longrightarrow H^+(aq) + NO_3{}^-(aq)$.

**4. Most acids react with most metals.** They form compounds called *salts*, releasing hydrogen gas. For example:

magnesium + hydrochloric acid (dilute) ⟶ magnesium chloride + hydrogen.

$$Mg(s) + 2HCl(aq) \longrightarrow MgCl_2(aq) + H_2(g).$$

Nitric acid does not do this reaction (see chapter 16) and the metals copper, mercury, silver and gold will not react. (See chapter 18.)

**5. Acids neutralise bases.** They form a salt and water. For example:

copper(II) oxide + sulphuric acid ⟶ copper(II) sulphate + water.

$$CuO(s) + H_2SO_4(aq) \longrightarrow CuSO_4(aq) + H_2O(l).$$

Alkalis are also part of the group called bases, so they too are neutralised by acids. For example:

sodium hydroxide + nitric acid ⟶ sodium nitrate + water

$$NaOH(aq) + HNO_3(aq) \longrightarrow NaNO_3(aq) + H_2O(l).$$

**6. Acids react with carbonates.** They form a salt, the gas carbon dioxide, and water. The reaction is fizzy, or *effervescent*. For example:

sodium carbonate + hydrochloric acid (dilute) ⟶ sodium chloride + carbon dioxide + water.

$$Na_2CO_3(s) + 2HCl(aq) \longrightarrow 2NaCl(aq) + CO_2(g) + H_2O(l).$$

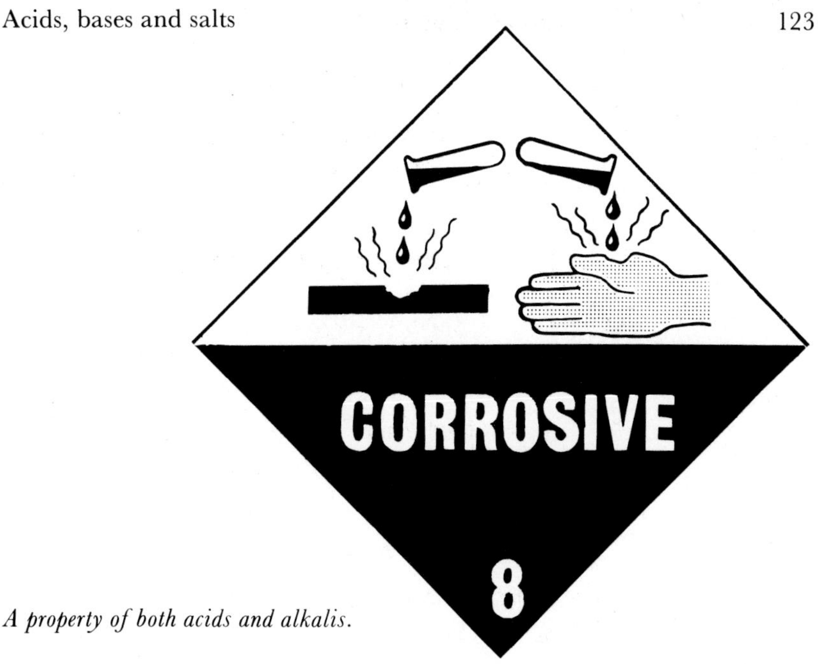


*A property of both acids and alkalis.*

9.6
**Properties of bases and alkalis**

**1. Alkalis are soapy to touch.** This is because they react with the natural oils in the skin, actually making soap.

**2. Alkalis turn litmus blue.**

**3. Bases contain oxide ions or hydroxide ions.** If they are oxides of metals, they contain oxide ions; if they are hydroxides of metals, they contain hydroxide ions. For example:

copper(II) *oxide* $CuO \longrightarrow Cu^{2+} + O^{2-}$
sodium *hydroxide* $NaOH \longrightarrow Na^{+} + OH^{-}$.

All alkalis are soluble in water and give solutions which contain hydroxide ions.

**4. Alkalis will react with most metal ions.** They form insoluble precipitates.
Whenever a solid substance is formed by the reaction of solutions, it is called a precipitate. This is what happens when copper sulphate is added to sodium hydroxide. This is really just a reaction between the copper ions and the hydroxide ions.

$Cu^{2+}(aq) + 2OH^{-}(aq) \longrightarrow Cu(OH)_2(s)$.

Because of this it is called an *ionic precipitation* reaction.

**5. Bases neutralise acids.** We have seen this reaction already in the previous section.
Whenever a base neutralises an acid it is called a neutralisation reaction, and water is always made.

## 9.7 Salts

Besides acids and alkalis, nearly all the compounds which you will meet in the laboratory are *salts*.

Salts are made whenever an acid is neutralised with a base or an alkali, and when an acid reacts with a metal or a carbonate. Salts are ionic compounds made up from two ions. The first part is positive metal ions or the ammonium ion. These come from a base. The second is negative non-metal ions, which have come from an acid. For example:
Sulphuric acid forms salts called sulphates.

| | |
|---|---|
| calcium sulphate | $CaSO_4$ |
| sodium sulphate | $Na_2SO_4$ |
| ammonium sulphate | $(NH_4)_2SO_4$. |

Hydrochloric acid forms salts called chlorides.

| | |
|---|---|
| copper(II) chloride | $CuCl_2$ |
| aluminium chloride | $AlCl_3$ |
| zinc chloride | $ZnCl_2$. |

Nitric acid forms salts called nitrates.

| | |
|---|---|
| magnesium nitrate | $Mg(NO_3)_2$ |
| potassium nitrate | $KNO_3$ |
| iron(II) nitrate | $Fe(NO_3)_2$. |

## 9.8 Water of crystallisation

Most salts contain a fixed percentage of water in their crystal lattice when they crystallise. It is essential to their shape and sometimes to their colour, and is called *water of crystallisation.* If it is removed, then the crystal changes form and colour and becomes dehydrated, or anhydrous. The number of moles of water per mole of salt is called the *degree of hydration*.

Figure 10 contains some salts which are hydrated and some which are not.

| **Name of salt** | **formula** |
|---|---|
| copper(II) sulphate-5-water | $CuSO_4.5H_2O$ |
| sodium sulphate-10-water | $Na_2SO_4.10H_2O$ |
| cobalt chloride-6-water | $CoCl_2.6H_2O$ |
| zinc sulphate-7-water | $ZnSO_4.7H_2O$ |
| iron(II) sulphate-7-water | $FeSO_4.7H_2O$ |
| sodium carbonate-10-water | $Na_2CO_3.10H_2O$ |
| sodium chloride | $NaCl$ |
| potassium nitrate | $KNO_3$ |
| potassium manganate(VII) | $KMnO_4$ |

**Figure 10**
*Hydrated and non-hydrated salts.*

If a salt loses its water of crystallisation it can often be replaced again. For example:
When copper(II) sulphate-5-water is heated, the blue crystals break down into a white powder and steam is given off. The process can be reversed by adding water to the powder. Heat is given back out and a blue solution is formed which may be evaporated to give crystals:

$$\underset{\text{blue}}{CuSO_4.5H_2O(s)} \underset{\text{heat out}}{\overset{\text{heat in}}{\rightleftharpoons}} \underset{\text{white}}{CuSO_4(s) + 5H_2O(g).}$$

Another similar substance is cobalt chloride-6-water. When its purple crystals are heated, steam is given off and a dark blue solid is formed. Addition of water gives a pink solution which may be evaporated to purple crystals:

$$\underset{\text{purple}}{CoCl_2.6H_2O(s)} \underset{\text{heat out}}{\overset{\text{heat in}}{\rightleftharpoons}} \underset{\text{blue}}{CoCl_2(s) + 6H_2O(g).}$$

Both of these anhydrous substances may be used to identify water. If water is added to anhydrous copper(II) sulphate it turns the white powder blue, and if water is added to anhydrous cobalt chloride, it turns the blue powder purple. The latter is often soaked into paper, which is blue when dry and pink when wet.

## 9.9 The solubility of salts

Different salts have different solubilities in water, but for convenience, they can be divided up into those which are obviously soluble in water, and those which are not.

You will find it very useful if you learn the following information.

**1** All nitrates are soluble in water.

**2** All sulphates are soluble in water except:
lead sulphate $PbSO_4$
barium sulphate $BaSO_4$
mercury(II) sulphate $HgSO_4$
calcium sulphate $CaSO_4$.

**3** All chlorides are soluble in water except:
silver chloride $AgCl$
mercury(II) chloride $HgCl_2$
lead chloride $PbCl_2$.

**4** All carbonates are *insoluble* in water except:
sodium carbonate $Na_2CO_3$
potassium carbonate $K_2CO_3$
ammonium carbonate $(NH_4)_2CO_3$.

## 9.10 Methods of preparing salts

If you want to prepare a salt, you must first see if it is soluble or insoluble, or if it is to be prepared in an anhydrous state. Then use the chart in figure 11 to choose the best method.

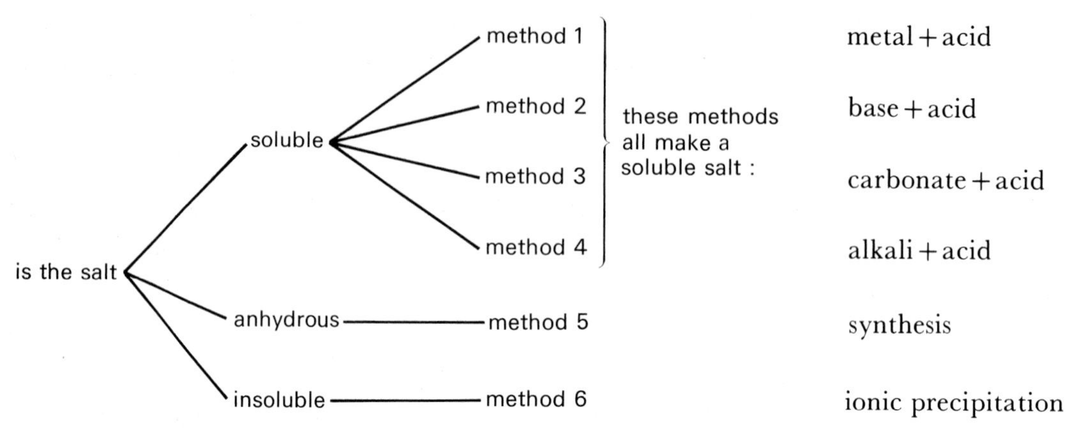


**Figure 11**
*Methods of making salts.*

**Metal + acid.** You have already seen this reaction as one of the properties of acids. In this method, the metal is dissolved in the acid, hydrogen is evolved, and a solution of the salt is left.
For example: the preparation of zinc sulphate-7-water.

$$Zn(s) + H_2SO_4(aq) \longrightarrow ZnSO_4(aq) + H_2(g).$$

**1** About $25\,cm^3$ of dilute sulphuric acid is placed in a small beaker and a few pieces of zinc are added. The zinc fizzes and hydrogen is evolved, so the reaction is best done in a fume-cupboard. If the zinc dissolves completely, more is added.

**2** When the zinc will not fizz any more, it means that all the hydrogen ions in the acid have been changed into hydrogen. The acid has been used up, leaving only a solution of zinc sulphate. It is filtered to remove the excess zinc.

**3** The salt solution is then carefully evaporated to remove some, but not all of the water. If it were completely dried it would lose its water of crystallisation.

**4** The partially evaporated solution is covered and left for several days to slowly crystallise. The crystals can be dried by blotting on filter paper. Figure 12 shows this procedure diagrammatically.

**Some exceptions.** This method is not suitable for very reactive metals such as sodium, potassium or calcium because the reaction would be too fast. On the other hand, it is too slow with unreactive metals such as lead. It will not work *at all* with copper, silver, gold and mercury. Nitric acid does not behave like other acids in its reactions with metals.

**Base + acid.** As in the first method, the base is added to the acid and it dissolves to form a solution of the salt. For example: the preparation of magnesium chloride-6-water.

$$MgO(s) + 2HCl(aq) \longrightarrow MgCl_2(aq) + H_2O(l).$$

**1** About 25 $cm^3$ of dilute hydrochloric acid are put into a small beaker and warmed over a bunsen flame. (Base + acid reactions usually need energy to start them.)

**2** White magnesium oxide is added to the warm acid, with stirring, until no more will dissolve. This means that all of the acid has been neutralised. It is filtered to remove the excess base.

**3, 4** Crystals of the salt are made by the same procedure as above. (See figure 12.)

**Carbonate + acid.** This method is very similar to the other two. The carbonate is added to the acid. It fizzes, evolving carbon dioxide leaving a solution of the salt. For example:
the preparation of copper(II) sulphate-5-water.

$$CuCO_3(s) + H_2SO_4(aq) \longrightarrow CuSO_4(aq) + H_2O(l) + CO_2(g).$$

**1** About 25 $cm^3$ of dilute sulphuric acid is placed in a small beaker, and solid copper(II) carbonate is added with stirring until no more will dissolve. This means that all the acid has been neutralised and that only the blue solution of copper(II) sulphate is left.

**2** The mixture is filtered to remove excess copper(II) carbonate.

**3, 4** Crystals of the salt are made by the same procedure as above. (See figure 12.)

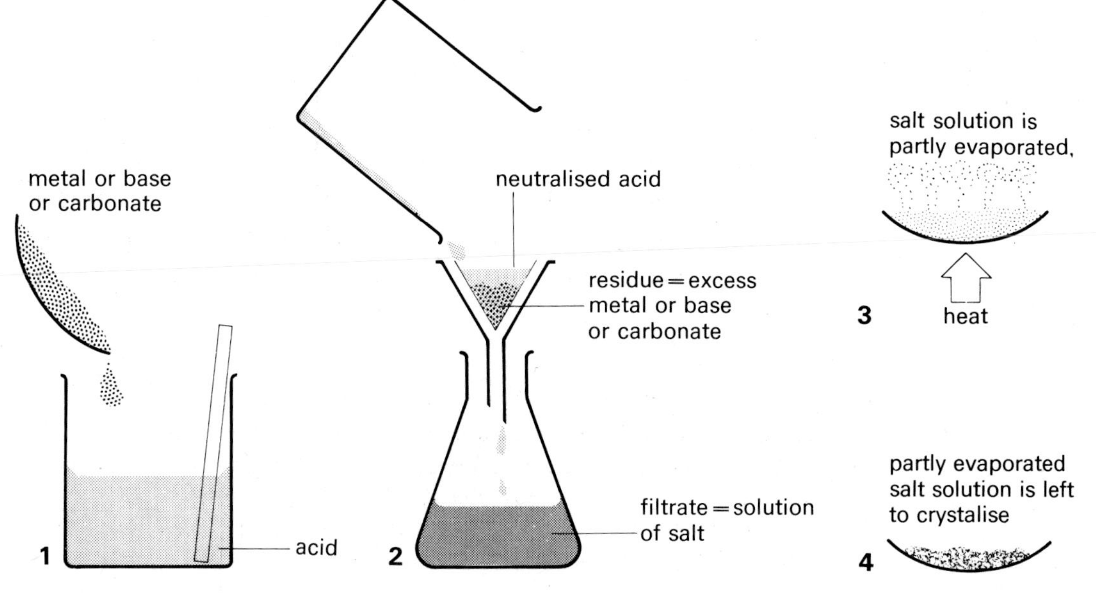


**Figure 12** *How to make crystals of any soluble salt.*

**Alkali + acid.** This method is a little different because alkalis are usually in the form of solutions, so it is difficult to know when the acid has been exactly neutralised. The acid and alkali must be reacted by a process called *titration*. Making sodium chloride from sodium hydroxide and hydrochloric acid is a typical example:

$$NaOH(aq) + HCl(aq) \longrightarrow NaCl(aq) + H_2O(l)$$

**1** Exactly 25 $cm^3$ of sodium hydroxide are put into a conical flask by means of a pipette. A pipette has only one gradation. When it is filled to that mark, it contains an exact quantity of liquid.

**2** Two or three drops of litmus solution are added to the alkali.

**3** A burette is filled with dilute hydrochloric acid, to above the zero mark. Liquid is run out of the tap into a spare conical flask until the level of liquid in the burette drops to the zero mark. Make sure that there are no air bubbles trapped in the burette.

**4** Acid is run from the burette into the conical flask containing alkali and litmus. The contents are swished to mix them. The litmus is at first blue, but when sufficient acid has been added to neutralise the alkali, it turns purple. The purple colour is often difficult to see, so the acid must be run in slowly so that one drop of acid past the end point turns the litmus pink.

**5** The conical flask now contains an almost neutral solution of sodium chloride, but it is coloured pink.

**6** So the titration is repeated using fresh acid and alkali, but no litmus. The quantity of acid which will be required to neutralize the alkali is known from the reading on the burette from the first titration. Finally, the colourless solution of sodium chloride can be evaporated to give crystals. All the water can be evaporated off at once, because sodium chloride contains no water of crystallisation.

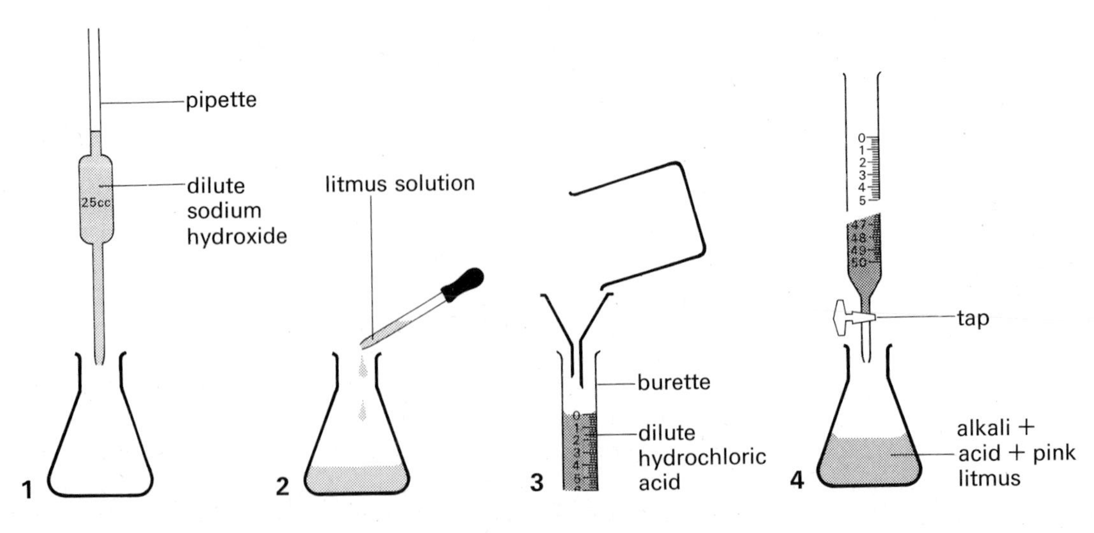


**Figure 13** *Stages of a titration.*

**Synthesis reactions.** If salts are required in an anhydrous form it is sometimes convenient to prepare them by direct combination or synthesis. In this method, the salt is made by reacting together the elements that make it up. For example: the preparation of iron(III) chloride.

$$2Fe(s) + 3Cl_2(g) \longrightarrow 2FeCl_3(s).$$

Figure 14 shows the apparatus that is used.

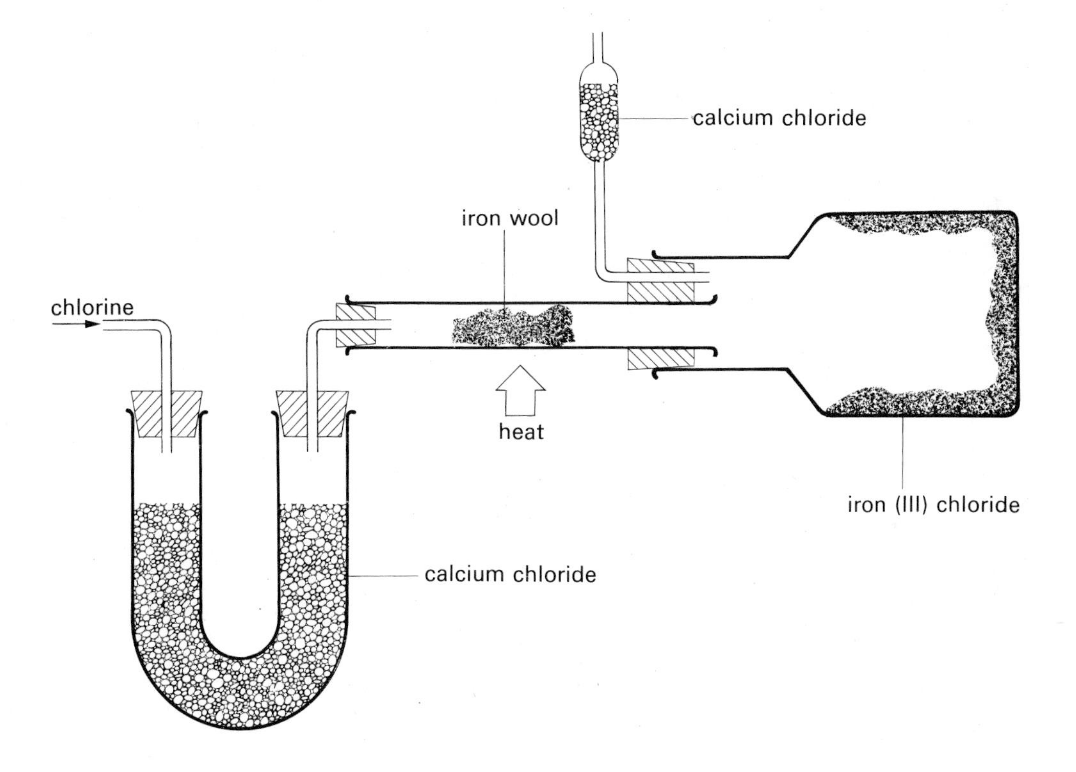


**Figure 14** *Direct synthesis of iron(III) chloride.*

The chlorine is dried before it is used. The iron is heated, because energy is needed to start the reaction, but after it has started, it is very exothermic and the iron glows brightly.

Anhydrous iron(III) chloride is formed as a vapour, but turns into a solid in the cold collecting jar. The calcium chloride tube allows excess chlorine to escape, but prevents moisture from entering the apparatus.

Anhydrous iron(III) chloride is black, but in its hydrated state it is yellow-brown.

**Ionic precipitation.** Insoluble salts cannot be made by any of the methods already described. Instead, they must be precipitated by adding solutions containing the correct ions. This method is called *ionic precipitation.*
For example: the preparation of lead iodide.
This may be made by adding lead ions to iodide ions:

$$Pb^{2+}(aq) + 2I^-(aq) \longrightarrow PbI_2(s)$$

But ions do not come on their own. We must choose two solutions which contain these ions. Two such solutions could be lead nitrate and potassium iodide:

$$Pb(NO_3)_2(aq) + 2KI(aq) \longrightarrow PbI_2(s) + 2KNO_3(aq).$$

**1** Put 25 $cm^3$ of the lead nitrate into a beaker and add a slightly larger quantity of potassium iodide solution of equal concentration. A bright yellow precipitate of lead iodide is formed.

**2** The mixture is filtered and the residue is retained.

**3** The residue is washed with water, and then carefully dried in an oven.

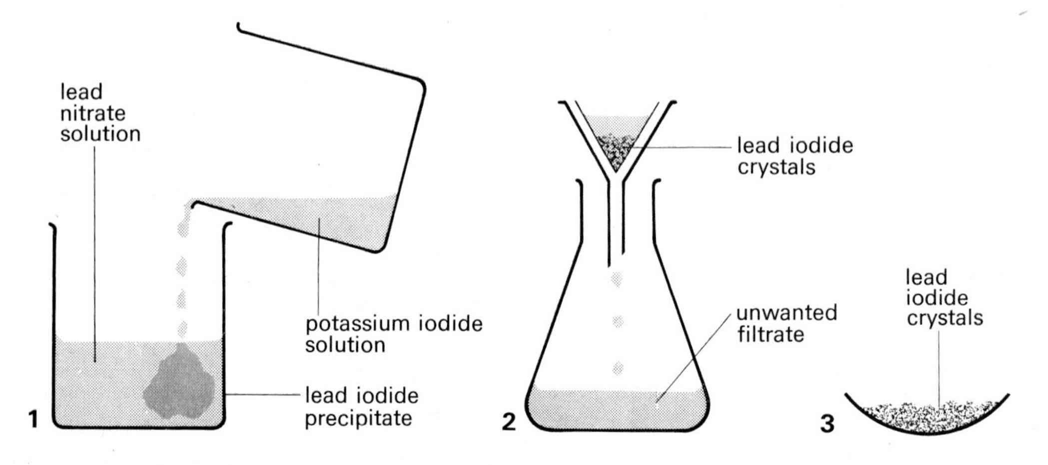


**Figure 15** *Preparation of lead iodide.*

## 9.11 Efflorescence

When left open to the air, some salts lose all or some of their water of crystallisation to the atmosphere, changing their appearance as they do so. This property is known as *efflorescence*, and the salts are said to effloresce. For example: sodium carbonate-10-water (commonly called washing soda), will lose 9 of its molecules of water over a few hours if left open to the air. In doing so, the crystals change from being transparent, to white and powdery.

$$Na_2CO_3.10H_2O(s) \longrightarrow Na_2CO_3.H_2O(s) + 9H_2O(g).$$

Sodium sulphate-10-water is another example of an efflorescent substance.

## 9.12 Deliquescence

When left open to the air, some salts do the opposite thing. They take in water from the atmosphere and dissolve themselves into a saturated solution. They are said to be *deliquescent*. For example: if blue copper(II) nitrate crystals are left open to the air for a few minutes, they quickly become covered with small drops of liquid, and soon turn into a sticky blue pool of liquid:

$$Cu(NO_3)_2(s) \longrightarrow Cu(NO_3)_2(aq).$$

Calcium chloride, zinc chloride-6-water and sodium hydroxide are other examples of deliquescent substances. Figure 16 shows some crystals which are stable.

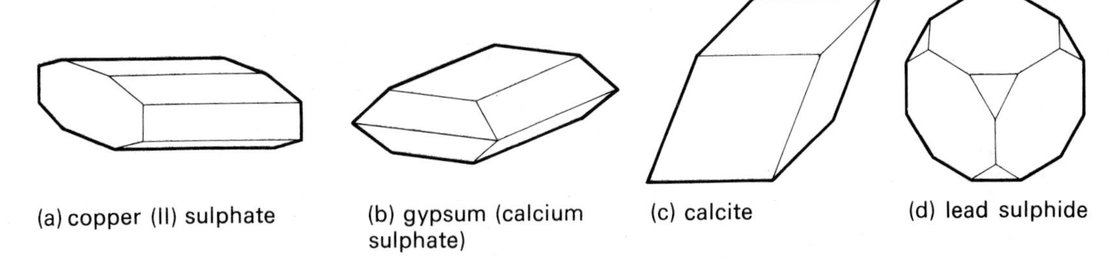


**Figure 16** *The shapes of some crystalline salts.*

## Summary

At the end of this chapter, you should be able to:

**1** Explain the term 'acid' and give examples of common acids.

**2** Describe the reactions of acids with indicators, metals, bases, alkalis and carbonates.

**3** Explain the difference between bases and alkalis.

**4** Explain what is meant by neutralisation.

**5** Explain what is meant by ionic precipitation.

**6** Explain how an indicator works.

**7** Distinguish 'strength' and 'concentration' for acids and alkalis.

**8** Describe the pH scale and its use with Universal Indicator.

**9** Say what a salt is.

**10** Explain the importance of water of crystallisation to some salts.

**11** Describe a test for water.

**12** Say whether or not any particular salt is soluble.

**13** Choose a suitable method for preparing a salt.

**14** Describe how a salt may be made by reacting: a metal with an acid; a base with an acid; an alkali with an acid; a carbonate with an acid; by synthesis and by ionic precipitation.

**15** Explain the terms efflorescence and deliquescence.

# Cleaning up clean water

*Even clean water is not pure. It contains ions which need to be removed before the water is any good for industrial processes such as electroplating, and for household uses such as washing. This process is known as de-ionisation and there are special chemicals which can do it. They are called ion exchange resins.*

Many modern ion exchange resins are fused onto a synthetic chemical called polystyrene which is made into small beads about 1 mm in diameter. If one of these beads were magnified about a million times, it would look rather like a dry bath sponge: the solid parts of the sponge being inter-linked chains, with sulphonic acid groups ($-SO_3^-H^+$) branching off them.

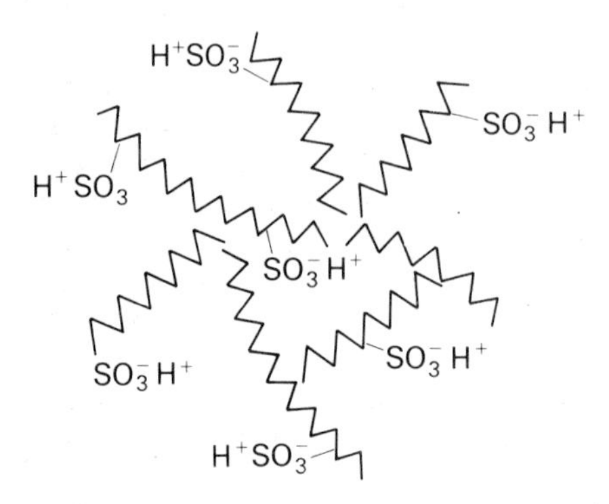


*The structure of ion exchange resin (greatly magnified).*

**Where have all the ions gone ?**

When impure water is passed through a container of the ion exchange resin beads, sulphonic acid groups swop their hydrogen ions for the unwanted metal ions in the water. These may be calcium or magnesium ions in hard water or heavier metal ions in industrial waste.

The water that emerges from the resin contains no metal ions, but more hydrogen ions. There are hydrogen ions in pure water anyway – so the water has been purified. It is said to have been *de-ionised.* If complete de-ionisation is required, the water can also be

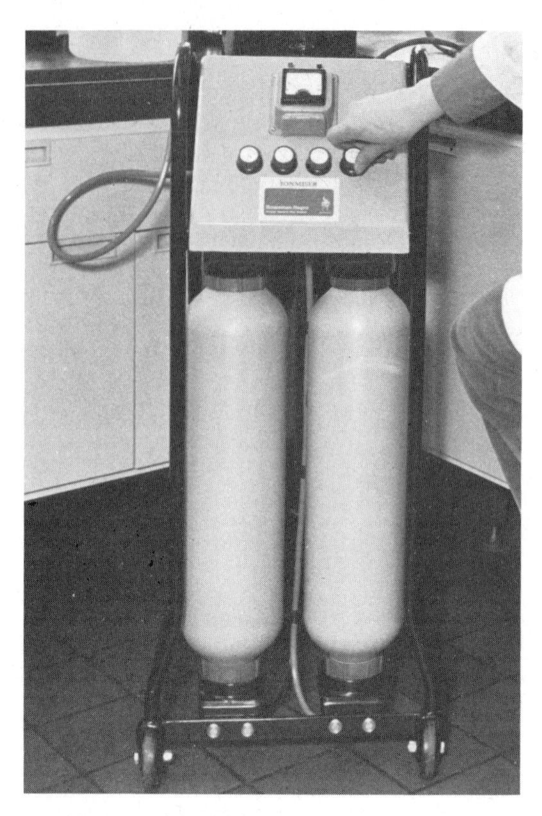

*Modern ion exchange equipment.*

passed through a slightly different resin which swops negative ions for hydroxide ions. Water that has passed through both these types of resin will have had all ions removed except hydrogen and hydroxide ions, and that is pure water. This diagram shows how hard water containing calcium sulphate might be de-ionised by such a method.

Very often, the two resins are mixed in the same container.

Obviously, the resins will run out of hydrogen and hydroxide ions after a time, and will no longer be able to de-ionise water. But when this happens, they can be regenerated. The hydrogen ions can be replaced by washing

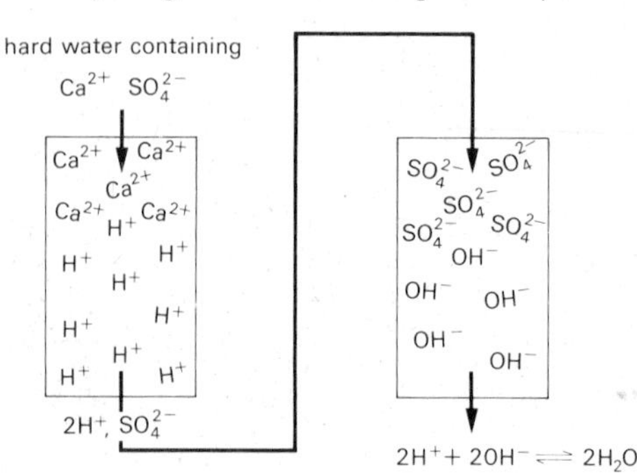

the resin with an acid solution and the hydroxide ions can be replaced by washing it with an alkaline solution.

It is a simple matter to find out when the resins are exhausted. Whilst the resins are still fresh, few ions are left in the water, because the hydrogen and hydroxide ions which are produced make up covalent water. In this state, it has a low electrical conductivity, whereas water containing a lot of ions conducts electricity well. Thus a simple conductivity meter placed in a pipe after the water has been treated will indicate a low reading while it is working efficiently. However, as soon as ions start to pass through unchanged, the meter reading rises sharply.

**Ion-free water at work**

The main application of ion exchange is in the treatment of water for boilers. In high performance boilers, such as those used in power stations, de-ionised water must be used.

Many industrial processes need water for rinsing. If water containing dissolved salts were used, a thin layer of solid might be left on the object. Transistors, for instance, are etched with acid or alkali and then washed with water since even tiny amounts of ionic impurities lower their electrical efficiency. De-ionised water is excellent for this purpose and, since it picks up only tiny quantities of impurities, it can be recycled through the de-ioniser and used again.

In one car manufacturing plant, when de-ionised water was introduced for washing the parts on the electroplating line, rejects dropped from 30% to zero. The washing water used in plating processes may pick up ions of metals, such as nickel and chromium, which are not only valuable but may exceed the quantities that may be legally discharged into rivers and sewage works. The 'swills' can be passed through a suitable resin which absorbs those ions – the purified water can then be reused.

Acids are used for etching steel before tin-plating or enamelling (to remove oxide impurities deposited on the steel while it is wet). The acid gradually becomes converted to the metal salt and so becomes useless. If it is passed through a hydrogen ion exchanger, the metal ions are replaced by hydrogen ions, so reforming the acid.

Computers need to work under strictly controlled conditions of temperature and humidity. Water is injected into the room in tiny droplets to raise the humidity. De-ionized water is ideal for this pupose, because if tap water were used, some would evaporate and deposit dissolved salts as a fine dust on the apparatus.

Ion exchange reains are also used in extracting metals from ores. For example, the residues from South African gold mines contain tiny quantities of uranium which are in very dilute solution. If this solution is passed through an ion exchange resin, the ions containing the uranium are absorbed. When the resin is regenerated, the ions are displaced as a concentrated solution from which the uranium can be obtained.

*De-ionised water used to clean car bodies before painting.*

**1** Put the following substances into three lists: one for acids, one for bases and one for salts.

| | | | |
|---|---|---|---|
| $NaNO_3$ | $H_3NSO_3$ | $CoSO_4$ | $FePO_4$ |
| $HClO_3$ | $(NH_4)_2HPO_4$ | $H_2S$ | $Na_2O$ |
| LiI | $H_2SiO_3$ | $Ba(OH)_2$ | $Fe(OH)_2$. |

**2** Write down the equations (words first, and then symbols) for any reactions that dilute hydrochloric acid might have with:

**a** iron **b** zinc oxide **c** potassium hydroxide
**d** sodium carbonate **e** copper.

**3** What is a base?
What happens when neutralisation takes place?

**4** What are alkalis?
Describe, with the equations where necessary, how potassium hydroxide will react with:

**a** litmus solution **b** dilute sulphuric acid
**c** cobalt chloride solution.

**5** Two acids both have the same concentration, but different strengths. How can this be? Explain how you would use universal indicator solution to distinguish between them.

Which of the three pH numbers 1, 5 and 13 corresponds to the stronger acid, and which to the weaker one?

**6** What would be the colour of:

**a** phenolphthalein in dilute hydrochloric acid?
**b** methyl orange in sodium hydroxide solution?
**c** bromothymol blue in dilute nitric acid?
**d** methyl orange in baking powder solution?
**e** phenolphthalein in health salts?
**f** bromothymol blue in orange juice?

Copper(II) sulphate solution is very slightly acidic. Why could you not use bromothymol blue indicator to distinguish it from sodium hydroxide solution?

**7** Aluminium nitrate is a hydrated salt whose formula can be written as $Al(NO_3)_3 . nH_2O$, where $n$ is a whole number.

**a** Calculate the relative molecular mass of the anhydrous salt.

**b** Calculate the relative molecular mass of water.

**c** If the relative molecular mass of hydrated aluminium nitrate is 375, calculate $n$.

**8** You are given a colourless liquid which might be water. You have no thermometer, but you do have some copper(II) sulphate-5-water crystals. Explain how you could test the liquid to see if it is water.

**9** State whether the following salts are soluble or insoluble in water:

calcium carbonate; sodium nitrate; copper(II) nitrate; ammonium carbonate; cobalt(II) sulphate; silver chloride; lead sulphate; barium carbonate; calcium chloride; zinc nitrate; potassium sulphate; iron(III) chloride.

**10** Suggest, without giving experimental details, the best methods for preparing the following salts:

zinc chloride; silver chloride; anhydrous aluminium chloride; lead sulphate; anhydrous zinc chloride; calcium carbonate; potassium chloride; sodium sulphate; copper(II) nitrate; calcium nitrate.

**11** Give full experimental details of how you would prepare:

**a** aluminium chloride from aluminium
**b** copper(II) nitrate from copper(II) oxide
**c** potassium sulphate from potassium hydroxide
**d** cobalt chloride from cobalt carbonate
**e** silver chloride from silver nitrate
**f** anhydrous iron(II) chloride from hydrogen chloride gas.

**13** Explain the meaning of the terms efflorescence and deliquescence. Can you suggest a part of the World where copper(II) nitrate might not be deliquescent?

Why is solid calcium chloride a good drying agent for damp gases?

# 10 Oxygen

H He
Li Be B C N O F Ne
Na Mg Al Si P S Cl Ar
K Ca Sc Ti V Cr Mn Fe Co Ni Cu Zn Ga Ge As Se Br Kr
Rb Sr Y Zr Nb Mo Tc Ru Rh Pd Ag Cd In Sn Sb Te I Xe

## 10.1 We've been here before

If you have read chapter 8, then you will know a lot about oxygen already. It includes sections on respiration, photosynthesis, combustion and the extraction of oxygen from the air. The uses of oxygen are mentioned as well. This chapter is about the different ways of preparing oxygen in the laboratory and the ways in which it reacts with other substances.

## 10.2 The laboratory preparation of oxygen

**Heating oxygen-rich compounds.** Many compounds contain oxygen which may be removed by heating. These compounds provide a simple means of obtaining a supply of oxygen.

**1** Heating lead(IV) oxide. This is a dark brown powder, which on heating gives off oxygen and turns yellow as it decomposes to leave lead(II) oxide:

$$\underset{\text{lead(IV) oxide}}{2PbO_2(s)} \longrightarrow \underset{\text{lead(II) oxide}}{2PbO(s)} + O_2(g).$$

**2** Heating dilead(II) lead(IV) oxide. This is the chemical name for a substance better known as red lead. It is a bright red powder, which when heated also turns into yellow lead(II) oxide, evolving oxygen as it does so:

$$\underset{\text{red lead}}{2Pb_3O_4(s)} \longrightarrow \underset{\text{lead(II) oxide}}{6PbO(s)} + O_2(g).$$

**3** Heating potassium nitrate. This substance is composed of colourless crystals which have a high melting point. Quite a high temperature is needed to decompose the crystals. When they melt, bubbles of oxygen are evolved, to leave a pale yellow salt called potassium nitrite:

$$\underset{\text{potassium nitrate}}{2KNO_3(s)} \longrightarrow \underset{\text{potassium nitrite}}{2KNO_2(s)} + O_2(g).$$

**4** Heating lead nitrate crystals. These are white in colour. When they are heated, they decompose noisily. Noisy decomposition is called *decrepitation*.

$$2Pb(NO_3)_2(s) \longrightarrow 2PbO(s) + 4NO_2(g) + O_2(g).$$

lead nitrate — lead(II) oxide — nitrogen(IV) oxide

Two gases are given off: oxygen, and brown nitrogen(IV) oxide which is very poisonous. This means that this is not a good laboratory method for making oxygen.

Most other nitrates behave in this way and you can read more about them in chapter 16.

**Heating potassium chlorate with a catalyst.** If potassium chlorate were to be heated on its own, a high temperature would be needed to decompose it.

Addition of a small quantity of manganese(IV) oxide catalyst, makes a lower temperature possible:

$$2KClO_3(s) \longrightarrow 2KCl(s) + 3O_2(g).$$

potassium chlorate — potassium chloride

Figure 1 shows the apparatus which could be used.

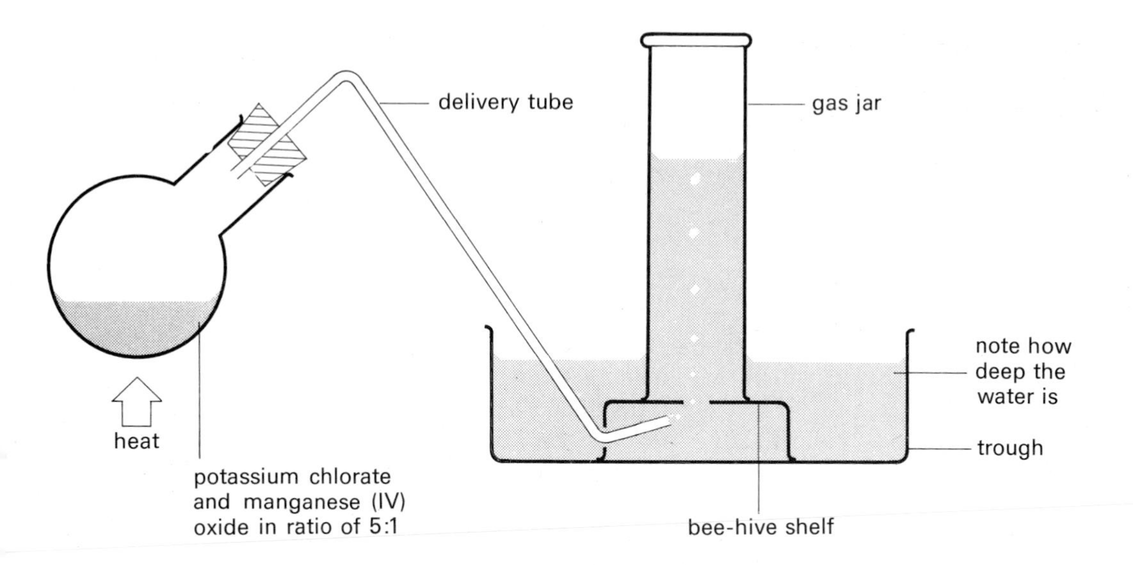


**Figure 1** *Preparing of oxygen from potassium chlorate.*

This method of collection is called downward displacement of water. Great care must be taken when heating potassium chlorate because it is a very powerful oxidising agent. This means that if it is allowed to mix with carbon, or any similar substance like fluff or dust – it can easily catch fire or explode.

**Decomposition of hydrogen peroxide.** Perhaps the most frequently used method of preparing oxygen in the laboratory is the decomposition of hydrogen peroxide using manganese(IV) oxide as a catalyst:

$$\underset{\text{hydrogen peroxide}}{2H_2O_2(l)} \longrightarrow 2H_2O(l) + O_2(g).$$

Hydrogen peroxide is a colourless liquid similar in appearance to water. When the catalyst is added, the hydrogen peroxide fizzes and decomposes, without any heating being necessary, leaving just water, and of course the catalyst. More hydrogen peroxide may be added to produce more oxygen. Figure 2 shows the apparatus that might be used.

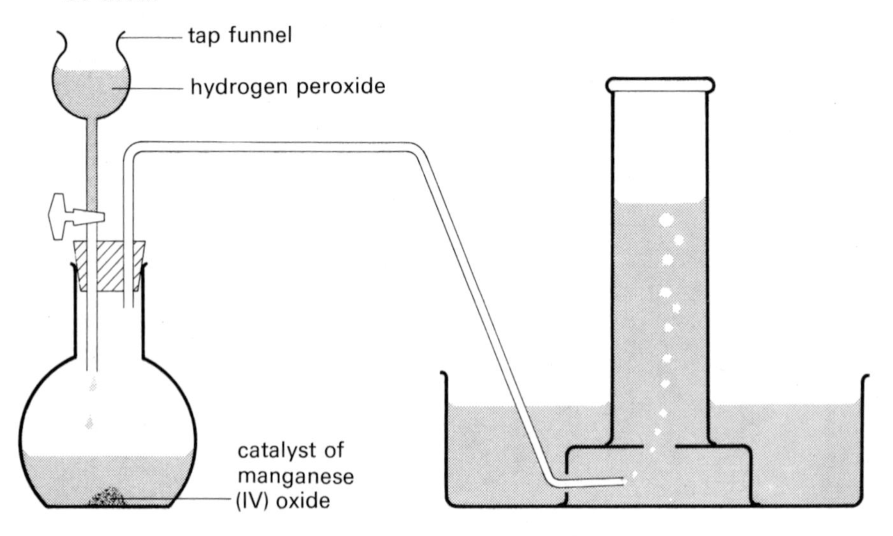


**Figure 2**

A tap funnel is used so that the amount of hydrogen peroxide which is added can be controlled. This makes it possible to control the speed at which the oxygen is made.

The oxygen produced by this method is wet, because it has been collected over water in the trough. If dry gas is required, then it must be bubbled through a wash bottle containing concentrated sulphuric acid, which is a drying agent. But, you cannot collect the gas over water after that because you would make it wet again! It can either be collected straight into a gas jar, or be drawn into a syringe.

## 10.3 Properties of Oxygen

The properties of gases can be divided into two sorts: physical properties, concerned with what they are like; and chemical properties, concerned with what they do. The following section deals with the physical and chemical properties of oxygen.

**Physical properties of oxygen.** Oxygen gas is: colourless, tasteless, odourless, and slightly soluble in water.

When water is heated, bubbles of air come out of it just before it boils. These bubbles are mainly oxygen, because nitrogen (the other component of air) is virtually insoluble in water. The slight solubility of oxygen in water is important for the fish who use the dissolved oxygen to breathe.

**Chemical properties of oxygen.** The following five sections deal with the various chemical properties of oxygen.

**Oxygen encourages combustion.** Substances which burn in air burn more brightly in oxygen. This is because there is no nitrogen to get in the way. The usual test for oxygen is to put a glowing wood splint into the gas. If it bursts into flame, the gas is oxygen.

## 10.4 Reaction of metals with oxygen

Oxygen reacts with most metals. To examine the reaction of a metal with oxygen, the metal often has to be in powder form since a solid piece will not heat up sufficiently to ignite. The material is placed in the cup of a combustion spoon, heated, and then lowered into a gas jar of oxygen. (See figure 3.)

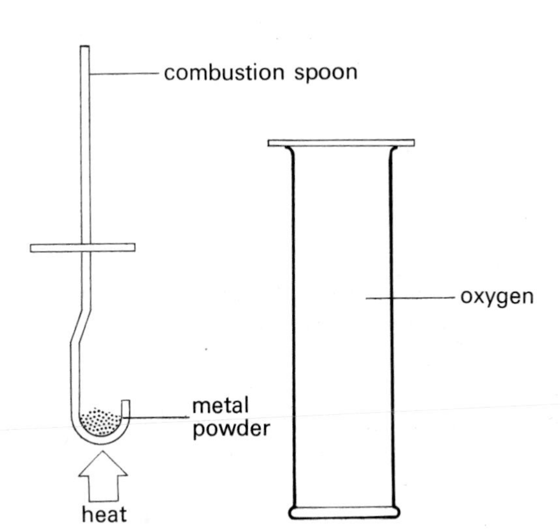


**Figure 3**
*Heating a sample of metal in oxygen.*

Figure 4 shows the reactions of some of the more common metals with oxygen. You can use it for reference.

| metal | description of reaction | equation |
|---|---|---|
| **potassium** | Potassium is a soft metal, kept under oil to try to protect it from air and moisture; when heated, it melts very easily and burns with a lilac flame to leave a white powder. | $K(s) + O_2(g) \longrightarrow KO_2(s)$. potassium oxide |
| **sodium** | This is very similar in appearance to potassium. When heated it melts at a slightly higher temperature, and burns vigorously with a yellow flame to leave a pale yellow solid. | $2Na(s) + O_2(g) \longrightarrow Na_2O_2(s)$. sodium oxide |
| **calcium** | This metal comes in small granules. It does corrode, but is not kept under oil. When heated it does not melt, but it burns with a brick red flame to leave a white solid. | $2Ca(s) + O_2(g) \longrightarrow 2CaO(s)$. calcium oxide |
| **magnesium** | This silver metal comes in ribbon form and corrodes slightly in air. When heated, it melts just before it burns. It does so with a blinding white flame, to leave a white ash. | $2Mg(s) + O_2(g) \longrightarrow 2MgO(s)$. magnesium oxide |
| **aluminium** | When silver aluminium powder is heated in oxygen it glows with a white flame which is less bright than that of magnesium, to leave a white powder. | $4Al(s) + 3O_2(g) \longrightarrow 2Al_2O_3(s)$. aluminium oxide |
| **zinc** | Grey zinc powder burns slowly with a dull red flame to produce wispy yellow/green flakes which are white when they cool. | $2Zn(s) + O_2(g) \longrightarrow 2ZnO(s)$. zinc oxide |
| **iron** | Grey iron filings sparkle when sprinkled into a flame. A pile of them will only just glow to leave a black solid. Steel wool burns with a bright sparkle leaving solid black lumps. | $3Fe(s) + 2O_2(g) \longrightarrow Fe_3O_4(s)$. iron(II) iron(III) oxide |

| | | |
|---|---|---|
| **copper** | Pink copper turns orange as it corrodes in the air. When heated it immediately becomes coated with black powder. | $2Cu(s) + O_2(g) \longrightarrow 2CuO(s)$. copper oxide |
| **mercury** | Silvery coloured liquid. Mercury becomes covered with red oxide; this coating disappears if it is heated more strongly. | $2Hg(l) + O_2(g) \longrightarrow 2HgO(s)$. mercury(II) oxide |
| **silver** | Does not oxidise. | |
| **gold** | Does not oxidise. | |

**Figure 4** *The reactions of metals with oxygen.*

Figure 5 shows what happens when solutions of the oxides formed by metals are tested with litmus solution.

| **metal oxide** | **solubility** | **effect on litmus solution** |
|---|---|---|
| potassium oxide | very soluble | turns blue |
| sodium oxide | very soluble | turns blue |
| calcium oxide | quite soluble, gets hot | turns blue |
| magnesium oxide | sparingly soluble | turns blue |
| aluminium oxide | sparingly soluble | turns blue |
| zinc oxide | very sparingly soluble | turns blue |
| iron(III) oxide | not soluble | none |
| copper(II) oxide | not soluble | none |
| mercury oxide | not soluble | none |

**Figure 5** *The solubility of metal oxides and their effect on litmus.*

**Metal oxides: bases or alkalis?** All metal oxides are bases. Remember that some of the metal oxides are soluble, and they have the special name, alkalis.

The oxides of potassium and sodium are very soluble, so their solutions are strong alkalis.

The oxides of calcium and magnesium are only slightly soluble – their solutions are weak alkalis.

The oxides of aluminium and zinc are only very slightly soluble – their solutions are just about alkaline.

The oxides of the other metals are not soluble. They are not alkaline at all.

You can see from figure 5 that the more reactive the metal, the more alkaline is its oxide.

**The Activity Table.** The metals may be placed in a 'league table' according to their reactivity with oxygen. (See figure 6.) This is called the Activity Table. There are more details of this in chapter 18.

| metal | reactivity |
|---|---|
| potassium | strongly reactive |
| sodium | |
| calcium | |
| magnesium | |
| aluminium | |
| zinc | |
| iron | |
| tin | |
| lead | |
| copper | |
| mercury | |
| silver | weakly |
| gold | reactive |

**Figure 6**
*The Activity Table.*

## 10.5 Reactions of non-metals with oxygen

Figure 7 shows the reactions of the more common non-metals with oxygen.

Figure 8 goes on to show the solubility of the oxides of the non-metals, and their effect on litmus solution. It shows that all non-metal oxides are soluble in water and form solutions of acids. This means that their solutions will turn litmus solution pink or red.

| non-metal | description of reaction | equation |
|---|---|---|
| **carbon** | This black solid burns slowly with a yellow-white flame to form a colourless gas. | $C(s) + O_2(g) \longrightarrow CO_2(g)$. carbon dioxide |
| **phosphorus** | This yellow solid, kept under water because it catches fire in air at room temperature, burns very brightly to produce clouds of white smoke which settle out as a white solid. | $P_4(s) + 5O_2(g) \longrightarrow P_4O_{10}(s)$. phosphorous(V) oxide |
| **sulphur** | This yellow solid melts and catches fire, burning with a blue flame. Produces misty gas with choking smell. | $S(s) + O_2(g) \longrightarrow SO_2(g)$. sulphur dioxide |

**Figure 7** *The reactions of non-metals with oxygen.*

| non-metal oxide | solubility | equation | effect on litmus |
|---|---|---|---|
| carbon dioxide | mildly soluble | $CO_2(g) + H_2O(l) \longrightarrow H_2CO_3(aq).$ | slowly turns pink |
| phosphorus(V) oxide | very soluble | $P_4O_{10}(s) + 6H_2O(l) \longrightarrow 4H_3PO_4(aq).$ | turns red |
| sulphur dioxide | very soluble | $SO_2(g) + H_2O(l) \longrightarrow H_2SO_3(aq).$ | turns red |

**Figure 8** *The solubility of non-metal oxides and their effect on litmus.*

## 10.6 Reaction of elements with oxygen

The preceding sections tell us that:

**1** All substances which burn in air burn much more brightly in oxygen. This is the basis of the glowing splint test for oxygen.

**2** Metals burn in oxygen to form oxides which are called basic oxides or more simply, bases. Some bases are soluble in water and are given the special name, alkalis.

**3** All non-metals which burn in oxygen form oxides called acidic oxides. All acidic oxides dissolve in water to form solutions of acids.

## Summary

At the end of this chapter you should be able to:

**1** Name several compounds which decompose on heating to give off oxygen, and write the equations for these reactions.

**2** Draw the apparatus, write the equation and describe the preparation of oxygen from potassium chlorate.

**3** Draw the apparatus, write the equation and describe the preparation of a dry sample of oxygen from hydrogen peroxide.

**4** Describe the physical properties of oxygen.

**5** Describe a test for oxygen.

**6** Describe the reactions of the common metals with oxygen, giving details of the solubility of their oxides and their effect on litmus.

**7** Describe the reactions of the common non-metals with oxygen giving details of the solubility of their oxides and their effect on litmus.

**8** Explain how metals can be placed in an Activity Table.

**9** Distinguish between basic and acidic oxides.

# Breathing out of bottles

*At high altitudes and below the sea, it is necessary to provide oxygen to support breathing. In the first article, Sir Edmund Hillary writes about some of the problems encountered 7 500 metres up Everest – the second article tells you about deep sea problems.*

**Time to breathe**

We prepared everything for the next day – loads were made up for the Sherpas and oxygen sets supplied with new bottles. The sun disappeared behind a distant peak and freezing cold descended. In order to guarantee ourselves a good night's sleep, we had our oxygen masks on. We could only use a low rate of flow but because of our inactivity at night even this had considerable effect.

I awoke feeling cold and uncomfortable. Our oxygen had run out and the whole camp was frozen and still. I glanced at my watch – it was very early, but I knew it would take a long time to prepare our breakfast and depart. Any task at high altitude takes a long time, as the brain and body, affected by the lack of oxygen, have little coordination or concentration.

We were taking with us three 'high altitude' Sherpas, and after a final check, we moved out of camp, across a large crevasse and into a series of steep and difficult ice walls.

We were moving steadily and rhythmically, taking no rests at all. My whole life was encompassed by the next step and the next breath. Forcing my lungs to the utmost, I sought to draw in the maximum of life-giving air – to relieve the dreadful feeling of deadness in my limbs and the pain in my chest. My whole body was crying out for a rest.

Suddenly I noticed two black dots on the snow, moving slowly and painfully down towards us. It must be John Hunt and Sherpa Da Namgyal! I asked Tenzing to prepare some hot drinks as quickly as possible, but knew this would take some time. Without bothering to use any oxygen, I started up the icy slopes.

Hunt and Da Namgyal were moving at a funereal pace. They seemed only able to walk fifty feet before slumping down on the ice for a rest. Spurred on by their obvious distress, I made rapid height, and was encouraged to find how quickly I could move without oxygen at this altitude. As the others came closer, I could tell by their stiff and clumsy movements just how tired they were. I reached them just as an exhausted John Hunt dropped to the ice once more. It was good to see them again and to feel I could help them. Hunt told me how he and Da Namgyal had carried their heavy loads up on to the South-east ridge as high as they possibly could. When they couldn't go a step farther and couldn't even crawl on their hands and knees, they left their loads in a little pile and struggled down.

I put John Hunt's arm over my shoulder and held him firmly around the waist and started down the ice-slope again. We were going very slowly and frequent rests were necessary. When John slipped down on to the ice once more, I realised that something more drastic was required. I left him sitting on the ice and raced back down to camp. I still had some oxygen left in my set, so I heaved it on to my back and returned. John was sitting slumped over on the ice and didn't seem to

*Tenzing on the summit of Everest.*

have moved. I put the oxygen mask on his face and turned on the maximum flow. It had an immediate effect. Soon he was able to drag himself to his feet and, with my assistance, move slowly down the ice-slope and finally up to the tents. We crawled inside, and as John lay resting on his sleeping-bag, I flooded the tent with oxygen. The door opened and Tenzing thrust in a hot drink to complete the recovery.

**Same problem beneath the sea**

At high altitudes above sea level it is impossible to do anything strenuous without extra oxygen.

Below sea level, the problem is much more immediate. Unlike fish, we are unable to extract the oxygen dissolved in the water. We breathe through lungs and must have our oxygen in gas form.

Divers can either have a long pipe trailing to the surface so that air may be pumped down to them. If they want to be more free to move, they must take air with them, compressed in bottles on their backs. The bottles are made of steel to withstand the pressure of the compressed air. They contain about one hour's supply of air.

However, there are other dangers. If a diver goes too deep the high pressure causes nitrogen as well as oxygen to dissolve in the blood. It is not used in any way and so it must come out again as the diver returns to the surface.

If the diver comes up slowly, the nitrogen is carried safely to the lungs where it is breathed out in the usual way. If the diver surfaces quickly, the nitrogen bubbles out of the blood while it is still in the blood vessels, causing the diver to get the 'bends'. This is very dangerous, often making the man unable to move because of the extreme pain. He may even become unconscious.

To prevent this from happening, and because it is dangerous to breathe pure oxygen at high pressures, divers breathe a mixture of 10% oxygen and 90% helium. The helium does not dissolve in the bloodstream, so they do not get the bends.

A curious effect is heard however if a diver speaks when he has his lungs full of this mixture. His voice sounds very high pitched and squeaky because the helium is a much lighter gas than nitrogen.

**Questions**

**1** Write short revision notes on: combustion, respiration, photosynthesis, the extraction of oxygen from the air, the uses of oxygen.

**2** Describe how you would prepare a dry sample of oxygen from lead(IV) oxide. Include a diagram of the apparatus you would use, and the equation for the reaction.

**3** Look at the equation for the effect of heat on lead nitrate. Nitrogen(IV) oxide is soluble in water. It also condenses to a liquid if it is dry, and cooled to about 0 °C. Can you devise an apparatus to get pure oxygen out of lead nitrate?

**4** Imagine you are a chemistry teacher. In tomorrow's lesson, you will show pupils how oxygen is made from hydrogen peroxide. You plan to collect the products from heating in oxygen the metals calcium, magnesium and iron, and the non-metals sulphur and carbon. After dissolving the oxides in water you are going to test them with litmus.

Draw up notes and diagrams to show exactly what you would do and what you would tell your pupils.

One clever-clogs in the class asks you if the acid made when sulphur dioxide dissolves in water is stronger than the acid formed when carbon dioxide dissolves in water. How would you answer his question?

**5** Four new elements have just been discovered and named after their discoverers. They are: Fredium (Fd), Bertium (Bt), Sidium (Sd), Gertrudium (Gt).
Look at this table of results, and answer the following questions.

| **element** | **description of burning in oxygen** | **solubility of oxide** | **pH of solution** |
|---|---|---|---|
| Fd | Powdered Fd burns quite brightly to form a white powder. | slightly soluble | 9 |
| Bt | A lump of Bt flares up brightly when heated slightly leaving a white solid. | very soluble | 14 |
| Sd | Does not appear to burn but turns grey on the surface. | not very soluble | 8 |
| Gt | Melts and burns quite easily to form a colourless gas. | quite soluble | 3 |

**a** Sort the new elements into metals and non-metals.
**b** Put those elements which are metals into the order in which they would appear in the Activity Table.
**c** Which of the oxides form alkalis in solution?
**d** Which elements form acidic oxides?
**e** Which of the oxides are bases?
**f** Which elements will probably react with acids to produce a salt and hydrogen?

# 11 Hydrogen

## 11.1 A simple atom

Hydrogen is the first element in the periodic table. It has an atomic number of 1 and also a mass number of 1. This is written as:

$^{1}_{1}H$.

This means that an atom of hydrogen consists of only one proton and one electron and no neutrons.

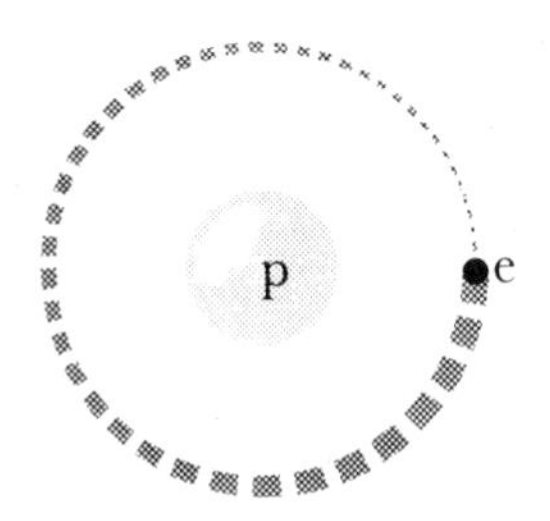


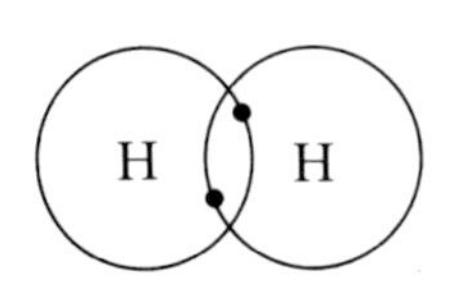


Hydrogen gas is made of molecules. In order to get full outside shells of electrons, two hydrogen atoms share their electrons and make a molecule. Remember that only 2 electrons are needed to fill the first shell.

## 11.2 Industrial sources of hydrogen

**Natural gas.** The majority of hydrogen used industrially is made from natural gas by a process called steam reforming. Natural gas – methane – is mixed with steam and passed over a catalyst of nickel at a high temperature and pressure. Carbon monoxide and hydrogen gases are produced:

$$CH_4(g) + H_2O(g) \longrightarrow 3H_2(g) + CO(g).$$

The temperature may be as high as 1000 °C and the pressure greater than 50 atmospheres. To remove the carbon monoxide, the gases are mixed with more steam, and passed over another catalyst of iron(III) oxide. The carbon monoxide is converted to carbon dioxide, and more hydrogen is made:

$$CO(g) + H_2O(g) \longrightarrow CO_2(g) + H_2(g).$$

Finally, the carbon dioxide is removed by dissolving it under pressure in water, so only the hydrogen is left.

**The Bosch process.** A small amount of hydrogen is still manufactured by the Bosch process. In this process steam is passed over white hot coke, and this endothermic reaction produces carbon monoxide and hydrogen:

$$C(s) + H_2O(g) \longrightarrow CO(g) + H_2(g).$$

This mixture of gases is called water gas. It was used as a fuel, since both gases burn. Hydrogen was separated from the carbon monoxide by the same method as in the previous section.

**As a by-product.** A lot of hydrogen is obtained as a by-product of other industrial processes. For example, when sodium hydroxide is made by the electrolysis of brine, hydrogen is made as well. (See chapter 17.)

## 11.3 Uses of hydrogen

**1** Probably the most important use of hydrogen is in the synthesis of ammonia. This is in turn made into fertilizers, nitric acid and explosives.

Hydrogen is mixed with nitrogen and heated to a temperature of 450 °C, at a pressure of 200 atmospheres, over a catalyst of iron. This produces ammonia:

$$N_2(g) + 3H_2(g) \xrightleftharpoons[\text{450 °C/200 atm.}]{\text{iron catalyst}} 2NH_3(g).$$

You can read more about this in chapter 16.

**2** Hydrogen is used in the manufacture of margarine.

**3** Hydrogen is used in the manufacture of methanol, which is used to make plastics and fertilizers. It is also used in the production of one of the starting materials for the synthesis of nylon.

## 11.4 Laboratory preparation of hydrogen

Hydrogen is produced whenever a metal reacts with a dilute acid:

metal + acid ⟶ salt + hydrogen.

Some metals react far too vigorously with acids, and some react very slowly, or not at all (chapter 9), so zinc is used because its rate of reaction is just right:

zinc + hydrochloric acid (dilute) ⟶ zinc chloride + hydrogen.

$$Zn(s) + 2HCl(aq) \longrightarrow ZnCl_2(aq) + H_2(g).$$

The apparatus for the preparation and collection of hydrogen is shown in figure 1.

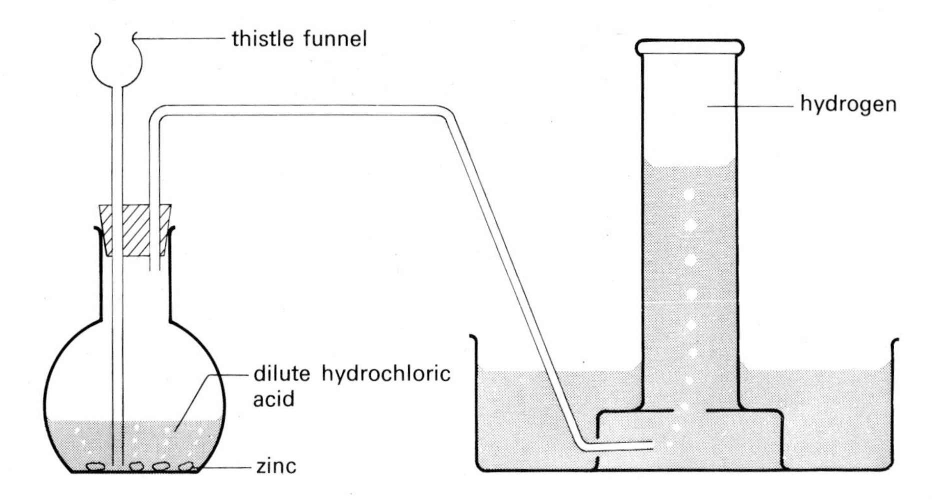


**Figure 1** *The preparation and collection of hydrogen.*

The zinc dissolves in the acid and bubbles of hydrogen are evolved. The reaction is quite exothermic. More acid might have to be added as the hydrogen ions are turned to hydrogen molecules.

If dry gas is required, then it must be bubbled through concentrated sulphuric acid, and collected in a syringe. However, because hydrogen is so light, it can easily be collected upwards into a gas jar, as in figure 2.

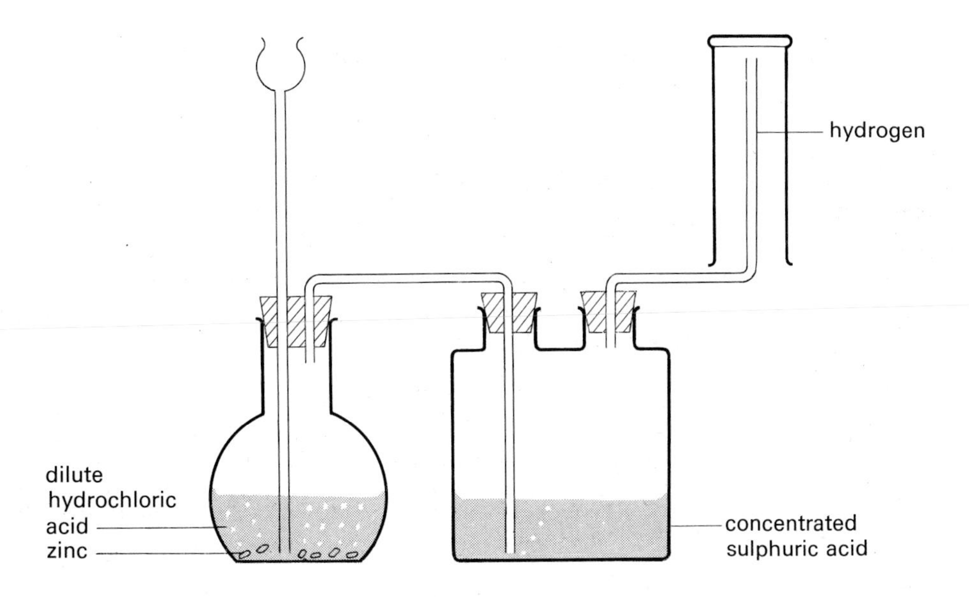


**Figure 2** *The collection of dry hydrogen.*

## 11.5 The physical properties of hydrogen

**1** It is odourless, tasteless and colourless.

**2** It is virtually insoluble in water.

**3** It is neither acid or basic.

**4** It is much lighter than air, in fact it is the lightest of all gases. This property could be illustrated in a number of ways.
**a** A balloon could be filled with hydrogen and allowed to float up to the ceiling.
**b** You could perform the gas jar trick. (See figure 3.)

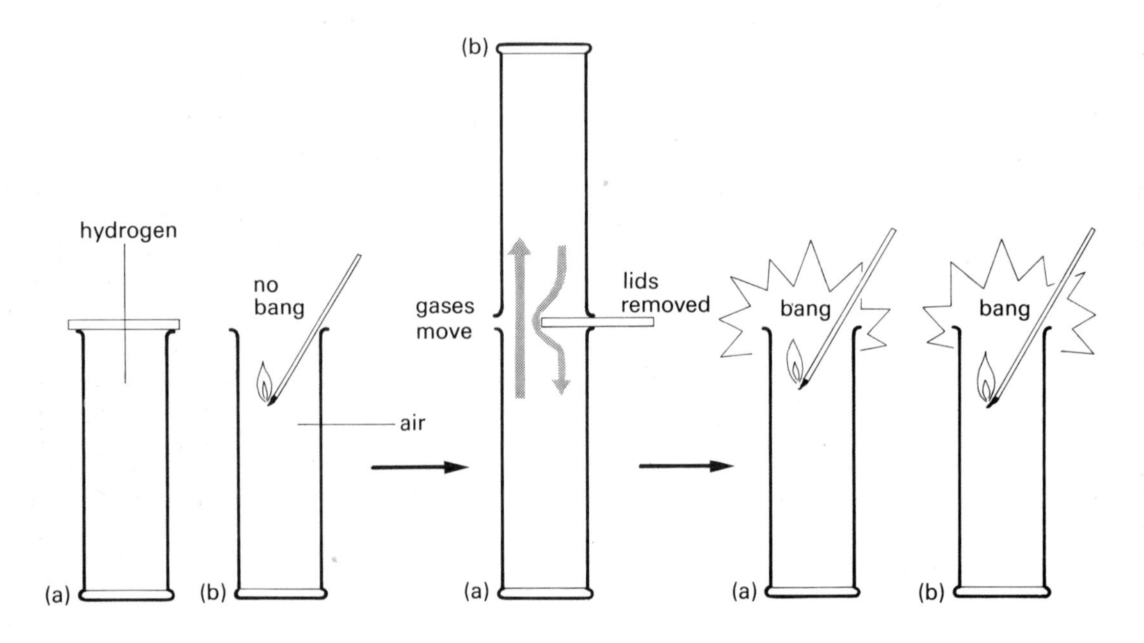


**Figure 3** *The gas jar trick.*

One gas jar is full of hydrogen. The other is full of air. Nothing happens when a lighted splint is put into the jar containing air, but an explosion would occur if it were put into the jar containing hydrogen. The air jar is placed on top of the hydrogen jar, and the glass lid is carefully removed for a few seconds and slid back into place. The gas jars are separated and each is tested with a lighted splint. Both explode!

This means that the hydrogen must have gone up and the air gone down quite quickly. The gases do not simply change places – they mix as well.

**5** Hydrogen molecules move much more quickly than air molecules, because they are much smaller and lighter. An experiment to demonstrate this is shown in figure 4.

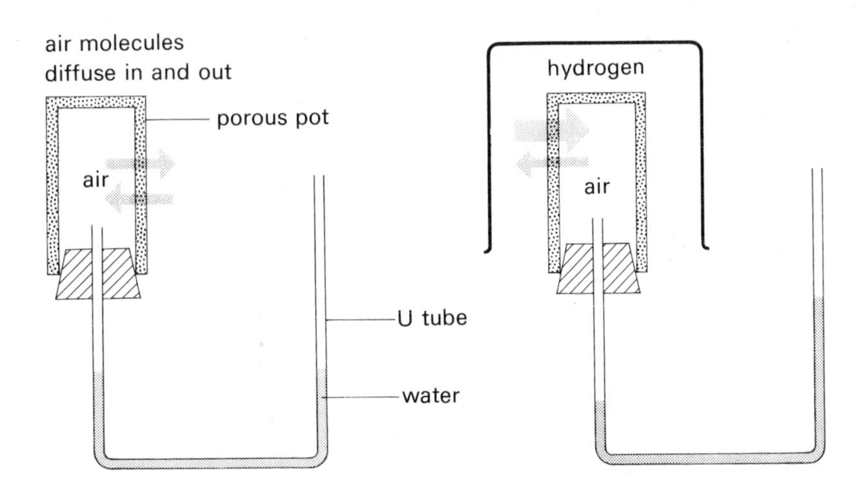


**Figure 4**
*The diffusion of hydrogen.*

The porous pot is made of unglazed porcelain – rather like flower pots (not the plastic ones). The porous pot contains air molecules and they are constantly moving, in and out of the pot through the millions of tiny holes that are in it. At first, the liquid levels in the U-tube are the same, because the pressure inside the porous pot is the same as that outside.

However, if hydrogen is put into the beaker around the porous pot, it too will start to pass through the little holes in the porous pot. Because the hydrogen molecules are smaller and faster than the air molecules, they can get into the pot quicker than the air molecules can get out. This increases the pressure in the pot, and the liquid in the U tube goes down on the left hand side.

This movement of gases is called *diffusion*.

*The surface of airships have to be specially treated to prevent the gas (nowadays helium) from leaking out by diffusion.*

## 11.6 The chemical reactions of hydrogen

**Hydrogen burns in air or oxygen.** If a jet of hydrogen is lit, it burns with a tiny blue flame and produces steam:

$$\text{hydrogen} + \text{oxygen} \longrightarrow \text{steam.}$$

$$2H_2(g) + O_2(g) \longrightarrow 2H_2O(g).$$

The steam may be condensed and collected in the apparatus shown in figure 5.

The hydrogen must flow through the apparatus for ten seconds before being lit, to make sure that all the air has been flushed out.

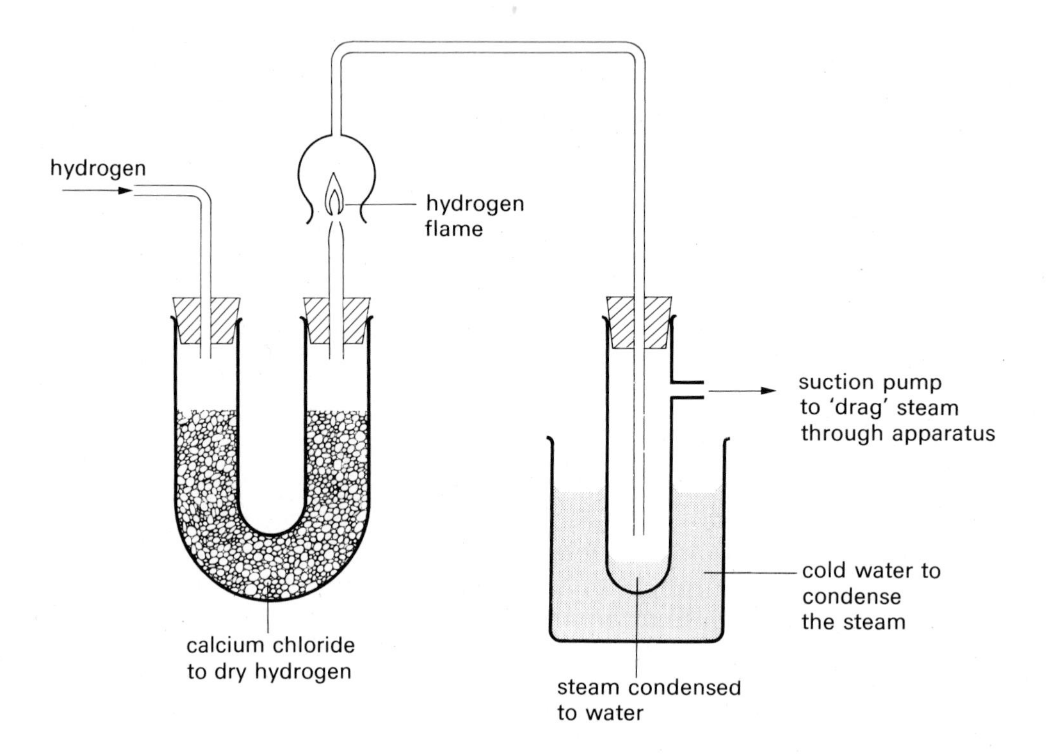


**Figure 5** *Burning hydrogen.*

A mixture of hydrogen and air (or better still oxygen), explodes when lit. This is the usual test for hydrogen. If you have a gas and you think it is hydrogen, collect a test tube of it and put a burning splint to its end. The hydrogen mixes with a small amount of air and explodes with a small squeaky pop.

The explosion is a very fast chemical reaction which produces a lot of energy in a short time.

**Hydrogen is a good reducing agent.** It will remove oxygen from metal oxides. Figure 6 shows the apparatus that might be used.

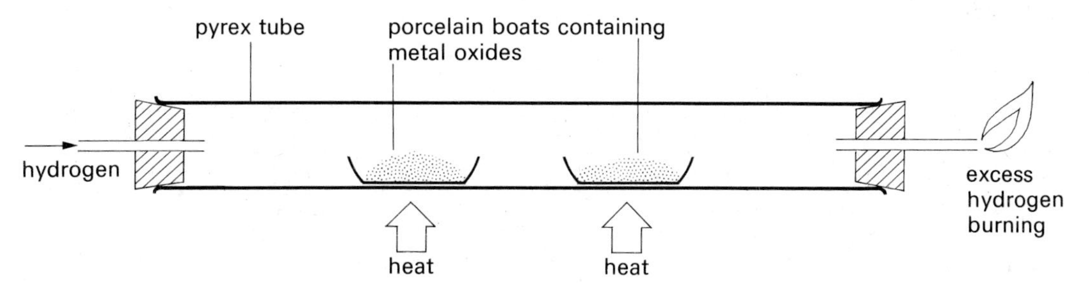


**Figure 6** *Hydrogen is a good reducing agent.*

The hydrogen is allowed to pass through the pyrex tube for ten seconds to flush out the air, and then the excess hydrogen is lit at the end of the tube.

Samples of metal oxides are placed in the porcelain boats and they are heated as the hydrogen flows over them. Lead(II) oxide and copper(II) oxide are easily reduced by this method. Oxides of more reactive metals cannot be reduced in this way. (See chapter 18.)

Suppose lead(II) oxide and copper(II) oxide were put into the apparatus. The black copper(II) oxide would be reduced to pink copper and the yellow lead(II) oxide would be reduced to silver coloured balls of lead:

$$CuO(s) + H_2(g) \longrightarrow Cu(s) + H_2O(g);$$

$$PbO(s) + H_2(g) \longrightarrow Pb(s) + H_2O(g).$$

The steam which is produced in each case is carried out in the hydrogen.

When the reaction is complete, the hydrogen is allowed to flow until the newly formed metals have cooled down, to prevent them from oxidising in the air.

**Summary**

At the end of the chapter you should be able to:

**1** Draw diagrams of a hydrogen atom and a hydrogen molecule, showing the arrangement of their electrons.

**2** Describe how hydrogen is made industrially, and state its main uses.

**3** Describe how hydrogen is made and collected in the laboratory.

**4** Describe the physical properties of hydrogen.

**5** Describe an experiment to demonstrate the lightness of hydrogen.

**6** Distinguish between burning and exploding of hydrogen.

**7** Give an example of hydrogen acting as a reducing agent.

# The Hindenburg

*Hydrogen has always been a very important chemical industrially. It was used for many years to build 'lighter than air' ships. The last one to be built for commercial use, was called 'The Hindenburg' . . .*

**The Hindenburg**

The Hindenburg was a huge German airship that was built in 1936 by the famous Zeppelin airship Corporation. Contrary to many people's ideas, this airship was not a balloon pumped full of gas, but was a vast framework of rings and girders, making a rigid structure nearly fifty metres high and 250 metres long. The framework was made of a copper and aluminium alloy called Duralumin, which was lighter than steel, but stronger than ordinary aluminium. The outside of the structure was covered with cotton fabric. This was stretched tight and sewn into place by riggers trained as sailmakers in the German Navy. It was painted with a coat of dope, and, then a coat of silver paint to protect it from the sun and rain.

The ship was hollow inside, save for the skeleton of metal girders. The hydrogen used to make the ship float was contained in 16 gas-bags which were secured inside the ship. They were made of cotton fabric lined with a gas-proof chemical, and between them they held almost 200 000 cubic metres of hydrogen.

Four powerful 16 cylinder diesel engines, attached to the outside of the ship could develop 5000 h.p. The diameter of the propellors was a gigantic seven metres.

With the ship fully laden, the hydrogen had to be able to provide lift for the ship's total

weight of 236 tonnes. It could do so with ease, and its engines could propel it along at a cruising speed of 140 kilometres per hour. The ship could stay in the air without stopping for fuel or supplies for 5 or 6 days, with a range of almost 20 000 kilometres.

The crew consisted of 15 stewards and cooks who cared for the 50 passengers. The ship was manned and piloted by about 40 engineers, navigators, riggers and officers.

Passengers on the Hindenburg spent their journey in hotel luxury. They had large staterooms with private bathrooms, a dining room, lounges, bars, and promenade areas where they could enjoy the view from windows.

This is why airships, and the Hindenburg in particular, were so popular for long voyages, such as trans-Atlantic crossings from Europe to the United States of America. Passengers could go in luxury but slowly by liner, or quickly but more uncomfortably by aeroplane. (Don't forget that it was 1937, and 'airliner' meant a 21-seat Douglas DC-3, which had two engines and had to stop every 2000 kilometres to refuel.) The Hindenburg allowed passengers to fly from Germany to America in four days, in old-fashioned comfort. Indeed, the journey was more comfortable in another sense. Unlike an aeroplane, an airship floated through the air rather than flying and it was not affected much by bad weather. It did not bump up and down in air currents, and did not pitch and roll like a boat on the sea. It was a very safe way of travelling, except for one thing. It contained hydrogen.

When hydrogen is mixed with air or oxygen, and ignited, it explodes violently. If it leaks from a container, it can burn as soon as it meets the air. But as long as it is inside something, away from air, it is quite safe. The hydrogen bags in the Hindenburg were carefully isolated from the engines and kitchens, and the area around them was ventilated so that any leaking gas would be carried out of the ship immediately. But to be absolutely certain, people were only allowed to smoke in 'safe' areas, and crew members who visited the hydrogen storage area had to wear special shoes with no metal parts on them so that there was no chance of a spark being made on the metal stairs.

At 8.00 p.m on Monday May 3rd 1937 the Hindenburg left Frankfurt in Germany for the U.S.A. It was a regular flight and as far as the passengers were concerned, perfectly ordinary. They did not know that three of their number were German Air Force officers in disguise and that the commanding officers of the ship had been warned that there might be a saboteur on board. In fact, the saboteur was thought to have been one of the crew who, angered by the German government's part in the Spanish Civil War, planted a small phosphorus incendiary bomb in one of the gas bags, timed to go off after the ship had landed and all the passengers and crew had departed. The destruction of the Hindenburg would have been a mighty gesture against the German government because Adolf Hitler had said that like the Third Reich, the Hindenburg was great and indestructible.

But the Hindenburg was delayed and the timing device went off prematurely.

The radio reporter, Herbert Morrison, who was there to record the arrival of the mighty ship, recorded the disaster instead. His recording for WLS, Chicago, had been going smoothly:

'Here it comes, ladies and gentlemen, and what a sight it is, a thrilling one, a marvellous sight . . . The sun is striking the windows of the observation deck on the westward side and sparkling like glittering jewels on the background of black velvet . . . Oh, oh, oh . . .! It's burst into flames . . . Get out of my way, please, oh my, this is terrible, oh my, get out of the way, please! It is burning, bursting into flames and is falling . . . Oh! This is one of the worst . . . Oh! It's a terrible sight . . . Oh! . . . and all the humanity! . . .'

The fire, slow at first, quickly expanded as the gas bags burned open and the hydrogen mixed with the air. As the great ship lost its buoyancy and sank to the ground the passengers and crew made desperate attempts to get away. Some jumped while they were still a hundred feet above the ground. Some waited until the framework had fallen and then fought their way through the white hot metal. Many were trapped and burned before they could get away; others were roasted as they ran under the burning gas or as flaming debris fell onto them.

After the hydrogen had burned away, the oil, metal and wood continued to burn for many hours until only a charred skeleton was left.

The scrap aluminium that was left was taken back to Germany and used to make fighter aircraft used in World War II.

## Questions

**1** Hydrogen forms a covalent compound with chlorine called hydrogen chloride. Draw a diagram of the structure of a molecule of hydrogen chloride showing the way in which you expect the electrons to be shared.

**2** Look up isotopes in the index and write revision notes about the isotopes of hydrogen.

**3** Find the chapter concerned with ammonia and read about the Haber process for the manufacture of ammonia.

**a** What is the catalyst? How much is used up in the reaction?
**b** Why does using a high temperature make the reaction go faster?
**c** Why does increasing the pressure make the reaction happen more easily?
**d** This is an equilibrium reaction. Try to explain what that means.
**e** Describe an important use for ammonia.

**4** Outline the way in which methane can be converted into hydrogen.

**5** Describe the way in which you would show that hydrogen produces water when it burns. Mention the way in which you would make the hydrogen, the way that you would collect the water formed, and the precautions that you would take in using hydrogen. How would you test the substance formed to see if it was actually water?

**6** If a balloon is filled with hydrogen and released, it floats up to the ceiling. A similar balloon filled to the same size with carbon dioxide sinks to the floor. By the next day, the hydrogen balloon has also fallen to the floor and it is much smaller than the carbon dioxide balloon. Why?

**7** How would you distinguish between three unlabelled gas jars containing hydrogen, oxygen and carbon dioxide?

**8** **a** What is a reducing agent?
**b** Name a substance which can be reduced by hydrogen.
**c** Name a substance which can oxidise hydrogen.
**d** Write an equation for a redox reaction involving hydrogen.
**e** Will hydrogen reduce sodium oxide?

**9** Write out a list of safety rules that you would issue for a factory or laboratory where a lot of hydrogen is going to be used.

Remember that the employees may well be smokers, have steel tipped shoes and wear nylon jumpers. They will switch on lights and work electrically operated machines. What special features will your factory or laboratory have?

# 12 Carbon

| | | | | | | | | | | | | | | | | | |
|---|---|---|---|---|---|---|---|---|---|---|---|---|---|---|---|---|---|
| | | | H | | | | | | | | | | | | | | He |
| Li | Be | | | | | | | | | | | B | C | N | O | F | Ne |
| Na | Mg | | | | | | | | | | | Al | Si | P | S | Cl | Ar |
| K | Ca | Sc | Ti | V | Cr | Mn | Fe | Co | Ni | Cu | Zn | Ga | Ge | As | Se | Br | Kr |
| Rb | Sr | Y | Zr | Nb | Mo | Tc | Ru | Rh | Pd | Ag | Cd | In | Sn | Sb | Te | I | Xe |

## 12.1 Graphite

Carbon exists in two different forms: graphite, and diamond. The graphite form is much more common. It can be soot, charcoal, or even carbon obtained from sugar or dehydrated bones. The molecular structure of all those substances is the same. (See figure 1**a**.)

The carbon atoms in graphite are arranged in rings of six, and they are joined together in large flat sheets. The sheets lie on top of one another and are held together by very weak forces. These are nowhere near as strong as chemical bonds, so the sheets can slide over one another. The properties of graphite are explained by its structure:

**1** Graphite is soft and greasy to touch. It is often used in lubricants, because the flat sheets are able to slide over each other.

**2** Graphite is black. The layers of rings are randomly arranged on top of each other, so carbon cannot transmit light. This gives it its black colour.

**3** Graphite conducts electricity. This is very unusual for a non-metal. Carbon has four electrons in its outside shell. But look at the structure of graphite. Each carbon atom has only three bonds – only three electrons are used in bonding. Each carbon atom has a free electron which is free to move and carry an electric current.

**4** Graphite burns slowly in air. In the form of coke it is often used as a fuel:

$$C(s) + O_2(g) \longrightarrow CO_2(g).$$

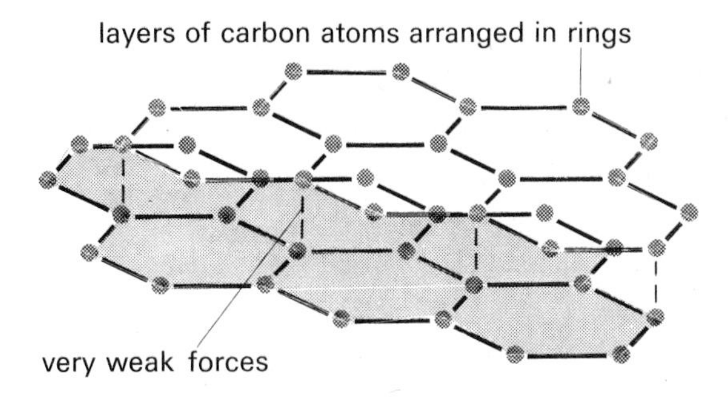


**Figure 1a**
*The structure of graphite.*

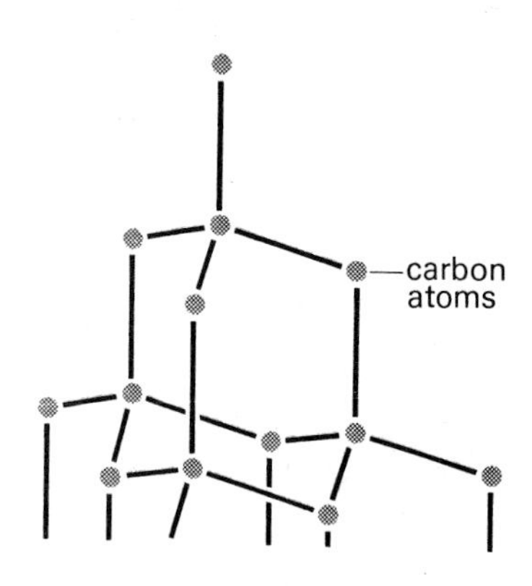


**1b**
*The structure of diamond.*

## 12.2 Diamond

The other form of carbon is called diamond. It is only rarely found. All natural diamonds were formed when graphite in the ground was subjected to extremely high temperatures and pressures. This is how synthetic diamonds are manufactured for industrial use. Figure 1**b** shows the structure of diamond.

You can see that the structure is quite different from that of graphite. The carbon atoms are joined together in a very strong tetrahedral arrangement. You can read a lot more about that in the next chapter. The properties of diamond can be explained by its structure:

**1** Diamond is the hardest natural substance known. The tetrahedral arrangement of the carbon atoms makes the structure very rigid and almost impossible to break. Diamonds can only be cut and shaped by other diamonds.

**2** Pure diamonds are colourless. They sparkle in light. This is because light entering the crystal is reflected from face to face inside the crystal before being reflected out again. Sometimes the light is split up into its component colours while it is being reflected, and this gives some diamonds characteristic colours. When cut and polished, diamonds are very beautiful stones, but their value lies more in their rarity than in their beauty.

**3** Diamond will not conduct electricity. All the electrons in the carbon atoms' outside shells are used in the bonding, so that none are left over to carry an electric current.

**4** Diamond will burn. It needs a slightly higher temperature than graphite to start burning, because of its stronger structure. Needless to say, it is not used as a fuel.

## 12.3 Allotropy

This property of carbon – its existence in two quite different forms – is known as *allotropy*.

**'When an element exists in two or more different forms without changing state (i.e. melting or boiling), simply by having different arrangements of its atoms or molecules, it shows the property of allotropy.'**

Graphite and diamond are allotropes of carbon.

## 12.4 Limestone

Carbon is rarely found uncombined with other elements. Huge quantities are found in combination with calcium and oxygen, in calcium carbonate.

Limestone, chalk, and marble are all forms of calcium carbonate. They are not allotropes however, because they are compounds and not an element. These were all made by different methods but they all started off as the shells of sea animals that lived millions of years ago, when most of the living creatures on Earth were in the sea.

## 12.5 Lime kilns

When limestone is heated, it decomposes into calcium oxide and carbon dioxide:

calcium carbonate $\longrightarrow$ calcium oxide + carbon dioxide.

$$CaCO_3(s) \longrightarrow CaO(s) + CO_2(g).$$

Industrially, calcium oxide is produced in a lime kiln. (See figure 2.)

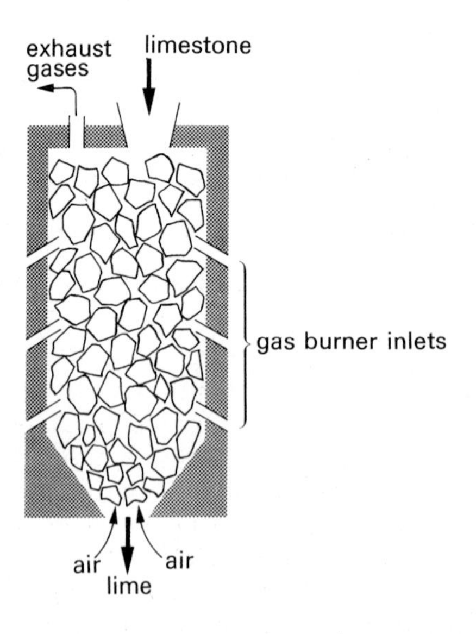


**Figure 2**
*A lime kiln.*

The limestone is heated with a gas flame. The degree of 'cooking' can be controlled, to produce calcium oxide of different qualities. Calcium oxide has been used for many years. In agriculture, it is put on the soil, to neutralise acidity. The kilns were once wood fired, and many can still be seen today, often on coastlines where the limestone was roasted as soon as it was unloaded from boats.

## 12.6 Laboratory preparation of calcium oxide and calcium hydroxide

If a piece of limestone or marble is heated strongly in a bunsen flame, it glows white hot. When it cools down, it is crumbly, and its surface is powdery. This powder is calcium oxide. If water is added to it, the powder hisses and steams as it is converted to calcium hydroxide.

Because the calcium oxide reacts so exothermically with water, its common name is quick lime. The work 'quick' comes from the latin meaning 'alive'. It is used as a drying or dehydrating agent. Calcium hydroxide is commonly called slaked lime. 'Slaked' means 'thirst quenched'.

calcium oxide + water $\longrightarrow$ calcium hydroxide

$$CaO(s) + H_2O(l) \longrightarrow Ca(OH)_2(s)$$

quick lime + water $\longrightarrow$ slaked lime

## 12.7 Limewater

Once made, calcium hydroxide is not very soluble in water, but it forms a weakly alkaline solution.

$$Ca(OH)_2(aq) \rightleftharpoons Ca^{2+}(aq) + 2OH^-(aq).$$

The common name for this solution is *limewater* and it is used as a test for carbon dioxide.

If carbon dioxide is bubbled through limewater, a white, cloudy substance is formed. The limewater is said to turn milky. The 'milkiness' is in fact due to minute particles of chalk:

$$Ca(OH)_2(aq) + CO_2(g) \longrightarrow CaCO_3(s) + H_2O(l).$$

Carbon dioxide is an acidic oxide and limewater is an alkali so when they react together, a salt is formed:

alkali + acid $\longrightarrow$ salt + water.

The salt in this case is calcium carbonate, which is insoluble in water. This is why the precipitate forms.

If the carbon dioxide is bubbled for a lot longer, the precipitate redissolves, leaving a clear solution once more. This is because the calcium carbonate, a normal salt, forms a new type of substance called an *acid salt* in the excess carbon dioxide. The acid salt is called calcium hydrogencarbonate, and it is soluble in water, so it re-dissolves:

$$CaCO_3(s) + CO_2(g) + H_2O(l) \longrightarrow Ca(HCO_3)_2(aq).$$

## 12.8 Laboratory preparation of carbon dioxide

The chapter on air mentioned how carbon dioxide is made as a product of burning. Some of it is also made by pupils in chemistry laboratories . . . there are three main ways of preparing carbon dioxide: heating carbonates or hydrogen carbonates, and by the action of dilute acids on carbonates.

**Action of heat on carbonates.** When carbonates are heated, carbon dioxide is evolved and an oxide is left:

$$\text{carbonate} \xrightarrow{\text{heat}} \text{oxide} + \text{carbon dioxide.}$$

Here are two examples:

$$\text{copper(II) carbonate} \xrightarrow{\text{heat}} \text{copper(II) oxide} + \text{carbon dioxide}$$

$$CuCO_3(s) \longrightarrow CuO(s) + CO_2(g);$$

$$\text{zinc carbonate} \xrightarrow{\text{heat}} \text{zinc oxide} + \text{carbon dioxide}$$

$$ZnCO_3(s) \longrightarrow ZnO(s) + CO_2(g).$$

Two exceptions are sodium carbonate and potassium carbonate which will not decompose when heated in a bunsen flame, because they are too stable.

**Action of heat on hydrogen carbonates.** When hydrogen carbonates are heated, carbon dioxide is evolved. A carbonate, and water are produced:

$$\text{hydrogen carbonate} \xrightarrow{\text{heat}} \text{carbonate} + \text{carbon dioxide} + \text{water}$$

For example:

$$\text{sodium hydrogen carbonate} \xrightarrow{\text{heat}} \text{sodium carbonate} + \text{carbon dioxide} + \text{water.}$$

$$2NaHCO_3(s) \longrightarrow Na_2CO_3(s) + CO_2(g) + H_2O(l).$$

Sodium hydrogen carbonate is used in baking powder. When a cake mixture containing baking powder is heated in the oven, the sodium hydrogen carbonate decomposes, and the carbon dioxide makes the cake rise.

**Action of dilute acids on carbonates.** When acids are added to carbonates, carbon dioxide is given off; a salt and water remain:

acid + carbonate $\longrightarrow$ salt + water + carbon dioxide.

This is shown in three examples. The third example is the usual laboratory preparation.

**1** Copper(II) carbonate + sulphuric acid (dilute) $\longrightarrow$ copper(II) sulphate + water + carbon dioxide.

$$CuCO_3(s) + H_2SO_4(aq) \longrightarrow CuSO_4(aq) + H_2O(l) + CO_2(g).$$

**2** Sodium carbonate + nitric acid (dilute) $\longrightarrow$ sodium nitrate + water + carbon dioxide.

$$Na_2CO_3(s) + 2HNO_3(aq) \longrightarrow 2NaNO_3(aq) + H_2O(l) + CO_2(g).$$

**3** **Usual laboratory preparation.** Carbon dioxide is usually prepared in the laboratory from marble chips and dilute hydrochloric acid. (See figure 3.)

Calcium carbonate + hydrochloric acid (dilute) $\longrightarrow$ calcium chloride + water + carbon dioxide.

$$CaCO_3(s) + 2HCl(aq) \longrightarrow CaCl_2(aq) + H_2O(l) \quad CO_2(g).$$

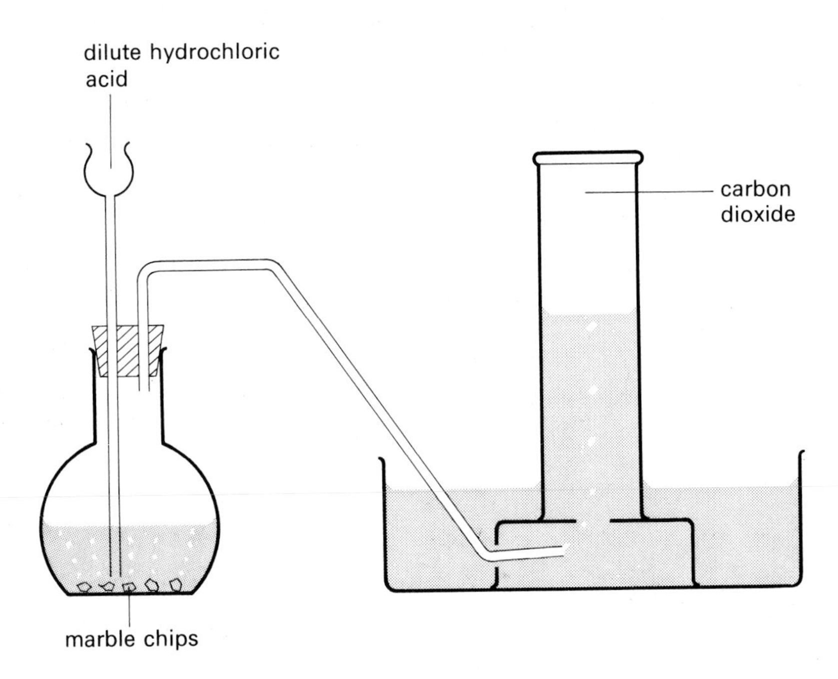


**Figure 3**
*The preparation and collection of carbon dioxide.*

When the acid is added through the thistle funnel, the marble chips fizz and dissolve. The carbon dioxide is collected over water.

## 12.9 Properties of carbon dioxide

**1** It is a colourless, odourless, tasteless gas.

**2** It is much heavier than air. Try the 'Invisible gas and candle trick'. (See figure 4.)

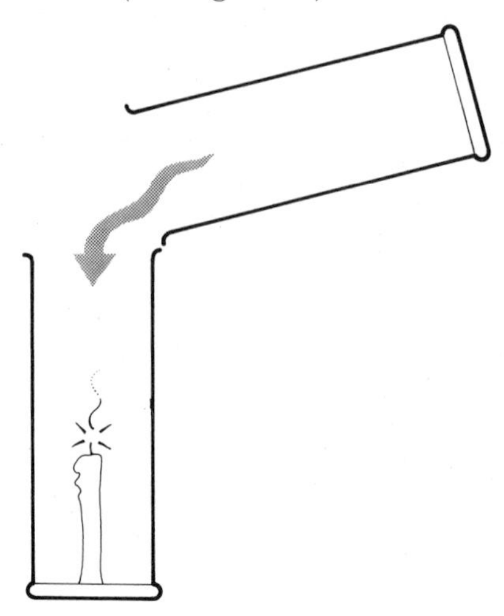

**Figure 4**
*The invisible gas and candle trick.*

**3** It is slightly soluble in water. Its solubility increases with pressure.

**4** If carbon dioxide is cooled and pressurised, it turns straight into a solid, which is called dry ice. Dry ice is used as a refrigerant. When it is heated, it does not melt, but sublimes to give a gas.

## 12.10 The reactions of carbon dioxide

**Carbon dioxide does not support combustion.** A lighted splint goes out when it is put into the gas.

**Carbon dioxide is an acidic oxide.** It forms a weakly acidic solution when added to water. This acid is called *carbonic acid*:

$$CO_2(g) + H_2O(l) \longrightarrow H_2CO_3(aq).$$

Carbonic acid is a weak acid and is not kept in bottles like other acids, because it easily decomposes back into carbon dioxide and water. However, its salts – carbonates and hydrogen carbonates are quite stable.

**Carbon dioxide can be reduced by burning magnesium.** If a piece of burning magnesium ribbon is lowered into a gas jar of carbon dioxide, a spluttering reaction takes place and white magnesium oxide is formed leaving black patches of carbon on the sides of the jar:

$$CO_2(g) + 2Mg(s) \longrightarrow 2MgO(s) + C(s).$$

## 12.11 The preparation of carbon monoxide

Carbon monoxide (CO) may not seem to be very different from carbon dioxide ($CO_2$). Its chemical properties are in fact very different indeed. Figure 5 shows the apparatus that may be used to prepare carbon monoxide in the laboratory.

The carbon dioxide is made from marble and dilute hydrochloric acid. It is passed through a pyrex tube which is loosely packed with charcoal. The tube must be heated to about 1000 °C for the carbon to reduce the carbon dioxide:

$$CO_2(g) + C(s) \longrightarrow 2CO(g).$$

The resulting gas is bubbled through potassium hydroxide solution to remove any unreacted carbon dioxide:

$$CO_2(g) + 2KOH(aq) \longrightarrow K_2CO_3(aq) + H_2O(l).$$

It is finally collected over water.

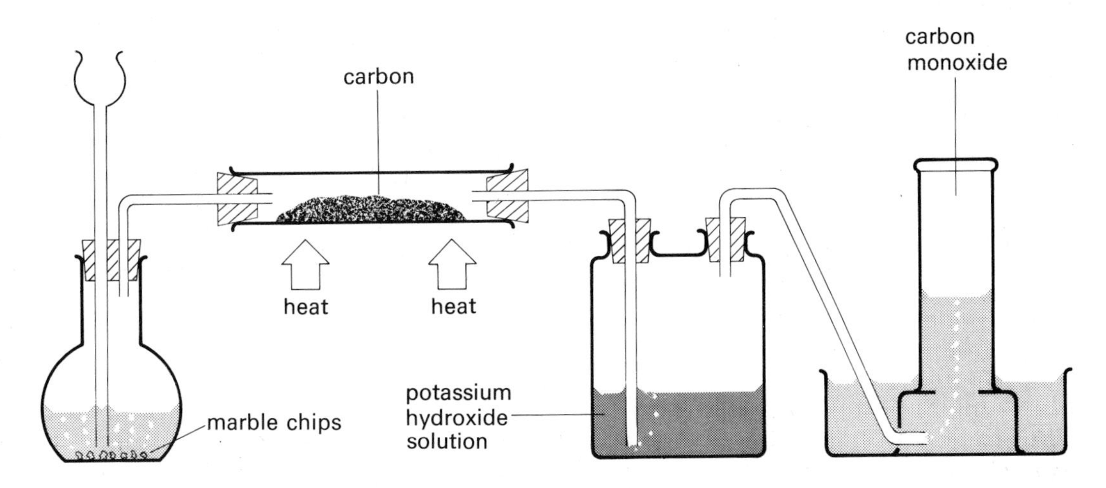


**Figure 5** *The preparation and collection of carbon monoxide.*

## 12.12 The properties of carbon monoxide

1 It is a colourless, tasteless, odourless gas.

2 It is insoluble in water.

3 It is very poisonous.

## 12.13 The reactions of carbon monoxide

**Carbon monoxide burns in air or oxygen.** It burns with a blue flame to form carbon dioxide.

$$2CO(g) + O_2(g) \longrightarrow 2CO_2(g).$$

This flame can often be seen on the top of a domestic coke fire. Figure 6 on the next page shows the reactions that take place.

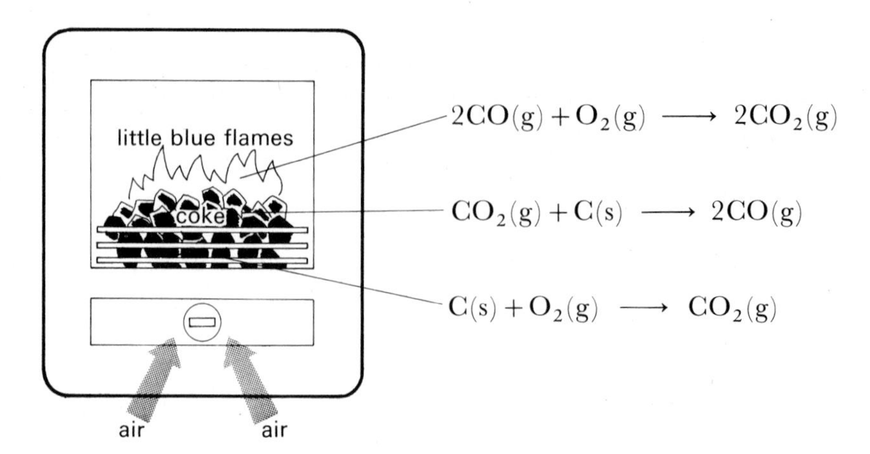


**Figure 6**
*An enclosed domestic coke fire.*

**Carbon monoxide is a reducing agent.** In the same way as hydrogen, it will reduce copper(II) oxide to copper, and lead(II) oxide to lead. Figure 7 shows an apparatus which could be used.

The carbon monoxide is allowed to flow through the apparatus for a few seconds at the start to flush out the air so that an explosive mixture of carbon monoxide and air is not formed. The excess carbon monoxide is burned at the jet because it is a poisonous gas. The carbon monoxide reduces the black copper(II) oxide to pink copper, and the yellow lead(II) oxide to silver coloured lead:

$$CuO(s) + CO(g) \longrightarrow Cu(s) + CO_2(g);$$

$$PbO(s) + CO(g) \longrightarrow Pb(s) + CO_2(g).$$

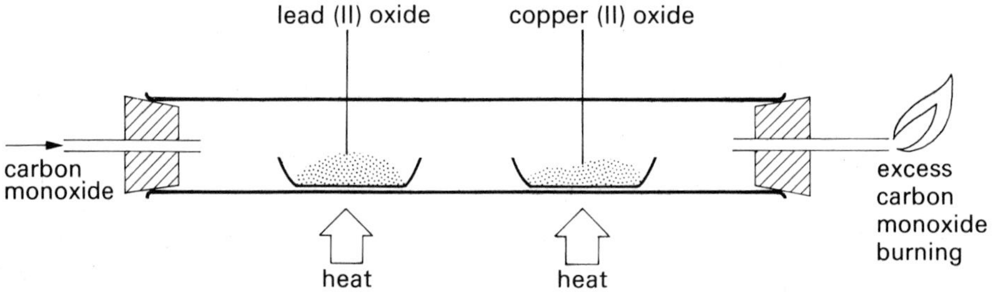


**Figure 7**
*Carbon monoxide is a reducing agent.*

## 12.14 The Carbon Cycle

Figure 8 shows a sequence of events which is called the *carbon cycle.*

**1** All green plants take in carbon dioxide in the process called photosynthesis. The carbon dioxide is combined with water to make carbohydrates.

**2** Animals eat plants, and people eat both of them, so that the carbon originally taken in by the plants gets into all animals.

**3** Animals give out carbon dioxide as they breathe.

**4** Dead animals and plants give out carbon dioxide as the respiration product of bacteria and fungi.

5 Factories and houses give out carbon dioxide as a result of combustion – even amateur beer and wine-makers take part in the Carbon Cycle because the process of fermentation gives out carbon dioxide. All these go on simultaneously – the carbon circulates.

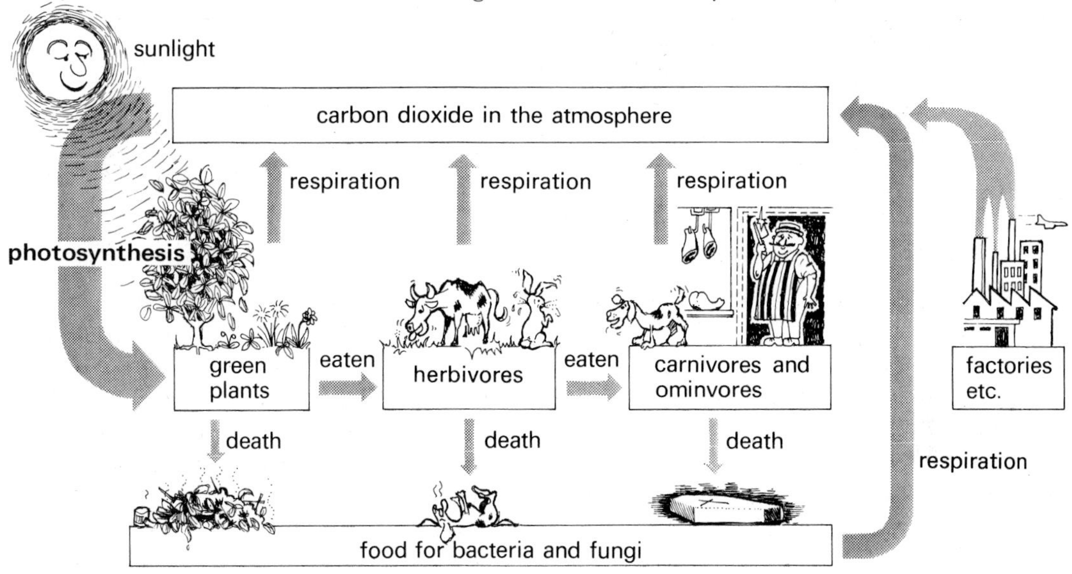


**Figure 8** *The Carbon Cycle.*

**Summary**

At the end of this chapter you should be able to:

**1** Explain the term Allotropy.

**2** Describe the difference in structure between diamond and graphite.

**3** Describe the properties of diamond and graphite and show how the properties of these two allotropes may be explained by their structures.

**4** Describe how quicklime and slaked lime are made from limestone.

**5** Explain the lime water test for carbon dioxide.

**6** Describe how carbon dioxide may be prepared by heating carbonates and hydrogen carbonates, and say how it is made in the laboratory from marble chips.

**7** Describe the physical properties of carbon dioxide.

**8** Explain the chemical reactions of carbon dioxide.

**9** Describe how carbon monoxide is prepared from carbon dioxide and discuss its reactions with oxygen and metal oxides.

**10** Give your version of the Carbon Cycle.

# Modern buildings–they're all concrete and glass

*We take many familiar materials for granted. Houses are made of bricks and concrete and their windows are made of glass. Sometimes whole buildings seem to be made of nothing but concrete and glass. And what about those plate glass windows shops have? How do they get them so flat? Where do cement and concrete come from? Read on.*

**Cement**

The first people known to have used cement were the Egyptians who used it to build their pyramids. The Romans used lots of it too; they had volcanic rocks that could be ground down and roasted into a powder rather like our modern cement.

In 1824, Joseph Aspdin patented his 'Portland Cement' which he said was 'equal to the best merchantable Portland stone in solidity and durability'.

Isambard Kingdom Brunel used cement when he built his famous tunnel under the Thames in 1838, and in 1851, a huge block of Portland cement was exhibited at the Great Exhibition in London.

The raw materials for cement are limestone, clay, water and coal. The clay and limestone are crushed and mixed with water, and this slurry is fed into a rotating kiln. Powdered coal is blown into the kiln at the same time and it is heated by oil or gas burners

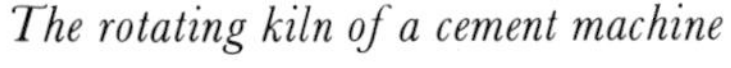

*The rotating kiln of a cement machine*

so that it burns at about 1500 °C. The mixture decomposes at this temperature into quicklime:

$$CaCO_3(s) \longrightarrow CaO(s) + CO_2(g).$$

The mixture is then cooled and crushed and finally mixed with powdered Gypsum (calcium sulphate) to make Portland Cement. When cement is mixed with sand and water mortar is formed. If gravel or small stones are added, it is called *concrete.* As the concrete dries, the quicklime, clay and gypsum in the mixture slowly crystallise and bind the sand and stones together in a strong framework. Sometimes iron bars or meshes are embedded in the concrete to make it even stronger. Recently, Pilkingtons, the glass manufacturers, have developed a form of concrete which has tiny fibres of glass mixed in it. This is called Glass Reinforced Concrete, and it can be cast into slabs, or sprayed and painted onto brickwork. It is very resistant to weathering and is very good at keeping the heat in.

## Glass

Like cement, glass is a very old material and many ancient civilizations made things with it. It is thought that glass was discovered by accident when a mixture of sand and soda was heated in a hot fire. Someone found that a strong transparent substance was left – glass has been discovered.

In a modern glass factory, sand (silicon dioxide) and sodium carbonate are mixed with scrap glass (called cullet) and this is heated in a big furnace or 'tank'. The tank is

*Milk-bottle manufacture*

heated to a temperature of 1500 °C by big gas or oil burners. They react to form sodium silicate, which is the glass:

$$SiO_2(s) + Na_2CO_3(s) \rightarrow Na_2SiO_3(s). + CO_2(g).$$

The glass is rather like treacle at this temperature and as it drains from the tank, it is chopped off by big metal shears into lumps called *gobs.* The gobs fall into moulds and are pressed by compressed air into the shapes of jam jars, milk bottles, beakers – almost anything – before being allowed to cool down very slowly over a period of many hours. Items like milk bottles that have to stand a great deal of scratching and bumping are sprayed with titanium chloride solution to make them tougher.

Plate glass for large windows is made by allowing the molten glass to flow onto a bed of molten tin so that it forms absolutely flat. This is known as float glass.

## Questions

**1** What does the term allotropy mean? Explain how the element carbon is capable not only of making marks on paper which can easily be removed with a rubber, but can also scratch marks on glass.

**2** Draw the apparatus you would use to:

**a** prepare a gas jar of dry carbon dioxide by heating copper(II) carbonate.

**b** show how carbon dioxide can oxidise magnesium.

**3** Read the section in Chapter 3 concerned with hardness of water. Write short revision notes on how hardness gets into water. There is a similarity here with a lime water test that goes on for too long? What is it?

**4** Read the section in Chapter 11 concerned with the lightness of hydrogen. Carbon dioxide molecules are much bigger than hydrogen molecules and therefore diffuse much more slowly. Think carefully and try to design an experiment to demonstrate that fact. You will need a porous pot as in chapter 11, but you will have to modify the apparatus.

**5** You are given two jars of gas. One contains hydrogen and the other contains carbon monoxide.

Both gases have several physical properties in common. What are they? How do they differ in their reactions with oxygen? How would you tell them apart?

**6** Describe how a molecule of carbon dioxide can start, and end up in the air, having made a journey through a blade of grass, a lamb, and a starving Chemistry teacher!

# 13 Organic chemistry

## 13.1 Organic and inorganic

Chapter 12 was about the element carbon and how it forms oxides and carbonates. Along with the chemistry of all the other elements, this is usually described as *inorganic chemistry*. The chemistry in this chapter is *organic carbon chemistry*. What is the difference?

Generally, organic chemistry deals with things that are living or have lived at one time. So one part of organic chemistry is concerned with plants and animals. The other part deals with anything that has been derived from coal or oil, because these had their origins in living things. This second part covers thousands of different types of chemicals from oils and solvents to plastics and drugs. After you have learned some organic chemistry you will see that organic chemicals fit into families, and that it becomes quite easy to distinguish organic chemicals from inorganic ones.

One thing that most organic chemicals have in common is that when they are heated, they burn or char. Sometimes their flames are smoky, and sometimes they leave deposits of carbon. Some good examples are wood and paper, plastics and foodstuffs. (Don't forget the toast!) These compounds not only contain carbon, but hydrogen and sometimes oxygen as well. When they burn, some of the carbon gets converted into carbon dioxide, and the hydrogen forms steam. Figure 1 shows an apparatus that could be used to detect these compounds.

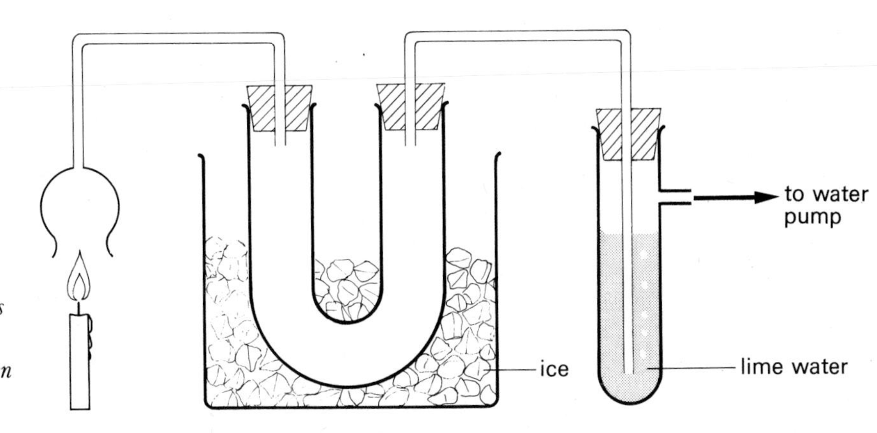


**Figure 1**
*Organic compounds produce steam and carbon dioxide when they burn.*

The candle wax is an organic substance. As it burns, the gases in its flame are sucked through the apparatus by the water pump. Any steam that is formed is condensed in the cold U-tube, and may be tested with cobalt chloride paper. Any carbon dioxide that is produced will turn the limewater milky as it bubbles through.

This section in the chapter has mentioned that organic chemicals fit into a series of families. The next few sections will introduce you to some of them.

## 13.2 Families of organic compounds: the Alkanes

The first group are called *alkanes*. This family of organic compounds are *hydrocarbons*. This means that they contain carbon and hydrogen only.

The simplest member of the family is a gas called methane. The molecular formula for methane is written as: $CH_4$.

When dealing with organic compounds, it is more helpful to write the *structural formula* for a compound because it shows much more clearly how the atoms are arranged in the molecule.

The structural formula for methane is:

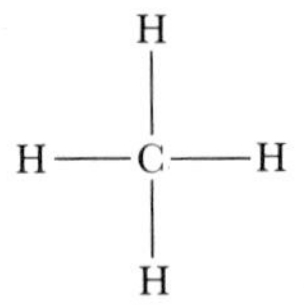

```
      H
      |
  H — C — H
      |
      H
```

The carbon is firmly bonded to the four other atoms. Alkanes are said to be *saturated* compounds because of this.

Even this structural formula does not truly represent the shape of methane molecules. In fact, the four bonds of the carbon atom are arranged in a tetrahedral (triangular pyramid) shape like this:

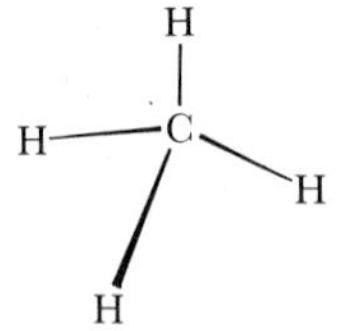

```
        H
        |
  H ——  C
       /  \
      /    H
     H
```

The molecule is completely symmetrical. It looks the same no matter which way you turn it.

Figure 2 shows the first four members of the family of alkanes, and gives some details about them.

A family of organic compounds is called a *homologous series*, because all the members of the family have a similar shape of molecule. You can see that each compound has one carbon and two hydrogen atoms more than the one before.

| name | molecular formula | structural formula | state | occurrence | use |
|---|---|---|---|---|---|
| methane | $CH_4$ | $\begin{array}{ccc} & H & \\ & \mid & \\ H- & C & -H \\ & \mid & \\ & H & \end{array}$ | gas | as natural gas and in rotting vegetation and 'fire damp' in mines | fuel; industrial preparation of hydrogen |
| ethane | $C_2H_6$ | $\begin{array}{cccc} & H & H & \\ & \mid & \mid & \\ H- & C- & C & -H \\ & \mid & \mid & \\ & H & H & \end{array}$ | gas | | fuel |
| propane | $C_3H_8$ | $\begin{array}{ccccc} & H & H & H & \\ & \mid & \mid & \mid & \\ H- & C- & C- & C & -H \\ & \mid & \mid & \mid & \\ & H & H & H & \end{array}$ | gas | obtained from oil | fuel |
| butane | $C_4H_{10}$ | $\begin{array}{cccccc} & H & H & H & H & \\ & \mid & \mid & \mid & \mid & \\ H- & C- & C- & C- & C & -H \\ & \mid & \mid & \mid & \mid & \\ & H & H & H & H & \end{array}$ | gas | | fuel |

**Figure 2** *The homologous series of alkanes.*

## 13.3 The properties of the alkanes

**The first four members of the series are gases.** As their molecular size goes up, their boiling point goes up. At room temperature, the next alkane pentane does not boil – it is a liquid. The next eleven alkanes are all liquids at room temperature. After this, their molecules are so large that they become solids.

**Methane is a good fuel.** We have abundant supplies of methane from the North Sea oil fields. With a good supply of air, methane burns with a clean flame:

$$CH_4(g) + 2O_2(g) \longrightarrow CO_2(g) + 2H_2O(g).$$

Ethane burns in a similar way. Both propane and butane can be easily liquefied by pressure and this makes them suitable for putting into containers for cigarette lighters and cylinders of camping gas. Calor gas is mainly butane.

## 13.4 Isomerism

Look once again at the structure of butane:

```
    H   H   H   H
    |   |   |   |
H — C — C — C — C — H
    |   |   |   |
    H   H   H   H
```

It is possible to draw another version of this molecule, with the same molecular formula, but with a different structural formula. The atoms are arranged in a different way. This arrangement is called 2-methyl propane.

```
    H   H   H
    |   |   |
H — C — C — C — H
    |   |   |
    H   |   H
    H — C — H
        |
        H
```

These compounds which have the same molecular formula, but have different structural formulae are called *isomers*. The general name for this is *isomerism*.

## 13.5 Alkenes and alkynes

Alkenes and alkynes are two other types of hydrocarbons, but they are *unsaturated*. This means that each carbon atom is bonded to less than four other atoms. Instead of having a single covalent bond between each of the carbon atoms, double and triple bonds exist in the molecule as well.

**Alkenes.** The first member of the homologous series of alkenes is ethene:

```
C2H4   H         H
         \     /
          C = C
         /     \
       H         H
```

Alkenes contain double bonds. Like the alkanes, ethene burns, but with a smoky flame. It needs a good supply of oxygen to burn completely.

$$C_2H_4(g) + 3O_2(g) \longrightarrow 2CO_2(g) + 2H_2O(g).$$

Ethene is used in the production of polyethene.

**Alkynes.** The first member of the homologous series of alkynes is ethyne:

$C_2H_2$ H—C≡C—H

Alkynes contain triple bonds.

Ethyne burns with a very sooty flame. It needs a very good supply of oxygen to burn completely:

$$2C_2H_2(g) + 5O_2(g) \longrightarrow 4CO_2(g) + 2H_2O(g).$$

Ethyne is used in the production of polyvinyl chloride (PVC). (See chapter 20.)

## 13.6 Alcohols

The word alcohol is the family name of another homologous series. Figure 3 gives details of the first four members.

| name | molecular formula | structural formula |
|---|---|---|
| methanol | $CH_3OH$ | 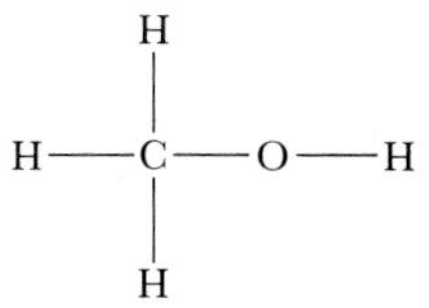 H—C(H)(H)—O—H |
| ethanol | $C_2H_5OH$ | 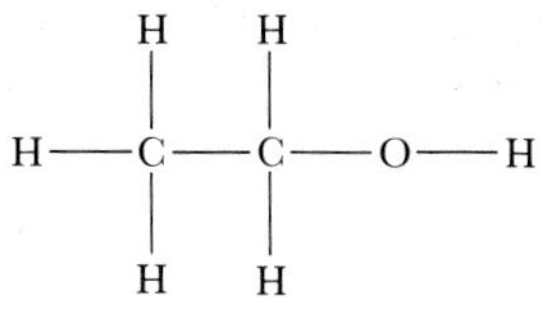 H—C(H)(H)—C(H)(H)—O—H |
| propanol | $C_3H_7OH$ | 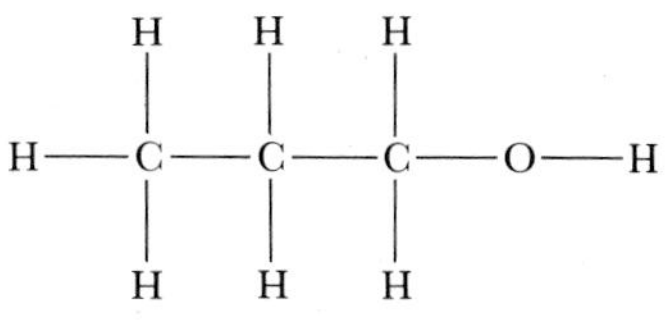 H—C(H)(H)—C(H)(H)—C(H)(H)—O—H |
| butanol | $C_4H_9OH$ | 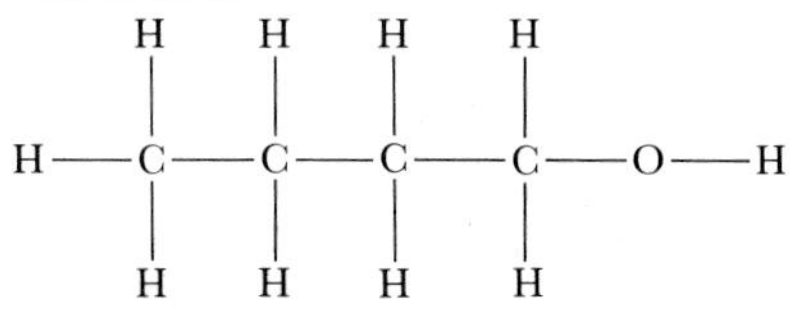 H—C(H)(H)—C(H)(H)—C(H)(H)—C(H)(H)—O—H |

**Figure 3**
*The homologous series of alcohols.*

Methanol is very poisonous. Ethanol is what is usually called 'alcohol'. It is made by the process of *fermentation*. Propanol and butanol are similar to ethanol in behaviour.

## 13.7 The formation of ethanol by fermentation

Beer is made from barley. Full details of how it is made are given in the extra-time section.

The essential part of the process is carried out by yeast. Yeast is a living substance, and as it works it multiplies itself. It contains chemicals called *enzymes*. They are organic catalysts. They help to break down malt from the barley into sugars. Eventually, glucose is formed which is converted into ethanol and carbon dioxide:

$$C_6H_{12}O_6(aq) \longrightarrow 2C_2H_5OH(aq) + 2CO_2(g).$$

This is the fermentation process. It needs a closely controlled temperature in the range 18–20 °C. At lower temperatures, the yeast becomes dormant and very slow acting. At higher temperatures, the yeast may be killed.

Many different sugars, and starch from many foods such as potatoes can be broken down by enzymes to make ethanol. Wine is made from grapes, which have their own enzymes on their skins. Wine can also be made from many fruits and berries.

Beers vary in the percentage of ethanol that they contain, but the fermentation is usually stopped when it has reached about 3–4%. Some beers are much stronger however! Wines have about 14–20% of ethanol. Fermentation stops naturally when all the sugar has run out, or when the ethanol has reached such a concentration that the yeast is poisoned. Drinks like sherry and port have extra ethanol added.

Spirits contain a much higher percentage of ethanol because they are distilled after fermentation.

Although it is quite legal to make your own wine and beer at home, you are not allowed to sell, or distil it.

Figure 4 shows a typical fermentation apparatus that might be used by an amateur wine maker. The air lock on top allows the carbon dioxide to get out, but prevents air from getting in.

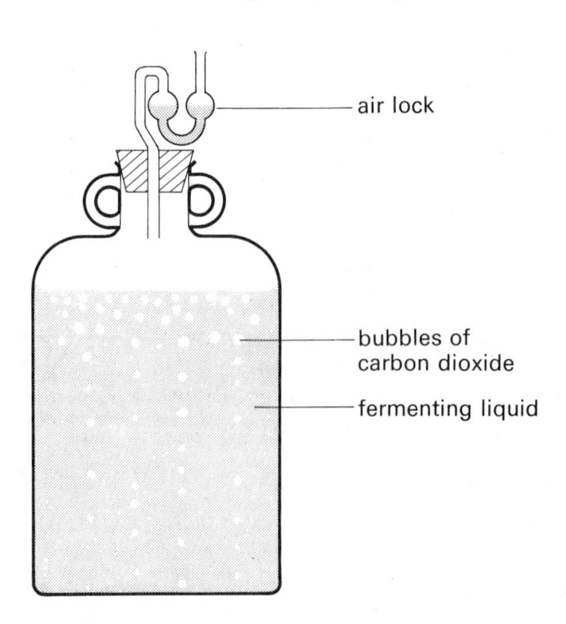


**Figure 4**
*A wine fermentation jar.*

## 13.8 The physical properties of ethanol

1. It is a colourless liquid with a sweet smell.
2. It has a boiling point of 78 °C.

## 13.9 The reactions of ethanol

**Ethanol is a fuel.** It burns with a fairly clean flame:

$$C_2H_5OH(l) + 3O_2(g) \longrightarrow 2CO_2(g) + 3H_2O(g).$$

Methylated spirits contains ethanol mixed with a small amount of methanol and water. Colouring and a taste are added to prevent people from drinking it by accident.

**Ethanol is a covalent liquid.** This means that it cannot be an alkali despite the fact that it contains an —OH group on the end of its molecule.

The —OH group is not an ion. Sodium hydroxide is an alkali because it contains an $OH^-$ ion:

$$C_2H_5OH \qquad Na^+OH^-.$$

Because ethanol does not contain hydroxide ions, it has no effect on litmus.

**Ethanol forms compounds called esters.** These are formed when it reacts with organic acids. You can read about this reaction in section 13.11 on the next page.

## 13.10 Organic acids

There is also a family or homologous series of organic acids. The first member is methanoic acid, which is found in stinging nettles and ants – but ethanoic acid is much more common. It is the acid in vinegar. It is made when ethanol is oxidised. This can happen to the ethanol in wine. It is attacked by bacteria who use air to oxidise it. Wine-makers use an air lock to prevent this happening:

$$\text{ethanol} \xrightarrow[\text{bacteria}]{\text{air}} \text{ethanoic acid.}$$

$$C_2H_5OH(l) \longrightarrow CH_3COOH(aq).$$

Figure 5 on the next page shows the first four members, their molecular and structural formulae.

| name | molecular formula | structural formula |
|---|---|---|
| methanoic acid | HCOOH | 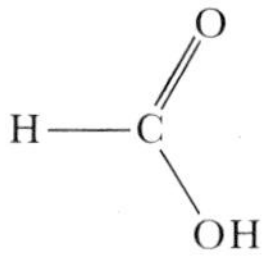 $H-C(=O)-OH$ |
| ethanoic acid | $CH_3COOH$ | 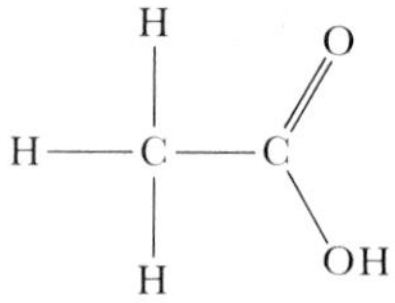 $H-C(H)(H)-C(=O)-OH$ |
| propanoic acid | $C_2H_5COOH$ | 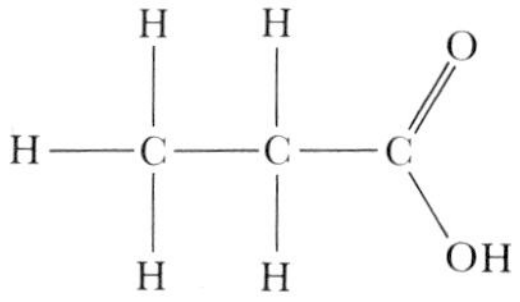 $H-C(H)(H)-C(H)(H)-C(=O)-OH$ |
| butanoic acid | $C_3H_7$ COOH | 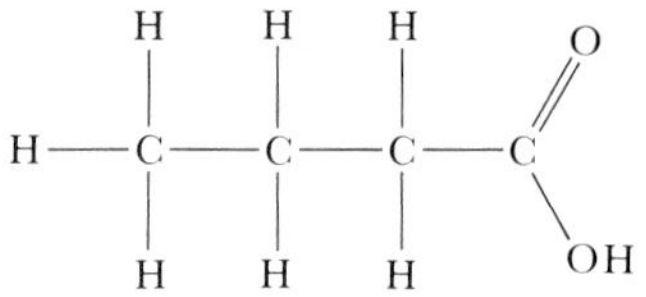 $H-C(H)(H)-C(H)(H)-C(H)(H)-C(=O)-OH$ |

**Figure 5**
*The first four organic acids.*

## 13.11 The properties of ethanoic acid

**It behaves like an ordinary weak acid.** It reacts with metals, to form a salt and hydrogen; with bases, to form a salt and water; with carbonates, to form a salt, carbon dioxide and water. The salts of ethanoic acid are called *ethanoates*. Ethanoic acid is a weak acid because it is only slightly ionised in solution.

**Ethanoic acid reacts with alcohols to form esters.** For example: if ethanol is warmed with ethanoic acid, in the presence of a few drops of concentrated sulphuric acid as a catalyst, a glue-like smell soon appears. This is due to the formation of an ester called ethyl ethanoate. The process is called *esterification*:

$$\text{ethanoic acid} + \text{ethanol} \rightleftharpoons \text{ethyl ethanoate} + \text{water.}$$
$$CH_3COOH(l) + C_2H_5OH(l) \rightleftharpoons CH_3COOC_2H_5(l) + H_2O(l).$$

Ethyl ethanoate has the structure:

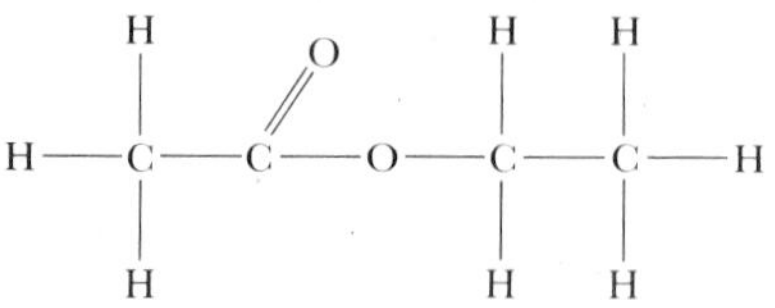

$$H-C(H)(H)-C(=O)-O-C(H)(H)-C(H)(H)-H$$

Note that the reaction has a two-way arrow in the middle. This is because the reaction is of a type called an *equilibrium reaction*. The ethanol and ethanoic acid will never be completely converted to ethyl ethanoate and water. The final mixture will contain some of all four chemicals. The ester must be removed by distillation. Esters are responsible for many of the odours and flavours in flowers and fruits. Because the reaction is reversible, the ester can be turned back into ethanol and ethanoic acid. The reverse process is called *hydrolysis* and it is done by boiling the ester with sodium hydroxide solution. Hydrolysis is used in the manufacture of soap from naturally occurring esters.

## 13.12 Soaps and detergents

One important group of compounds based on organic molecules are *detergents*. Detergents are cleaning agents, and soap is one of the oldest of these. Soap is made when animal fats or plant oils are boiled with concentrated sodium hydroxide solution. This process is called *saponification* (soap-making). The fats and oils contain esters which are made up from an alcohol called glycerol, and big organic acids. A typical ester might be glyceryl stearate, and this is its structural formula:

$$\begin{array}{l} C_{17}H_{35}\text{—}C(=O)\text{—}O\text{—}CH_2 \\ C_{17}H_{35}\text{—}C(=O)\text{—}O\text{—}CH \\ C_{17}H_{35}\text{—}C(=O)\text{—}O\text{—}CH_2 \end{array}$$

When glyceryl stearate is boiled with sodium hydroxide solution, it is hydrolysed. Glycerol and the sodium salt of stearic acid are formed:

$$\begin{array}{l} C_{17}H_{35}\text{—}C(=O)\text{—}O\text{—}CH_2 \\ C_{17}H_{35}\text{—}C(=O)\text{—}O\text{—}CH \\ C_{17}H_{35}\text{—}C(=O)\text{—}O\text{—}CH_2 \end{array} + 3NaOH \longrightarrow \underset{\text{sodium stearate}}{3C_{17}H_{35}COO^-Na^+} + \begin{array}{l} HO\text{—}CH_2 \\ HO\text{—}CH \\ HO\text{—}CH_2 \end{array}$$

The salt – sodium stearate – is one form of soap. It is precipitated from solution by the addition of brine. After purification, perfume and colouring are added before the blocks of soap are cut and wrapped. Chapter 3 explains the washing action of soap.

Modern detergents are called soapless detergents – they do not form a scum with hard water. The way they work is similar to that of soap. Note that they usually end with a sulphonic acid group. A typical soapless detergent might be:

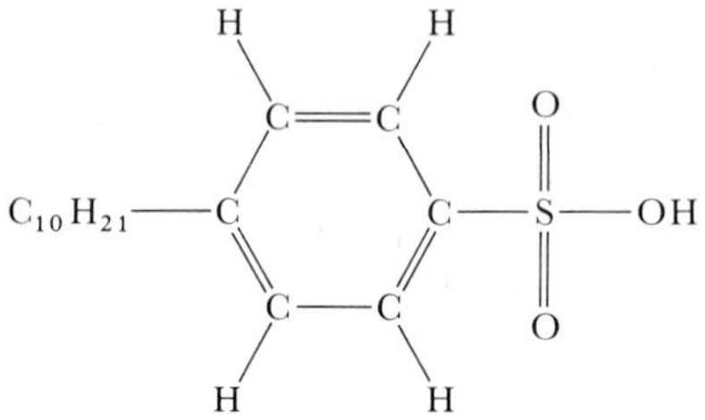

```
              H         H
               \       /
                C═════C        O
               /       \       ║
C10H21 ──── C           C ──── S ──── OH
               \\     //       ║
                C ─── C        O
               /       \
              H         H
```

Soapless detergents are made by heating vegetable oils such as olive oil, palm oil or oils from petroleum – with concentrated sulphuric acid. They have the same sort of washing action as soap.

## Summary

At the end of this chapter you should be able to:

**1** Identify an organic compound by the way it occurs or is made.

**2** Describe an experiment to show that most organic compounds contain carbon and hydrogen.

**3** Explain what is meant by the term homologous series.

**4** Distinguish between a molecular formula and a structural formula.

**5** Name the first four members of the homologous series of alkanes, and give their structural formulae.

**6** Suggest uses for the alkanes.

**7** Write a combustion equation for an alkane.

**8** Explain the term isomerism, using butane as an example.

**9** Draw the structure of ethene and ethyne.

**10** Distinguish between a saturated hydrocarbon and an unsaturated hydrocarbon.

**11** Write the structural formulae of methanol and ethanol.

**12** Explain what is meant by the term fermentation.

**13** Explain why ethanol is not an alkali.

**14** Draw the structure of ethanoic acid.

**15** Explain why ethanoic acid is a weak acid.

**16** Write equations for the reactions of ethanoic acid with a metal, a base and a carbonate.

**17** Write the equation for the reaction between ethanol and ethanoic acid.

**18** Explain the terms esterification and hydrolysis.

**19** Describe one advantage of a soapless detergent over soap.

**20** Briefly outline the preparation of soap.

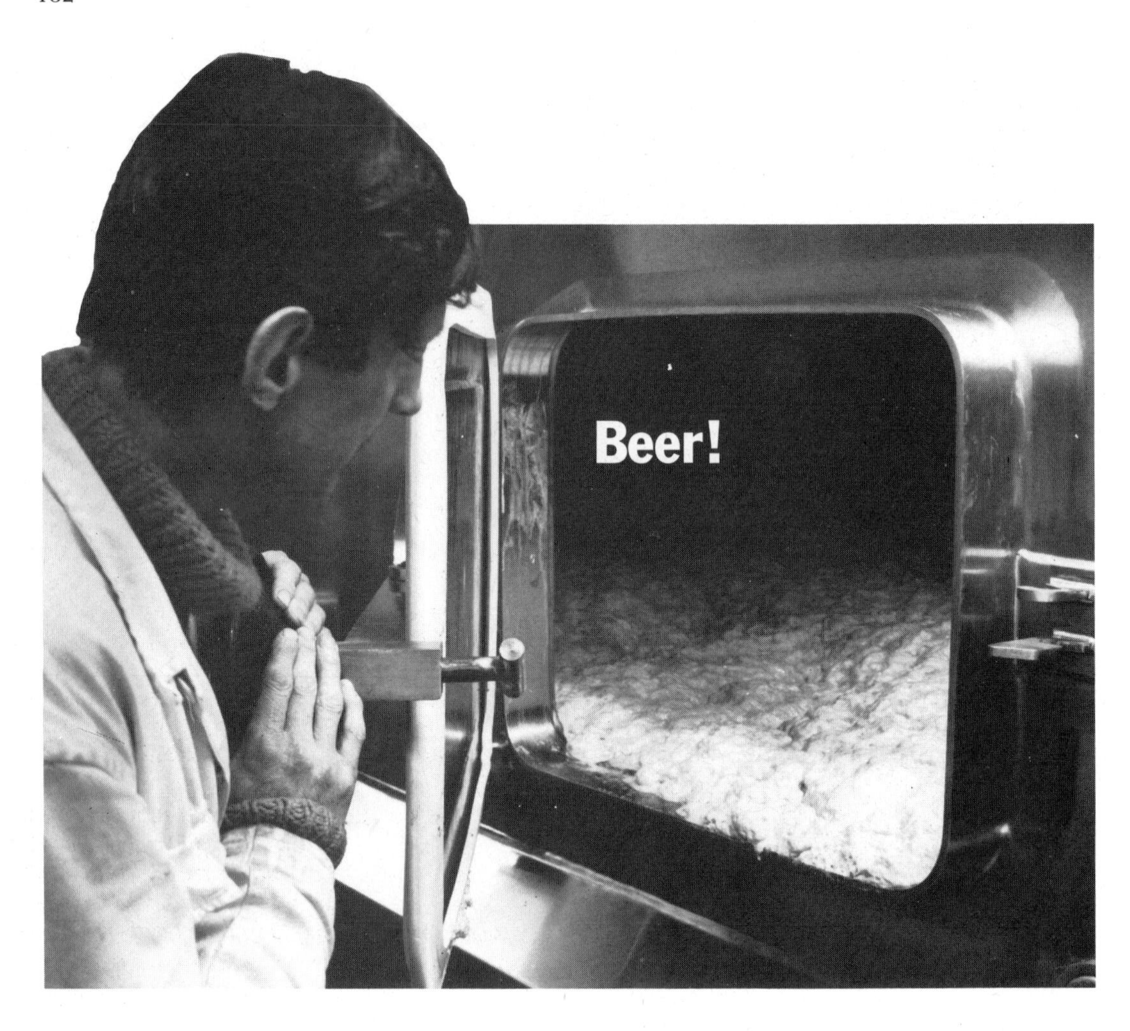


*The ancient Egyptians discovered it – today, it is still big business. How is it made? Read on.*

**Breweries swallowed up**

For as long as books have been written in this country, beer seems to have been a traditional drink. In the last century, each inn brewed its own beer for its customers, making a few gallons each week. Then larger breweries were built to cater for the thirsts of whole towns. Beers of quite different sorts were to be had in different parts of the country. Today, the small breweries have been swallowed up by the well known big companies. Nevertheless, the beer is still brewed by the same methods.

**Mash your grist and sparge your wort!**

It all starts with barley, a plant like wheat which when fully grown produces ears full of seeds or grains. These grains are harvested, and then they are *malted*. This involves keeping the grains moist and warm until they just begin to germinate and small roots grow out of them. At this stage, enzymes in the grains have converted some of the starch into sugars. If the grains were allowed to go on growing, they would use up the sugars themselves, so the growing process is stopped by gently roasting them in kilns. The sugars are preserved and the grains have become *malted barley*. If you were to chew some of the grains at this stage they would taste sweet.

When the malted barley is needed, it is cracked and split between rollers in a mill to form *grist*. The grist (crushed malt) is put into a large copper vessel called a *mash-tun* and hot water (at 65 °C) is added to extract the sugars from the grist. In brewing language, the water is called *liquor*, and after the husks and waste matter from the grain have been removed, the sugar extract that is left is called the *wort*. To make sure that all the sugar has been extracted, the husks and residues are sprayed with more hot water. This is called *sparging*.

*Adding the hops, at Norwich brewery.*

The wort is then put into another vessel and more sugar is added while it is boiled. Hops are added to give the beer a slightly bitter flavour (they used to be added to help preserve the beer), and the sugary wort is boiled for several hours to infuse the hops and sterilise everything. Next it is strained from the vessel, and suddenly, the tax man appears.

**Gauger sticks to the law**

All ethanol that is produced is taxed by the Government. (This is where they get a lot of their money for running the Country.) The Excise man measures the specific gravity of the wort to see how much sugar it has in it. The more sugar there is in the wort, the more ethanol will be made. He charges the brewers an appropriate sum of money, and they get it back from us when we buy the beer.

Taxing beer hasn't always been a simple matter of putting a hydrometer into the wort. In the 17th Century, a man called a *gauger* tasted the beer, and then, wearing leather trousers, sat in a puddle of the beer for 30 minutes. If he stuck to the seat afterwards, he charged a higher rate of tax than if he didn't. Sometimes the excise man tested spirits like gin by pouring them over gunpowder. By trying to light the powder afterwards, they could measure how much flammable ethanol was in the gin.

**Fermentation**

Once the sugar content has been worked out, the beer is ready for fermentation. Yeast is added and this works on the sugar producing ethanol and carbon dioxide. A thick froth is formed on top of the fermenting beer and this helps to keep bacteria out and stops the beer from going bad and turning into vinegar. After about three days, the fermentation is complete. Some of the yeast is skimmed off (to be used for more brewing, or to be sold for yeast extracts and cattle food), and substances called *finings* are added to clear the beer by making all the suspended matter collect together and fall to the bottom of the fermentation vessel.

At one time, all beer was sold 'live'. The yeast cells in the beers were not dead, and fermentation had only stopped because the sugar had all been used up, or the ethanol content had got too high. The beer was bottled or drawn off into wooden casks to be hand pumped to glasses when it was drunk. Now-a-days, most beer is pasteurised rather like milk. It is carefully heated so that the yeast is killed and the fermentation is stopped. This makes the beer keep much better and different batches of beer always come out the same. This beer is put into metal casks and is drawn out by carbon dioxide under pressure.

There are still lots of different types of beers. It all depends on how long the barley was roasted, the sort of water that was used, the hops, the sugar, the fermentation temperature and the speed of fermentation.

**Do it yourself?**

You don't have to be too much of an expert to make beer. You can buy malted barley and hops in the shops, and beer kits even extract the sugars from the malt for you and put them in a tin. But care needs to be taken that everything is kept very clean. Some skill in judging when and how much of the various ingredients must be added, before a really good quality brew can be made.

**Questions**

**1** Explain why nylon, whose ingredients come from coal and oil, is thought of as an organic substance, whilst the ore zinc carbonate, is said to be inorganic.

**2** Write a list of 10 everyday substances which go black and char when they are heated. Are they all organic substances?

**3** The following apparatus was set up to show that custard powder gives off carbon dioxide and steam when heated.

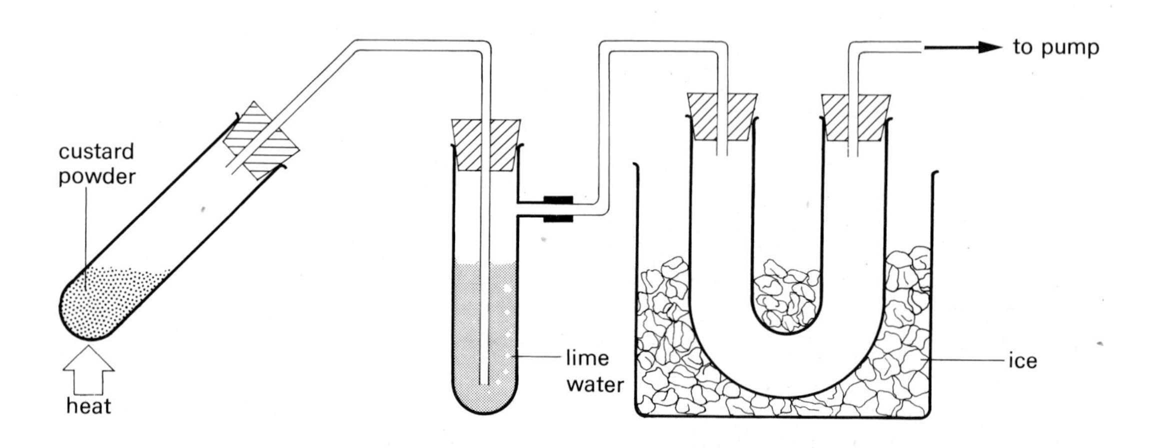


The custard powder charred and went black and the limewater went milky. However, very little condensation appeared in the U-tube. Where had the steam gone? How would you modify the apparatus?

**4** Write down the molecular formula of hexane and draw its structure and formula. Is it a liquid or gas?

**5** Pentane has the molecular formula $C_5H_{12}$. Draw its structure and formula. Pentane has three isomers. Can you draw them? Your teacher may use models to discuss your answer with you.

**6** Write down the structures of ethanol and methanoic acid. Now try to write the esterification reaction that occurs between them. What is the ester called?

**7** If you had a mixture of ethanol and water draw the apparatus that you might use to obtain ethanol from it.

Why isn't wine or beer bottled before it has finished fermenting?

Sometimes home made beer that has been left open to the air after it has finished fermenting tastes of vinegar. What has happened to it? What chemical has been formed?

**8** Explain why 2-methyl-3-ethylpentane is a saturated hydrocarbon, but 2-methylpen-3-ene is an unsaturated hydrocarbon. Their structure and formulae are given below.

```
              H
              |
          H — C — H
              |
          H — C — H                     H   H   H       H
  H   H       |   H   H                 |   |   |       |
  |   |       |   |   |             H — C — C — C = C — C — H
H—C — C ————— C — C — C — H             |   |       |   |
  |   |       |   |   |                 H   |       H   H
  H   |       H   H   H                 H — C — H
  H — C — H                                 |
      |                                     H
      H
```

2-methyl 3-ethyl pentane 2-methyl pent-3-ene

Which one will burn with the more sooty flame?

**9** Compare and explain the reactions of ethanol and sodium hydroxide solution with: **a**, litmus, **b** oxygen, **c** ethanoic acid and **d** an electrical test circuit like the one below.

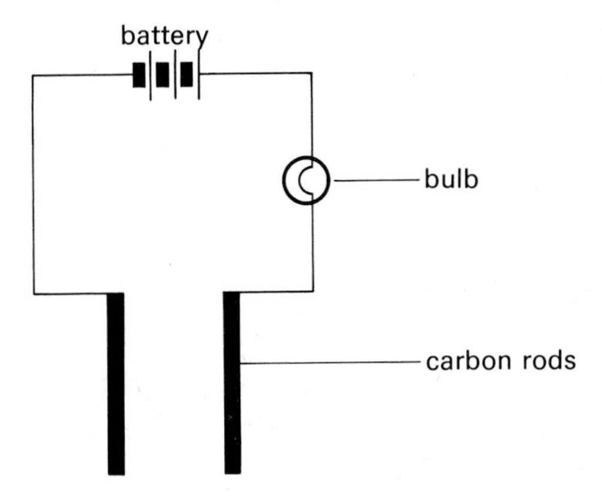


**10** Read chapter 3 and write revision notes about the washing action of soap, and the way that it is affected by hardness of water.

# 14 Chlorine

| | | | H | | | | | | | | | | | | | | He |
|---|---|---|---|---|---|---|---|---|---|---|---|---|---|---|---|---|---|
| Li | Be | | | | | | | | | | | B | C | N | O | F | Ne |
| Na | Mg | | | | | | | | | | | Al | Si | P | S | Cl | Ar |
| K | Ca | Sc | Ti | V | Cr | Mn | Fe | Co | Ni | Cu | Zn | Ga | Ge | As | Se | Br | Kr |
| Rb | Sr | Y | Zr | Nb | Mo | Tc | Ru | Rh | Pd | Ag | Cd | In | Sn | Sb | Te | I | Xe |

## 14.1 Group VII The halogens

Chlorine is a member of Group VII of the periodic table, and the family name of this group is *the halogens*. (See the diagram above.) Halogen is an Ancient Greek word which means 'Salt producer'.

Each of the elements in Group VII has seven outside shell electrons. Because of this they all react in the same sort of way.

A halogen atom needs one more electron to fill its outside shell with 8 electrons, so when it reacts, it does so to take in one electron and form a negatively charged ion:

| | | |
|---|---|---|
| fluorine atom | $F + e^- \longrightarrow F^-$ | fluoride ion |
| chlorine atom | $Cl + e^- \longrightarrow Cl^-$ | chloride ion |
| bromine atom | $Br + e^- \longrightarrow Br^-$ | bromide ion |
| iodine atom | $I + e^- \longrightarrow I^-$ | iodide ion. |

Because of this, the halogens are good oxidising agents. They take electrons away from other atoms or molecules. Remember that when something is oxidised, it can either have gained oxygen, but it can also have lost electrons. (See chapter 7.)

The structure of a halogen atom also dictates the form that molecules of the element takes. Two atoms share electrons to make a covalent, diatomic molecule.

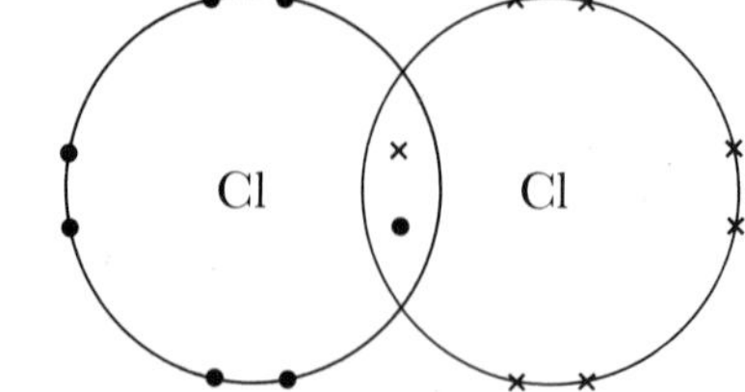


**Figure 2**
*Chlorine is a diatomic molecule.*

The halogens are also very similar in their physical properties. Figure 3 compares the physical properties of chlorine, bromine and iodine.

| halogen | state | colour | effect of heat | smell | danger |
|---|---|---|---|---|---|
| chlorine | gas | green/ yellow | | choking | very poisonous gas. |
| bromine | liquid | deep red | boils easily to a brown vapour | burns nose and throat | liquid causes burns and vapour is poisonous. |
| iodine | solid | silver/ black | melts and boils to form a violet vapour which sublimes back to the solid on a cold surface. | vapour affects nose and throat. | A solution of the solid in water is used as an antiseptic, but vapour is poisonous. |

**Figure 3** *The physical properties of the halogens.*

## 14.2 The industrial production of chlorine

The majority of chlorine is obtained from the electrolysis of salt. It is broken down to give both sodium, and chlorine gas. You can read more about this in chapter 17.

### The uses of chlorine.

**1** The majority of chlorine is used in the production of chemicals such as solvents for dry cleaning; anaesthetics; and petrol additives.

**2** A large amount is used in the production of hydrochloric acid, which in turn is used to make other substances such as polyvinyl chloride.

**3** Some of it is used to make bleaches such as sodium chlorate(I).

**4** Chlorine is used in the production of antiseptics, and directly as a germ-killer in swimming pools.

## 14.3 The laboratory preparation of chlorine

Chlorine can be made by the oxidation of hydrochloric acid:

$$2HCl(aq) + \underset{\text{oxidising agent}}{O} \longrightarrow Cl_2(g) + H_2O(l).$$

The oxidising agents that are usually used are manganese(IV) oxide (formula $MnO_2$) and potassium manganate(VII) (formula $KMnO_4$). The next two paragraphs give details.

**Manganese(IV) oxide as oxidising agent.** Concentrated hydrochloric acid is warmed with black manganese(IV) oxide powder, and chlorine gas is evolved. The reaction is difficult to control once it has started, so it must be done in a fume cupboard:

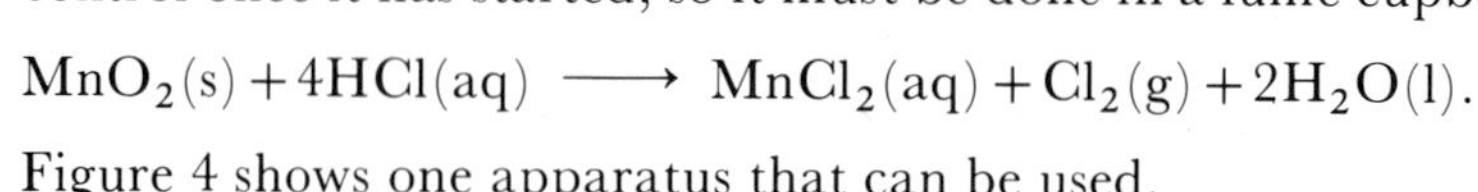

$$MnO_2(s) + 4HCl(aq) \longrightarrow MnCl_2(aq) + Cl_2(g) + 2H_2O(l).$$

Figure 4 shows one apparatus that can be used.

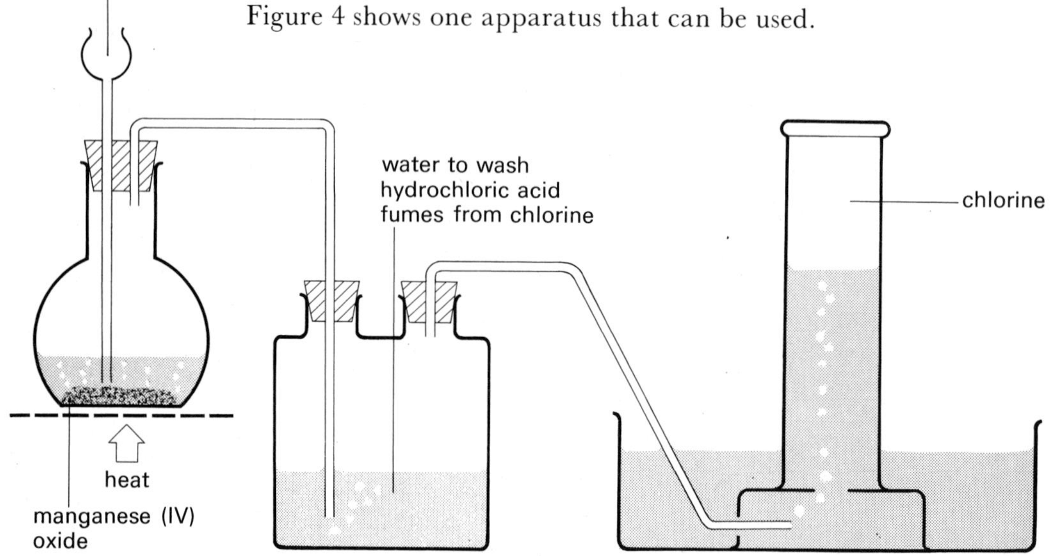


**Figure 4** *The preparation of chlorine using manganese(IV) oxide.*

**Potassium manganate(VII) as oxidising agent.** This reacts with concentrated hydrochloric acid without heating. If the acid is dripped into the solid from a tap funnel, the chlorine is evolved in small puffs of greenish yellow gas. Figure 5 shows the apparatus that may be used for collecting a gas jar of dry chlorine.

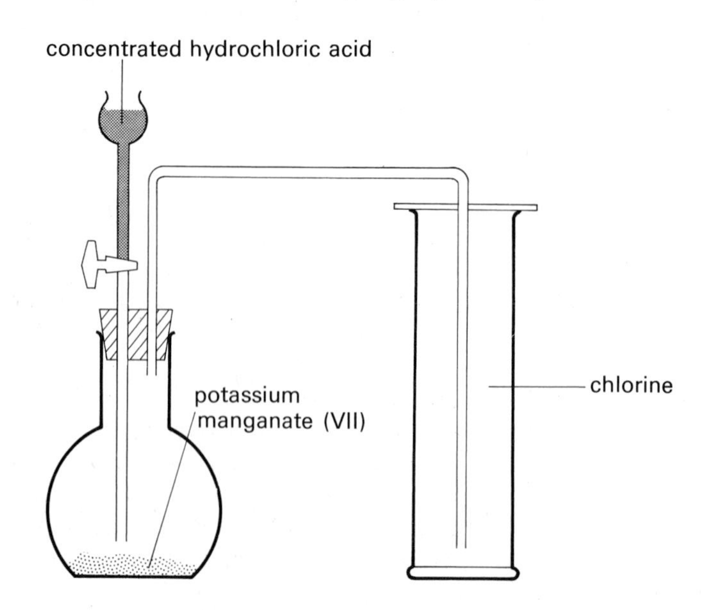


**Figure 5** *The preparation of chlorine using potassium manganate(VII).*

## 14.4 The physical properties of chlorine

**1** Chlorine is a greenish yellow coloured gas.

**2** It is much heavier than air.

**3** It has a choking smell and is very dangerous if breathed, even in small quantities.

**4** It is moderately soluble in water.

## 14.5 The chemical reactions of chlorine

Chlorine is very reactive. This section shows five of its main reactions.

**Chlorine dissolves in water.** It reacts to form two acids:

$$Cl_2(g) + H_2O(l) \longrightarrow HCl(aq) + HOCl(aq).$$

hydrochloric acid (dilute) chloric acid(I) (dilute)

This solution is known as chlorine water.

Chloric acid(I) is unstable and decomposes in light to form hydrochloric acid and oxygen:

$$2HOCl(aq) \longrightarrow 2HCl(aq) + O_2(g).$$

Look at figure 6. The tube containing chlorine water is left near a window for a few days, and a bubble of gas is formed at the top of the tube. If a glowing splint is put in the gas it will relight, showing that the gas is oxygen.

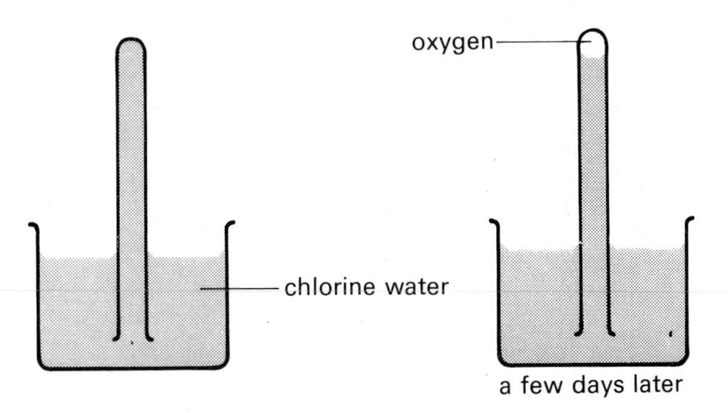


**Figure 6**
*Chloric acid(I) is unstable in light.*

This unstable acid is responsible for the bleaching action of chlorine. If wet litmus paper is put into a gas jar of chlorine, it is immediately bleached white. Chlorine water is formed in the water on the litmus paper; the chloric acid(I) in it oxidises the litmus dye. Since wet chlorine always bleaches by oxidation, it will only bleach those substances which turn white when oxidized.

**Chlorine is a powerful oxidising agent.** If iron wool is heated until it is red hot, and then chlorine is passed over it, the iron wool glows brightly as a very exothermic reaction takes place. Iron(III) chloride is formed as a brown smoke. This reactions is called a synthesis and the apparatus used may be found in chapter 7. It is an example of a redox reaction. The chlorine oxidises the iron and the iron reduces the chlorine:

$$3Cl_2(g) + 2Fe(s) \xrightarrow{\text{redox}} 2FeCl_3(s).$$

The chlorine is the oxidising agent because it takes electrons away from the iron. The iron is oxidised because it loses electrons.

**Hydrogen gas will burn in chlorine.** If a hydrogen flame is put into a gas jar containing chlorine, the greenish yellow colour quickly disappears and white misty fumes of hydrogen chloride gas are formed. This is shown in figure 7.

This is one way in which hydrogen chloride gas is manufactured.

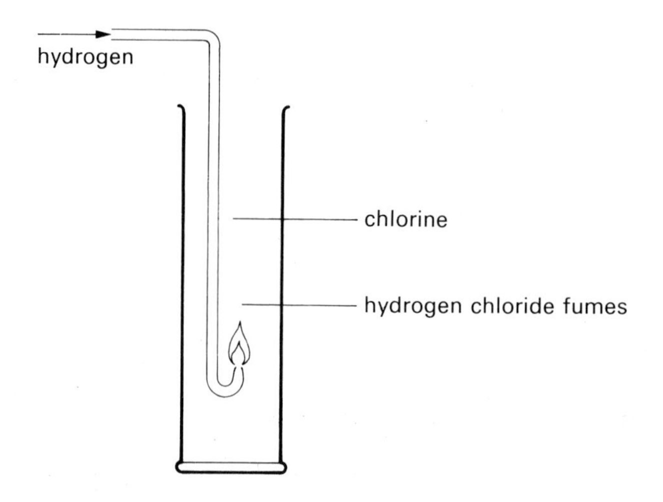


**Figure 7**
*The formation of hydrogen chloride.*

**Displacement: another redox reaction.** If chlorine is bubbled through a colourless solution of potassium bromide, a red colour of bromine is immediately produced. The chlorine has oxidised the bromide ions to bromine:

$$Cl_2(g) + 2KBr(aq) \longrightarrow Br_2(aq) + 2KCl(aq).$$

The same thing happens when chlorine is bubbled through a colourless solution of potassium iodide:

$$Cl_2(g) + 2KI(aq) \longrightarrow \underset{\text{red colour turning to a black solid}}{I_2(s)} + 2KCl(aq).$$

## 14.6 Manufacture of hydrogen chloride

Hydrogen chloride is a covalent gas which is mainly used to make hydrochloric acid. It is manufactured from hydrogen (a by-product of the petrol industry) and chlorine. This is shown in figure 8.

The chlorine is burned in an atmosphere of hydrogen, and the hydrogen chloride gas formed is immediately dissolved in water in absorption towers:

$$H_2(g) + Cl_2(g) \longrightarrow 2HCl(g).$$

$$HCl(g) \longrightarrow H^+(aq) + Cl^-(aq).$$

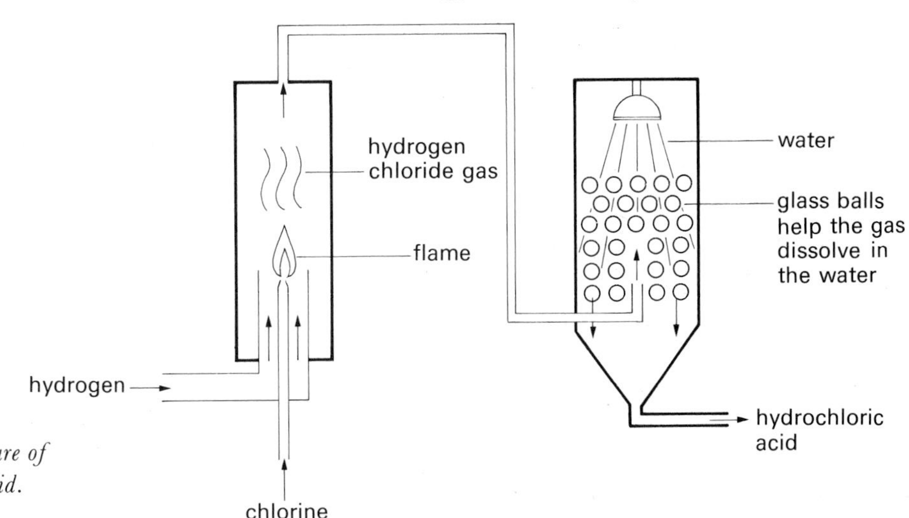


**Figure 8**
*The manufacture of hydrochloric acid.*

## 14.7 Laboratory preparation of hydrogen chloride

Hydrogen chloride is formed whenever a chloride is reacted with concentrated sulphuric acid. For example:

$$NaCl(s) + H_2SO_4(l) \longrightarrow HCl(g) + NaHSO_4(aq).$$

sodium hydrogen sulphate

This is shown in figure 9. Great care must be taken because of the concentrated acid. The HCl(g) is also very choking and corrosive. The hydrogen chloride is dried by bubbling it through concentrated sulphuric acid and then collected downwards into a gas jar.

concentrated sulphuric acid
hydrogen chloride gas
concentrated sulphuric acid to dry gas
sodium chloride

**Figure 9** *The preparation of dry hydrogen chloride.*

## 14.8 Physical properties of hydrogen chloride gas

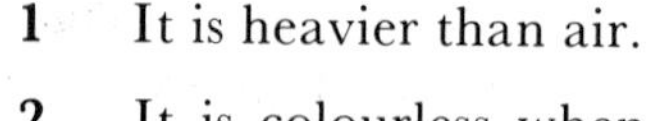

**1** It is heavier than air.

**2** It is colourless when dry, but in damp air it fumes. This is because it combines with the moisture in the air to form tiny droplets of hydrochloric acid which appear as a mist.

**3** Hydrogen chloride has a choking smell.

**4** Damp litmus turns red when added to the gas, because the gas dissolves in the water in the paper to form hydrochloric acid.

**5** Hydrogen chloride is very soluble in water. This may be demonstrated by the fountain experiment. (See figure 10.)

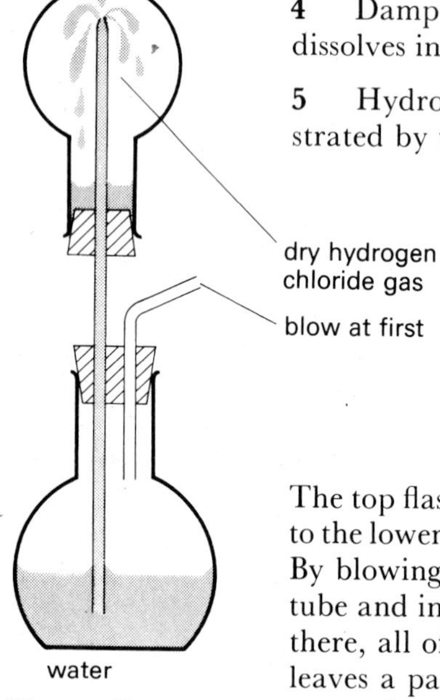


**Figure 10**
*The fountain experiment.*

The top flask is filled with dry hydrogen chloride gas. It is connected to the lower flask of water by a narrow tube at the top of which is a jet. By blowing down the side tube, the water is forced up the narrow tube and into the upper flask. As soon as the first drop of water gets there, all of the hydrogen chloride dissolves in that one drop. This leaves a partial vacuum in the upper flask and the pressure of the atmosphere forces the rest of the water up the tube in a fountain to fill the vacuum. The fountain continues until the pressure in the upper flask is again normal.

## 14.9 The chemical reactions of hydrogen chloride

**Hydrogen chloride gas reacts with ammonia gas.** White ammonium chloride is formed:

$$HCl(g) + NH_3(g) \longrightarrow NH_4Cl(s).$$

This happens when a gas jar of hydrogen chloride is opened next to a gas jar of ammonia, or even when bottles of concentrated hydrochloric acid (hydrogen chloride solution) and ammonia solution are opened near one another. The ammonium chloride is formed by sublimation because a solid is formed from two gases, without the liquid state being involved.

**Hydrogen chloride gas dissolves in water.** Concentrated hydrochloric acid is a saturated solution of the gas in water. It contains about 35% of the gas by weight. When a bottle of the concentrated acid is open in a warm room, it fumes as hydrogen chloride gas escapes from solution. The gas may be identified with damp blue litmus paper, or with the ammonia solution bottle. (See previous section.) Hydrochloric acid is a typical acid – it reacts with metals, to form a salt plus hydrogen, bases, to form a salt plus water, and carbonates, to form a salt plus carbon dioxide plus water. In each case the salt formed is a chloride.

## 14.10 Tests for chlorides

**1** Any chloride will react with concentrated sulphuric acid to form hydrogen chloride gas. For example:

$$NaCl(s) + H_2SO_4(l) \longrightarrow HCl(g) + NaHSO_4(aq).$$

**2** Adding silver ions will identify chloride ions in solution:

$$Cl^-(aq) + Ag^+(aq) \longrightarrow \underset{\text{(white precipitate)}}{AgCl(s)}.$$

The silver ions are obtained from silver nitrate solution. The suspect chloride is first acidified with a few drops of nitric acid. This prevents other compounds from being precipitated by the silver ions. Then a few drops of silver nitrate solution are added. A white precipitate of silver chloride confirms the presence of chloride ions. As a further test, the silver chloride will redissolve in dilute ammonia solution.

## Summary

At the end of this chapter, you should be able to:

**1** Describe the position of the halogens in the periodic table.

**2** Draw the structure of the chlorine molecule.

**3** Compare the physical properties of chlorine, bromine and iodine.

**4** Explain how chlorine is made industrially.

**5** Describe the main uses of chlorine.

**6** Give details of the laboratory preparation of chlorine.

**7** Describe the reactions of chlorine as a bleaching agent, and as an oxidising agent with metals and hydrogen.

**8** Describe the manufacture of hydrogen chloride and hydrochloric acid.

**9** Show how hydrogen chloride is made in the laboratory, and describe its reaction with water and ammonia.

**10** Describe an experiment to show that hydrogen chloride is very soluble in water.

**11** Describe reactions which show that hydrochloric acid is a typical acid.

**12** Describe a test for chloride ions.

# The halogens–use and abuse

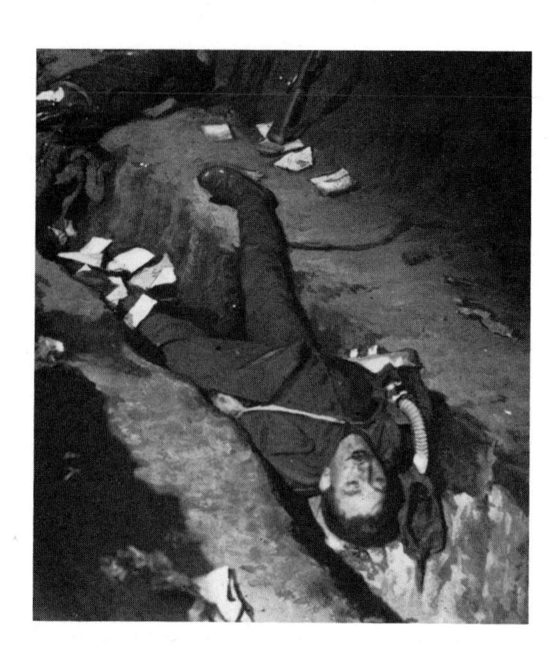

*Most chemicals can be used in a variety of different ways. The Halogens present an interesting example of this – the two articles that follow show you just how differently they can be used.*

**Chemical warfare**

The idea of using choking fumes as an offensive weapon in war is very old indeed. One of the earliest instances to be recorded occurred in 428 BC, when the Spartans burnt wood saturated with pitch and sulphur under the city walls of Athens in order to subdue the defending citizens. But the events that occurred at Ypres on 22nd April 1915, were a turning point in the development of chemical warfare.

**Green, deadly 'mist'**

The day was a fine one in the immediate vicinity of Ypres. At 5 p.m., men in forward positions heard a slight hissing in the direction of the German trenches. Within 15 minutes, no less than 170 tonnes of chlorine had been released along a front extending about six kilometres. A light wind of 5–6 kilometres per hour bore the wall-like cloud towards trenches manned by the British, Canadian, French, and Algerian troops. Distant observers spoke of a low greenish mist 'such as is seen over water meadows on a frosty night'. The deadly gas brought horror and confusion into the ranks of the Allies. Those still capable fled, only to run the risk of being shot by their comrades unaware of the advancing terror. More chlorine was discharged on an adjoining sector of the front on 24th April, this time against Canadian troops.

No-one knew exactly how many people were affected by the gas. The most common figure is that there were 15 000 casualties, of whom 5000 died during or soon after the attacks.

**Doctors could do little to help**

As with other poisonous gases, the effects of chlorine are related to the duration and amount of exposure. Victims suffering a high dose experience feelings of intense suffocation, fall to the ground, struggle for a few moments, and die. Those exposed to lower doses suffer from a burning throat and feelings of suffocation, and cough repeatedly. Breathing becomes intensely difficult, and death sometimes follows within two days. The main effect of the gas is to cause the secretion of massive quantities of a frothy fluid into the airspaces of the lungs. As fluid builds up, lack of oxygen induces feelings of weakness, fatigue, and headache.

Doctors at Ypres were able to detect bubbling noises in the chests of victims. The body temperature would tend to fall, and the features become blueish. When slightly tilted in a head downwards position, the patients sometimes produced over a litre of frothy liquid. Gased individuals surviving for 36 hours generally developed bronchitis. The fluid coughed up became greenish and purulent; the pulse became weak and rapid, while respirations became shallow and fast. Headache and debility would develop, persisting sometimes for several weeks. Those who lasted out the initial two days usually made some kind of recovery.

## Fluoride: the halogen which replaces dentists

It has been estimated that about two-thirds of all children between the ages of 5 and 15 need treatment for holes in their teeth. Worse than this is the fact that about one-third of the adults in this country haven't got any teeth at all, because they've been taken out. Teeth rot because we eat too many sweets and sticky buns and biscuits and gooey puddings – you know the list as well as I do. On top of this, we don't look after our teeth properly. Do you brush your teeth for two or three minutes after each meal? Do you brush your teeth at night? Have you even got a toothbrush? Even if you do clean your teeth regularly, you probably don't remove all of the plaque, which is the decaying substance that builds up on your teeth after meals.

Dentists and scientists are very aware of the problem. As well as encouraging children and adults to clean and look after their teeth properly, many of them want to put a chemical into our water supply which will help fight tooth decay.

### Fluoride stops the rot

The chemical contains the fluoride ion, and tests show that in areas where fluoride has been put in the water, or where the natural fluoride content of the water is high already, children have less need of fillings because their teeth are more resistant to decay.

The obvious answer would be to add fluoride to the whole country's water supply. The amount of fluoride that would need to be added is very small – about one part for every million parts of water. Some areas of the country have double that amount naturally. The trouble with putting fluoride in the water, even in such small quantities, is that everyone has to drink it, whether they want to or not.

### Poisonous when concentrated

Too much fluoride can cause teeth to drop out. What happens if there is a long hot summer and everyone drinks a lot? Could we suffer from fluoride poisoning? Just as there are a lot of people who want fluoride put into the water because of its good effect on teeth, there are many who don't want their water to be 'poisoned'. What will be the *next* additive, they argue?

### Any other ideas ?

Alternatives to putting it in the water supply are not very satisfactory. It could be given in tablet form – but people forget, or take overdoses by mistake. It could be put in special milk for children, but this would be very difficult to distribute. The best answer so far is to put it in toothpaste, which many manufacturers now do. Presumably *they* aren't too worried about people's teeth dropping out . . .

Everything considered, it seems that fluoride in water seems to be the best answer. Everyone drinks it regularly so that the concentration of the chemical need only be very small because it is absorbed continually. But there are still plenty of people who think that it is quite wrong to force them to take in intentional impurities against their will, which is what it amounts to.

Perhaps the argument is going on in your town at the moment. If it is, you had better clean your teeth – with fluoride toothpaste if you want to – until they make up their minds.

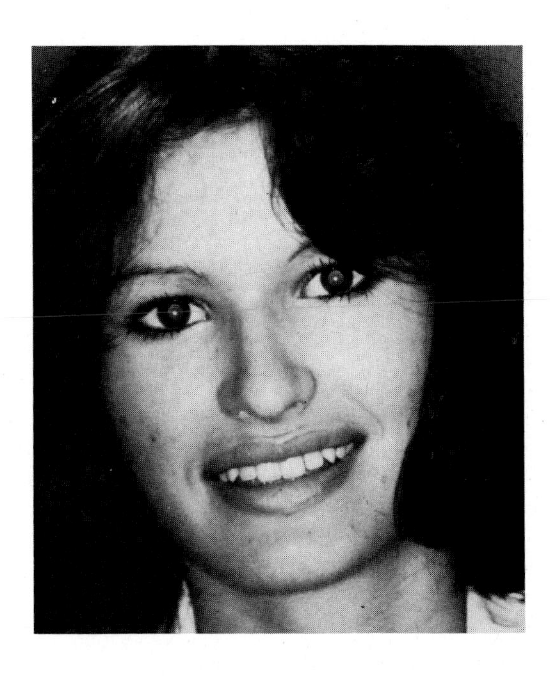

**Questions**

**1** Draw an outline of the periodic table and write each of the following in the correct position:

sodium, which has two full shells, and one electron in the third;
neon, which has three full shells;
aluminium, which has two full shells and three electrons in the third;
chlorine, which has two full shells and seven electrons in the third;
carbon, which has one full shell and four electrons in the second;
magnesium, which has two full shells, and two electrons in the third.

**2** Draw the structure of a bromine molecule, showing how the outside shell electrons are shared.

**3** Describe a method by which you might get some pure dry iodine from a mixture of iodine and sand.

**4** Use the index to find out how polyvinyl chloride (which is a plastic) is made. Trace the route that chlorine follows in the manufacture of this plastic, from its elemental state, to the final product.

**5** Describe how you would make and collect a gas jar of pure, dry chlorine starting with manganese(IV) oxide and concentrated hydrochloric acid. Draw the apparatus you would use, and stress the precautions you would take.

**6** Read the section on the bleaching action of chlorine. Why do you think that items bleached in chlorine have to be washed afterwards?

A certain coloured material was put into a gas jar of chlorine but it stayed coloured. Can you suggest two reasons why it was not affected?

**7** You are asked to prepare a sample of anhydrous zinc chloride. Draw the apparatus which you would use for its preparation, and write the equation for the reaction. Describe what would happen.

**8** You are given two unlabelled gas jars. One contains chlorine and the other contains hydrogen chloride. Give one physical difference, and two chemical reactions that would enable you to distinguish the two gases. Describe the tests, giving full details.

**9** Explain why a bottle of concentrated hydrochloric acid fumes when it is opened in a warm room, yet a bottle of dilute hydrochloric acid does not.

What will you see if a bottle of concentrated ammonia solution is opened near the concentrated hydrochloric acid bottle? Explain what is happening.

**10** A white crystalline solid is thought to be sodium chloride. Describe two tests that you could perform to confirm that it does contain chloride ions. Say what reagents you would use and describe what you would see. Write equations for the reactions that take place.

# 15 Sulphur

| | | | | | | | | | | | | | | | | | |
|---|---|---|---|---|---|---|---|---|---|---|---|---|---|---|---|---|---|
| | | | H | | | | | | | | | | | | | | He |
| Li | Be | | | | | | | | | | | B | C | N | O | F | Ne |
| Na | Mg | | | | | | | | | | | Al | Si | P | **S** | Cl | Ar |
| K | Ca | Sc | Ti | V | Cr | Mn | Fe | Co | Ni | Cu | Zn | Ga | Ge | As | Se | Br | Kr |
| Rb | Sr | Y | Zr | Nb | Mo | Tc | Ru | Rh | Pd | Ag | Cd | In | Sn | Sb | Te | I | Xe |

## 15.1 Where does it come from?

Some sulphur is found on the surface of the Earth in volcanic areas such as Italy and Japan. In parts of France and Canada, sulphur compounds are removed from natural gas and crude oil, but by far the largest amount of sulphur is extracted from underground deposits in Texas, in the U.S.A. 90% of the World's sulphur is obtained in this way, and that amounts to millions of tonnes each year.

The deposits of sulphur are several hundred metres below the surface of the ground. A hole is drilled into the sulphur bed, and a pipe which is about 20 cm in diameter is put down. This pipe not only strengthens the hole, but it contains a piece of engineering apparatus called a Frasch sulphur pump. Figure 1 shows what a Frasch pump looks like.

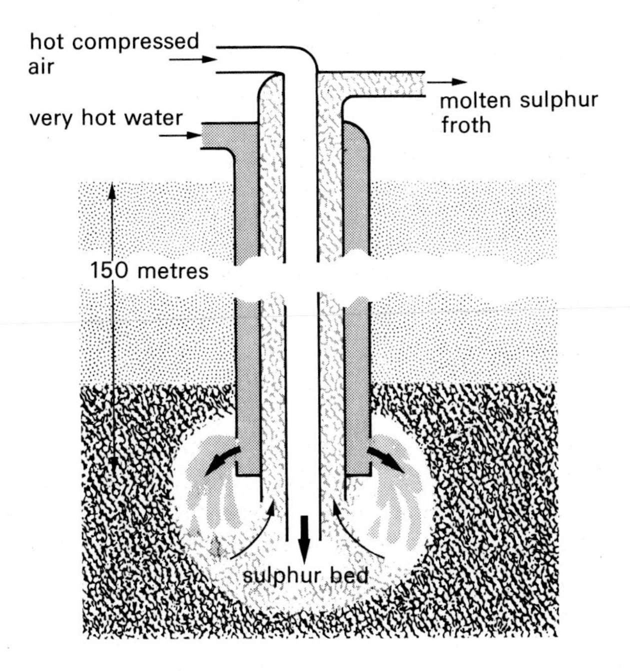


**Figure 1**
*The Frasch sulphur pump.*

The pump consists of three tubes – a small one in the middle, surrounded by a larger one, which is surrounded by a larger one still. Compressed air is forced down the smallest tube; super heated water, at 170 °C is pumped down the largest tube: and the sulphur comes out of the middle tube.

At the bottom of the hole, the sulphur and the water meet each other.

The melting point of sulphur is 115 °C so the super heated water melts it. The compressed air then forces it up the middle tube, all the way to the surface. Because this pipe is always surrounded by the hot water pipe, the molten sulphur does not solidify again.

On the surface, the sulphur is stored in huge piles, and needs very little purification because it is so pure already. It is usually remelted and poured into moulds to produce sticks of sulphur 'rock' called roll sulphur.

## 15.2 The uses of sulphur

Sulphur is a very important element. Most of it is made into sulphuric acid. The rest is used to make chemicals for various industries. Sulphur compounds are used for bleaching wood pulp in the paper industry and for preserving fruit (you can often smell the sulphur dioxide when you open a new bottle of orange squash concentrate). At one time it was extensively used for vulcanising rubber and is still used for making carbon disulphide which is used as a solvent.

## 15.3 The allotropy of sulphur

Sulphur can exist in two different solid forms. This property is known as allotropy (remember carbon in chapter 12).

Below 96 °C, sulphur exists as the allotrope called *rhombic sulphur*. Above 96 °C, it exists in a form called *monoclinic sulphur*. The allotropes are identical chemically, but they have different crystal shapes.

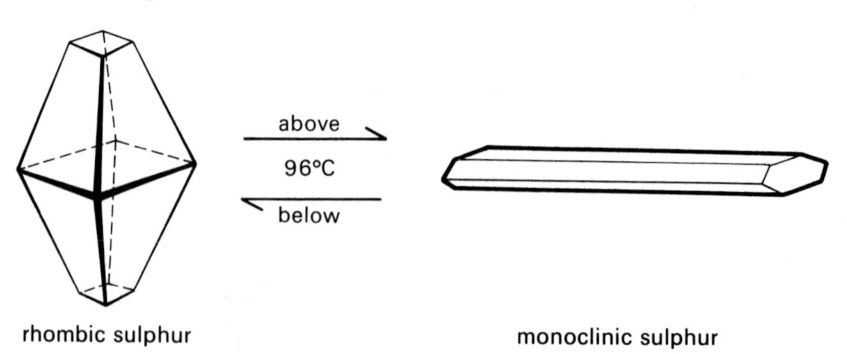


**Figure 2** *The shapes of rhombic and monoclinic sulphur.*

96 °C is called the *transition temperature*. Above the transition temperature, only monoclinic sulphur is stable. Below the transition temperature, only rhombic sulphur is stable.

A single molecule of sulphur consists of eight atoms arranged in a ring. Figure 3 shows the shape of a sulphur molecule.

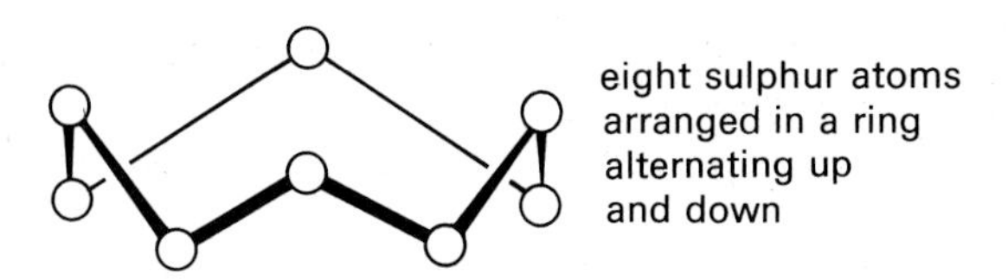


**Figure 3**
*The shape of a sulphur molecule.*

When the $S_8$ molecules pack together in a crystal lattice, they take up two different arrangements, according to their temperature. (See figure 4.)

**Figure 4**
*The molecule arrangement in rhombic and monoclinic sulphur.*

## 15.4 The preparation of rhombic sulphur

Rhombic crystals are formed whenever a solution of sulphur is allowed to evaporate at a temperature below 96 °C.

Sulphur is not soluble in water, but it does dissolve in organic solvents such as methylbenzene.

Sulphur powder is stirred into warm methylbenzene (below 96 °C) until a saturated solution is formed.

The solution is then quickly filtered.

Finally, it is allowed to evaporate slowly in an evaporating basin covered with a filter paper. In this way, large, yellow crystals may be formed. The procedure for this preparation is shown in figure 5.

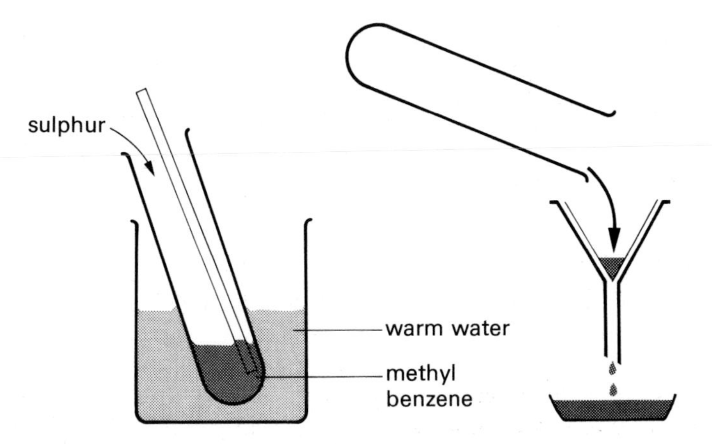


**Figure 5** *The preparation of rhombic sulphur.*

## 15.5 The preparation of monoclinic sulphur

Monoclinic crystals of sulphur are formed when a solution of sulphur is allowed to crystallise *above* 96 °C.

The same solvent as was used for rhombic sulphur may be used again, but it must be heated nearly to its boiling point (110 °C) so that a hot, saturated solution is formed.

This is then allowed to crystallise while the test tube containing the solution is kept in boiling water, so the crystals form above the transition temperature of 96 °C. Long, pale yellow crystals of monoclinic sulphur are formed. If the test tube of solution is removed from the boiling water, and allowed to cool down, the long crystals quite quickly break down into smaller, rhombic crystals.

The details of this preparation are shown in figure 6.

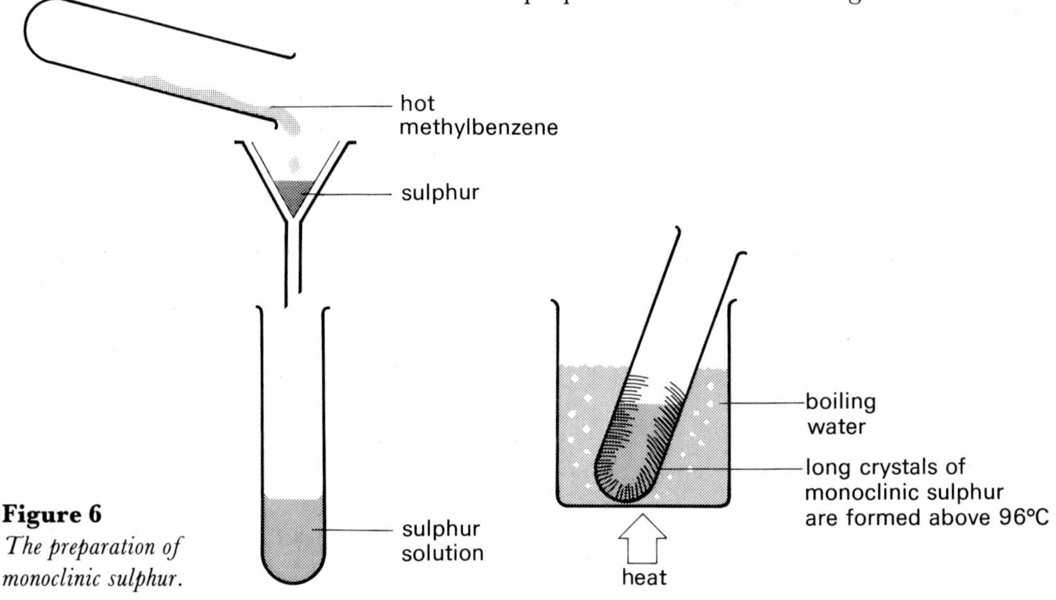


**Figure 6**
*The preparation of monoclinic sulphur.*

## 15.6 The effect of heat on sulphur

Sulphur melts at 115 °C. As the temperature increases beyond this, it turns into a runny, yellow liquid, but this soon becomes like treacle in texture and darker in colour. As the temperature rises further, the liquid sulphur becomes runny again, but the colour darkens even more and becomes almost black. At 444 °C, the sulphur boils and gives off a dark brown vapour. This sometimes catches fire, and burns to form sulphur dioxide.

These changes in the form of liquid sulphur may be understood by looking at how its structure changes with temperature. These are shown in figure 7.

If hot, runny liquid sulphur is poured into cold water, it forms a soft pliable substance, rather like chewing gum, which is called *plastic sulphur*. This still contains the long chains that are in liquid sulphur. However, the plastic sulphur soon goes hard and brittle as its crystalline form changes to that of rhombic sulphur, which is the only stable form of sulphur at that temperature.

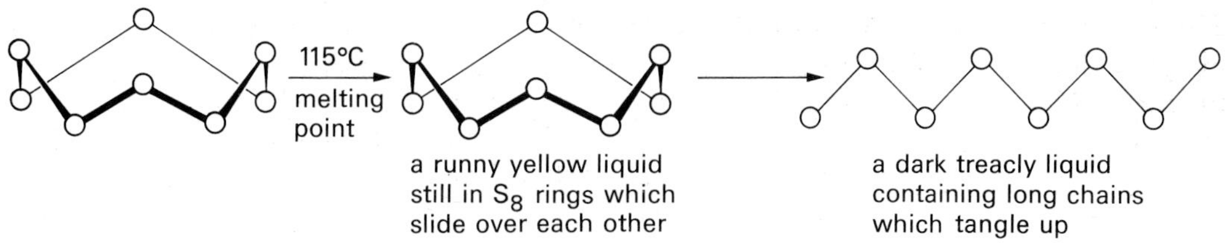

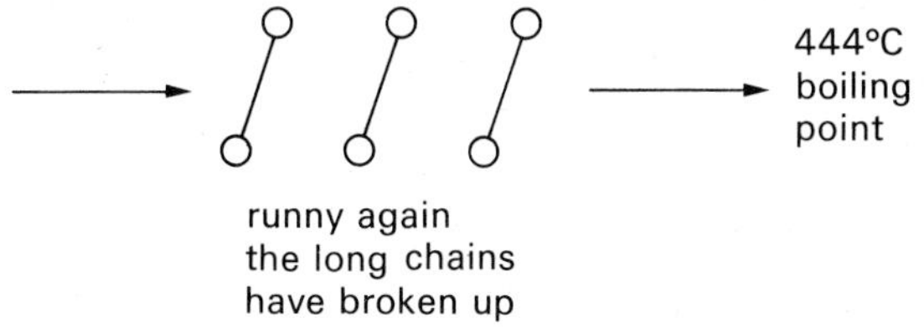

**Figure 7**
*The changes in liquid sulphur.*

## 15.7 The chemical reactions of sulphur

**Sulphur will burn when heated.** First, it melts to a dark red liquid, and then catches fire, burning with a small dark blue flame. Sulphur dioxide gas is formed:

$$S(s) + O_2(g) \longrightarrow SO_2(g)$$

**Sulphur reacts with metals.** When heated with any metal, it forms a salt called a *sulphide*. For example:

$$\begin{array}{ccccc} \text{magnesium powder} & + & \text{sulphur powder} & \longrightarrow & \text{magnesium sulphide.} \\ Mg(s) & + & S(g) & \longrightarrow & MgS(s) + \text{much heat and light.} \end{array}$$

Because magnesium is quite a reactive metal, the reaction is extremely exothermic. With a less reactive metal such as iron, the reaction is less vigorous. Nevertheless, it is still exothermic. If sulphur powder is added to iron filings and the mixture is heated, a glow spreads through the mixture, and grey, brittle iron(II) sulphide is left:

$$Fe(s) + S(s) \longrightarrow FeS(s).$$

## 15.8 Sulphur dioxide

Sulphur dioxide is a heavy, colourless gas with a choking smell. It has a boiling point of $-10\,°C$, so it can be easily liquefied by pressurising it to about three times atmospheric pressure. Metal canisters of liquid sulphur dioxide are a convenient source of the gas. The compound is a liquid while it is under pressure in the canister. When the tap is opened and the pressure is released, the liquid boils, and the gas forms.

Sulphur dioxide is used as a preservative for fruit and jam. But its most important use is in the manufacture of sulphuric acid.

## 15.9 The laboratory preparation of sulphur dioxide

**Burning sulphur in air.** This is probably the simplest way of making sulphur dioxide. A lot of sulphur dioxide for industry is made this way:

$$S(s) + O_2(g) \longrightarrow SO_2(g).$$

**Roasting sulphur-containing metal ores in air.** This method is used in industry as a source of metal oxides and sulphur dioxide. The heating of iron pyrites is typical:

$$4FeS_2(s) + 11O_2(g) \longrightarrow 2Fe_2O_3(s) + 8SO_2(g).$$

The iron oxide formed is used to make iron.

**Sulphite + acid.** Sulphur dioxide is made whenever sulphite salts are added to a dilute acid:

$$\text{sulphite} + \text{acid} \longrightarrow \text{salt} + \text{sulphur dioxide} + \text{water}.$$

For example:

$$\begin{matrix}\text{sodium} \\ \text{sulphite}\end{matrix} + \begin{matrix}\text{hydrochloric} \\ \text{acid (dilute)}\end{matrix} \longrightarrow \begin{matrix}\text{sodium} \\ \text{chloride}\end{matrix} + \begin{matrix}\text{sulphur} \\ \text{dioxide}\end{matrix} + \text{water}.$$

$$Na_2SO_3(s) + 2HCl(aq) \longrightarrow 2NaCl(aq) + SO_2(g) + H_2O(l).$$

Compare this with the reaction between sodium carbonate and hydrochloric acid. The apparatus for the preparation and collection of sulphur dioxide is shown in figure 8.

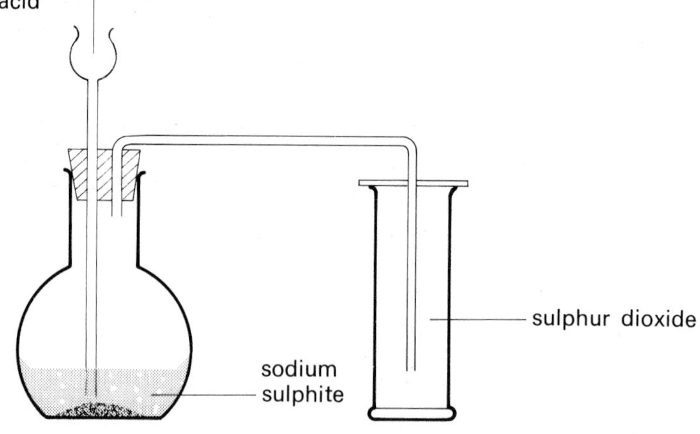


**Figure 8**
*The preparation and collection of sulphur dioxide.*

## 15.10 The reactions of sulphur dioxide

**Sulphur dioxide is quite soluble in water.** It turns damp blue litmus paper red. This is because it forms a solution of *sulphurous acid*. Sulphurous acid is a weak acid, and easily decomposes into sulphur dioxide and water, so it is not kept as an ordinary laboratory acid. Its salts, however are kept, since they are stable.

$$SO_2(g) + H_2O(l) \longrightarrow H_2SO_3(aq).$$

**Sulphur dioxide reacts with alkalis.** As soon as sulphur dioxide is bubbled through a solution of sodium hydroxide, the gas dissolves in the water of the solution and forms sulphurous acid. This then neutralises the alkali, so that a salt plus water are formed.

sulphur dioxide + sodium hydroxide ⟶ sodium sulphite + water.

$$SO_2(g) + 2NaOH(aq) \longrightarrow Na_2SO_3(aq) + H_2O(l).$$

**Sulphur dioxide is a strong reducing agent.** This is true only when it is wet or in solution. In this state, it is really in the form of sulphite ions:

$$SO_2(g) + H_2O(l) \longrightarrow 2H^+(aq) + SO_3{}^{2-}(aq).$$

These sulphite ions readily take in oxygen atoms to form more stable sulphate ions, and this is how the sulphite ion reduces – by removing oxygen:

$$SO_3{}^{2-}(aq) + O \longrightarrow SO_4{}^{2-}(aq).$$

Certain coloured substances may be bleached because of this reaction. Any substance which changes from being coloured, to being white when it has oxygen removed, will be bleached by wet sulphur dioxide. Such substances as straw, paper and even blue flowers like bluebells, will react in this way.

**Change of colour as a test for sulphur dioxide.** Wet sulphur dioxide will reduce certain compounds, and in doing so, will change their colours quite noticeably:

**1** If sulphur dioxide is bubbled through acidified potassium manganate(VII) solution, then the potassium manganate is reduced. It changes from a purple colour, to colourless.

$$\text{purple} \xrightarrow[\text{dioxide}]{\text{sulphur}} \text{colourless}$$

**2** Similarly, if sulphur dioxide is bubbled through acidified potassium dichromate(VI) solution, it is reduced from an orange colour to a green one:

$$\text{orange} \xrightarrow[\text{dioxide}]{\text{sulphur}} \text{green}$$

Either of these colour changes can be used as a test for sulphur dioxide. A special version of the second is used in the 'breathalyser'.

## 15.11 Sulphuric acid

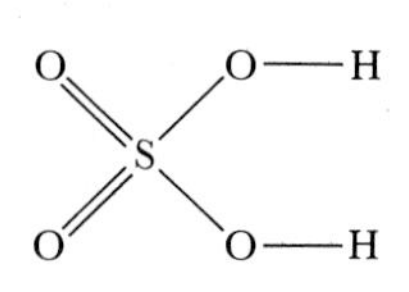


Concentrated sulphuric acid is a heavy, oily, colourless liquid, which has a covalent structure.

It is a dangerous chemical and can cause very severe burns and blisters. But it is an extremely important chemical industrially – each year, many millions of tonnes are made in Britain. About a million tonnes a year are used in the production of fertilizers such as ammonium sulphate. Almost as much is used in the production of paints and pigments, and man-made fibres. The rest is used to make detergents and soaps, plastics, and as battery acid. Some is used in the preparation of steel for electroplating and in the refining of oil.

## 15.12 The manufacture of sulphuric acid: the Contact Process

In this process, sulphur dioxide and oxygen are the starting points for a complicated reaction. The sulphur dioxide may be made in a number of different ways; either by burning sulphur, or by heating ores which contain sulphur. Air is used as the oxygen source.

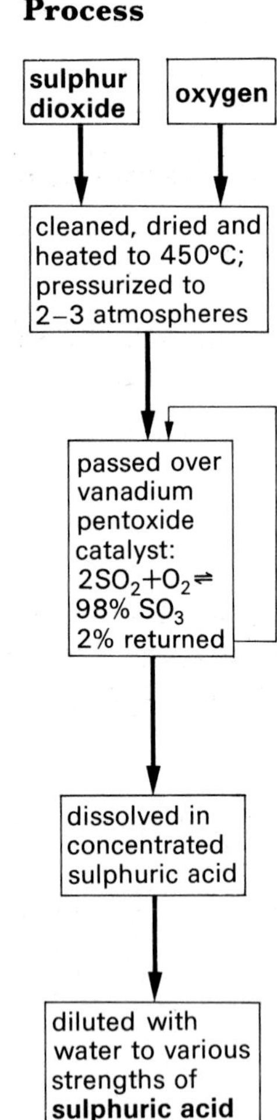


**Figure 9**
*The Contact Process.*

The gases are cleaned, dried and heated to a reaction temperature of 450 °C. They are then pressurised to a pressure of 2–3 times atmospheric pressure.

The mixture is then fed into a reaction chamber containing a catalyst of vanadium (V) oxide, making sulphur trioxide:

$$2SO_2(g) + O_2(g) \rightleftharpoons 2SO_3(g).$$

The reaction is a reversible one, but because of the carefully chosen conditions of temperature, pressure and the catalyst, about 98% conversion to sulphur trioxide is achieved. The sulphur trioxide must not be allowed to come into contact with water, otherwise it would form a mist of tiny droplets of sulphuric acid, which besides being corrosive, are very difficult to remove.

To prevent this happening, the sulphur trioxide is dissolved in concentrated sulphuric acid to form oleum:

$$SO_3(g) + H_2SO_4(l) \longrightarrow \underset{\text{oleum}}{H_2SO_4.SO_3(l)}.$$

This mixture can then be carefully diluted to the required strength of acid. Ordinary concentrated sulphuric acid contains about 98% of the acid with water.

## 15.13 The reactions of sulphuric acid

**Concentrated sulphuric acid is not an acid.** This is because its structure is covalent. It does not contain hydrogen ions. When it is diluted its covalent structure breaks down and it ionises, evolving a great deal of heat as it does so:

$$H_2SO_4(l) \longrightarrow 2H^+(aq) + SO_4^{2-}(aq).$$

Remember, when you dilute concentrated sulphuric acid,

**ADD ACID** TO WATER.

Even when this precaution is taken, the diluted solution may boil as the acid is being added, so always wear gloves and safety glasses. Once sulphuric acid is dilute, it adopts its acidic characteristics:

**1** It turns blue litmus red;

**2** It reacts with most metals to give a salt plus hydrogen;

**3** It reacts with bases to give a salt plus water;

**4** It reacts with carbonates to give a salt plus carbon dioxide plus water.

(See chapter 9 for details of these reactions.)

Because it is a dibasic acid, it can form both normal salts called sulphates, and acid salts, called hydrogensulphates
sodium sulphate $Na_2SO_4$;
calcium sulphate $CaSO_4$;
sodium hydrogensulphate $NaHSO_4$.

**Hot concentrated sulphuric acid is an oxidising agent.** It will oxidise metals to form salts, but instead of hydrogen being formed, it is reduced by the metals to form sulphur dioxide and water. For example:

copper + hot concentrated sulphuric acid ⟶ copper sulphate + sulphur dioxide + water.

$$Cu(s) + 2H_2SO_4(l) \longrightarrow CuSO_4(aq) + SO_2(g) + 2H_2O(l).$$

**Cold concentrated sulphuric acid is a dehydrating agent.** This is shown by two examples:

**1** When cold concentrated sulphuric acid is added to blue copper(II) sulphate crystals, they slowly turn white as their water of crystallisation is removed by the acid:

$$\underset{\text{blue}}{CuSO_4.5H_2O(s)} \xrightarrow[\text{sulphuric acid}]{\text{concentrated}} \underset{\text{white}}{CuSO_4(s)} + 5H_2O(g).$$

**2** When cold concentrated sulphuric acid is added to sucrose, it goes black. Sucrose is a carbohydrate, and when the water elements are removed from its molecules, only carbon is left:

$$C_{12}H_{22}O_{11} \xrightarrow[\text{sulphuric acid}]{\text{concentrated}} 12C + \text{steam} + \text{decomposition products of the acid.}$$

15.14
**Tests for sulphates**

There are two tests for sulphate salts – the effect of heat, and using barium chloride solution.

**1** Some sulphates decompose when they are heated, and give off white fumes of sulphur trioxide. For example:

$$\underset{\text{blue crystals}}{CuSO_4.5H_2O(s)} \xrightarrow{\text{heat}} \underset{\text{white solid}}{CuSO_4(s)} + 5H_2O(g);$$

$$CuSO_4(s) \xrightarrow{\text{heat}} \underset{\text{black}}{CuO(s)} + \underset{\text{white fumes}}{SO_3(g)}$$

Iron(II) sulphate behaves in a similar way:

$$\underset{\text{green crystals}}{FeSO_4.7H_2O(s)} \xrightarrow{\text{heat}} \underset{\text{white powder}}{FeSO_4(s)} + 7H_2O(g);$$

$$2FeSO_4(s) \xrightarrow{\text{heat}} \underset{\text{red powder}}{Fe_2O_3(s)} + \underset{\text{white fumes}}{SO_3(g) + SO_2(g)}.$$

Notice that the iron(II) sulphate ends up as Iron(III) oxide.

**2** Sulphates may be detected with barium chloride solution. If a solution of a sulphate is added to a solution of barium chloride, the sulphate ions react with the barium ions to form an insoluble white precipitate of barium sulphate. For example:

$$Na_2SO_4(aq) + BaCl_2(aq) \longrightarrow BaSO_4(s) + 2NaCl(aq).$$

Ionically:

$$SO_4^{2-}(aq) + Ba^{2+}(aq) \longrightarrow BaSO_4(s).$$

Unfortunately, sulphites do this reaction in a similar way. They form insoluble precipitates of barium sulphite with barium ions. For example:

$$Na_2SO_3(aq) + BaCl_2(aq) \longrightarrow BaSO_3(s) + 2NaCl(aq).$$

Ionically:

$$SO_3^{2-}(aq) + Ba^{2+}(aq) \longrightarrow BaSO_3(s).$$

But remember, sulphites dissolve in acids, so that if the suspect sulphate is first acidified with dilute hydrochloric acid before the barium chloride is added, only a sulphate should give a white precipitate.

## Summary

At the end of this chapter you should be able to:

**1** Say where the majority of the world's sulphur comes from.

**2** Describe the action of the Frasch pump for mining sulphur.

**3** List the uses of sulphur.

**4** Discuss the allotropy of sulphur.

**5** Describe the preparation of rhombic and monoclinic sulphur.

**6** Describe the effect of heat on sulphur.

**7** Describe the reaction of sulphur with air, and with metals.

**8** Give the physical properties of sulphur dioxide.

**9** Describe the preparation of sulphur dioxide from sulphites.

**10** Describe the reactions of sulphur dioxide with water and with alkalis; and give an account of its bleaching action and its reducing reactions with potassium manganate(VII) and potassium dichromate(VI).

**11** Describe the physical properties of concentrated sulphuric acid.

**12** Give an account of the Contact Process for manufacturing sulphuric acid.

**13** Describe the properties of sulphuric acid as an oxidising agent, a dehydrating agent, and as an acid.

**14** Explain what happens when concentrated sulphuric acid is diluted.

**15** Describe what happens when certain sulphates are heated.

**16** Describe the barium chloride test for sulphates.

# The foul case of the acid bath murderer

*Haigh leaving the court.*

*This chapter has dealt with sulphur, and sulphuric acid. You already know how to make salts by the action of acids on metals or carbonates. This section deals with a man who made some complex salts by using concentrated sulphuric acid, to react with a more valuable material than either metals or carbonates. He paid a very high price for doing so . . . his life . . . .*

The year is 1949. The scene, is the County Court, Lewes, where John George Haigh is being tried for murder. There is a big crowd (mainly of women) outside the courtroom, for this case has caught the horrified imagination of the British Public. This is not altogether surprising – because John George Haigh has admitted to murder, but at the same time, claims that no charges can be brought against him! He maintains that if no trace of the body can be found, then there can be no charge. There is no trace of the body, because it has been dissolved in sulphuric acid . . .

**Unpaid bill**

The story opens in South Kensington, in a hotel which was the home of two elderly, respectable ladies – Mrs Lane, and her friend, Mrs Durand-Deacon. A middle aged business-man, whose business he would only very vaguely describe, also lived there: his name was John Haigh. The management were becoming increasingly interested in the subject of his business, because his bill had not been paid for some time, and it was growing large.

One day, Mrs Durand-Deacon disappeared. Mrs Lane, beginning to grow anxious, started to make a few enquiries round the hotel. Mr Haigh quickly appeared, and expressed concern too. Mrs Lane and Mr Haigh agreed to go to the police, for Mr Haigh had some information that suggested there was cause for concern.

As he told the police, he had been trying to interest Mrs Deacon in a business venture which would manufacture 'stick-on' plastic fingernails. They had driven into London, where he had dropped her off to do some shopping. She had not returned to their arranged meeting spot, and so after a while he had returned to the Hotel without her.

**Criminal record**

For a while, the police worked on this information, but they could find no further clues to suggest what had happened to her. Enquiries were made into Mr Haigh's background, and it turned out that he had been in prison twice before, on charges of forgery and fraud. The police were now suspicious. Mr Haigh was interviewed again; a cleaner's receipt in his possession led them to a shop where they found Mrs Deacon's fur coat, being cleaned. Haigh was arrested. He then made a statement, which, when published, shocked Britain.

**Evidence 'disappeared'**

The police inspector questioning him had no idea that he would make the following statement.

'If I tell you the truth you would not believe it; It sounds too fantastic for belief...' At this point he was cautioned again. But he went on:

'I will tell you all about it. Mrs Durand-Deacon no longer exists. She has disappeared completely and no trace of her can ever be found again.'

'I have destroyed her with acid. You will find the sludge that remains at Leopold Road. Every trace has gone. How can you prove murder if there is no body?'

But Haigh was wrong on two counts. Firstly, murder *can* be proven if no trace of the body exists. Secondly, it was not the case that the body had been completely dissolved. Some traces did remain . . .

**Forensic science team called in**

This was a job for the forensic science team, and it was quite a job that they had to do. The police had uncovered some hundredweights of greasy sludge, and they now had the task of working out whether these were the remains of a human person.

Sulphuric acid had been used. You will have been warned in the lab to quickly wash off any spillages. The acid will react with flesh. Bones are tougher, being a mixture of organic and inorganic compounds – but both will dissolve eventually. There are very few things that will *not* react with concentrated acid, to produce either a simple, or a complex salt.

But some things will not – and they provided a major source of evidence against Haigh. The handle of a red plastic bag, and a plastic denture – all of which had belonged to Mrs Deacon – remained intact and identifiable. The police could even work out that she had suffered from a gall-bladder complaint...

**Haigh pays the price**

Haigh realized that the evidence against him was very serious indeed. Having made the statement mentioned above, he then began a string of increasingly bizarre statements hoping to make people believe that he was insane. If this could be demonstrated, then instead of being hung, he would be committed (for life) to Broadmoor Prison. He even asked the inspector in charge of his case:

'Has anyone ever been released from Broadmoor?'

To which the inspector replied, 'Before you start worrying how to get out of Broadmoor, start worrying how you get in!'

A leading Harley Street doctor examined him, and became convinced that Haigh was insane. He spoke in Haigh's defence at his trial.

But the jury was not convinced. It took them only fifteen minutes to come to their verdict, 'Guilty'. He was hanged seventeen days later at Wandsworth prison. Haigh paid the price for performing what must be the most gruesome experiment to produce salts, of all time.

**Questions**

**1** Describe how the molecules in a piece of sulphur change their arrangement as the sulphur is slowly heated from room temperature to 450 °C. How does the appearance of the sulphur change?

**2** What is meant by the term 'Allotropy', and how does it apply to sulphur?

Describe how you would prepare crystals of monoclinic sulphur. What would happen to them if they were left in the laboratory overnight?

What is plastic sulphur? How is it made? Why isn't plastic sulphur another allotrope like rhombic and monoclinic sulphur?

**3** Some powdered sulphur was put on the end of a combustion spoon and it was heated in a bunsen flame until the sulphur started to burn.

**a** Write the equation for this reaction and describe what you would see.

The spoon is then lowered into a gas jar of oxygen.

**b** Do some revision and find out how the flame would change.

After the sulphur had stopped burning, a few $cm^3$ of water were added to the jar, and the lid was put on to allow the gas in the jar to dissolve.

**c** Name the solution that would be formed and write the equation for the reaction which would take place.

A few drops of litmus solution were added to the solution.

**d** What colour would the solution become?

Sodium hydroxide solution was then added drop by drop until the litmus changed colour.

**e** What would the new colour be?

**f** What type of reaction would have taken place between the sodium hydroxide and the solution in the jar?

**g** Write the equation for the reaction.

**4** 14 g of iron powder was added to 8 g of sulphur powder and the two solids were thoroughly mixed. Suggest two ways of separating this mixture:

**a** to get the iron out and leave the sulphur,

**b** to get the sulphur out and leave the iron.

The mixture was then heated and a compound was made. No iron or sulphur remained.

Work out the empirical formula of this compound. The relative atomic masses of sulphur and iron are 32 and 56 respectively. (Remember the index if you are stuck.)

**5** Write word equations and chemical equations for the following reactions:

**a** The roasting of zinc blende ore (ZnS) in air to get sulphur dioxide.

**b** The reaction of magnesium with dilute sulphuric acid.

**c** The reaction of potassium sulphite with dilute hydrochloric acid.

**d** The reaction of barium chloride solution with zinc sulphate solution.

**6** Explain why concentrated sulphuric acid gets hot when it is diluted. What safety precautions would you take when diluting it?

One of the most important compounds made from sulphuric acid is ammonium sulphate – a fertilizer. Describe how you would make a sample of this compound in the laboratory starting with dilute sulphuric acid and ammonium hydroxide solution.

**7** Summarise the Contact Process for the manufacture of sulphuric acid in the following way:

**a** name the starting materials and say how they are made.

**b** Describe the conditions under which the Contact Process reaction is carried out and write the equation for it.

**c** Describe the product of the reaction and say how it is finally made into dilute sulphuric acid.

**8** Describe one reaction of sulphuric acid for each of the following set of conditions.

**a** hot, concentrated acid on a metal.

**b** cold, dilute acid with a carbonate.

**c** cold, concentrated acid with a carbohydrate.

In each case write the equation and say what you would see.

**9** The labels have come off two chemical bottles. They both contain white powders. One is sodium sulphite and the other is sodium sulphate, but the Head of the Chemistry Department doesn't know which is which. Luckily, two pupils solved the problem. They both used a different method.

Both hydrochloric acid and barium nitrate solution were available in the laboratory. Explain how both the pupils did it. Don't forget the equations.

# 16 Ammonia and nitric acid

## 16.1 Ammonia?

There are several possible accounts of how the gas ammonia got its name. The most likely seems to be that it came from the Greek word Ammoniac, meaning 'salt of the sand'. This was a reference to ammonium chloride, a naturally occurring substance that would yield the gas.

Other stories say that the name came from Armenia, a country in which a lot of the ammonium chloride was mined. The least likely explanation is the one concerning the Egyptians. They kept the temple of the god Amun warm by putting fresh camel dung under the floor. The pong that followed was referred to as 'Amunia'.

Ammonia is an extremely important chemical. Most of it is made into fertilizers such as ammonium sulphate (see Extra Time), and nitric acid (see section 16.8). Some of it is used in the preparation of plastics such as Nylon.

## 16.2 The Industrial manufacture of ammonia

Ammonia is manufactured by the *Haber process*. The whole process is shown as a flow diagram in figure 1.

The hydrogen needed for the process is made from methane by the process of steam reforming. (See chapter 11). The nitrogen is taken from the air, either by fractional distillation (see chapter 8), or more usually by burning some hydrogen in air. This removes the oxygen as steam which may be liquefied, thus leaving the nitrogen gas:

$$2H_2(g) + O_2(g) + N_2(g) \longrightarrow 2H_2O(l) + N_2(g).$$

The nitrogen and hydrogen are mixed together in the proportions of 1:3 by volume. The gases are dried, compressed to about 350 times atmospheric pressure, and heated to a temperature of 450 °C.

The mixture is then passed into a reaction chamber containing a catalyst of iron mixed with various chemicals such as aluminium oxide and potassium hydroxide. These additional chemicals are called promoters, and they increase the efficiency of the iron catalyst. The reaction that takes place is:

$$N_2(g) + 3H_2(g) \rightleftharpoons 2NH_3(g).$$

As you can see from the equation, the reaction is reversible and the gases emerging from the reaction chamber contain unreacted nitrogen and hydrogen.

After the ammonia has been removed by reducing the temperature and causing it to liquefy, the other gases are recycled.

Ammonia, although a gas at normal temperatures and pressures, can be kept as a liquid under a pressure of about 10 atmospheres. Alternatively, it is dissolved in water to make ammonium hydroxide.

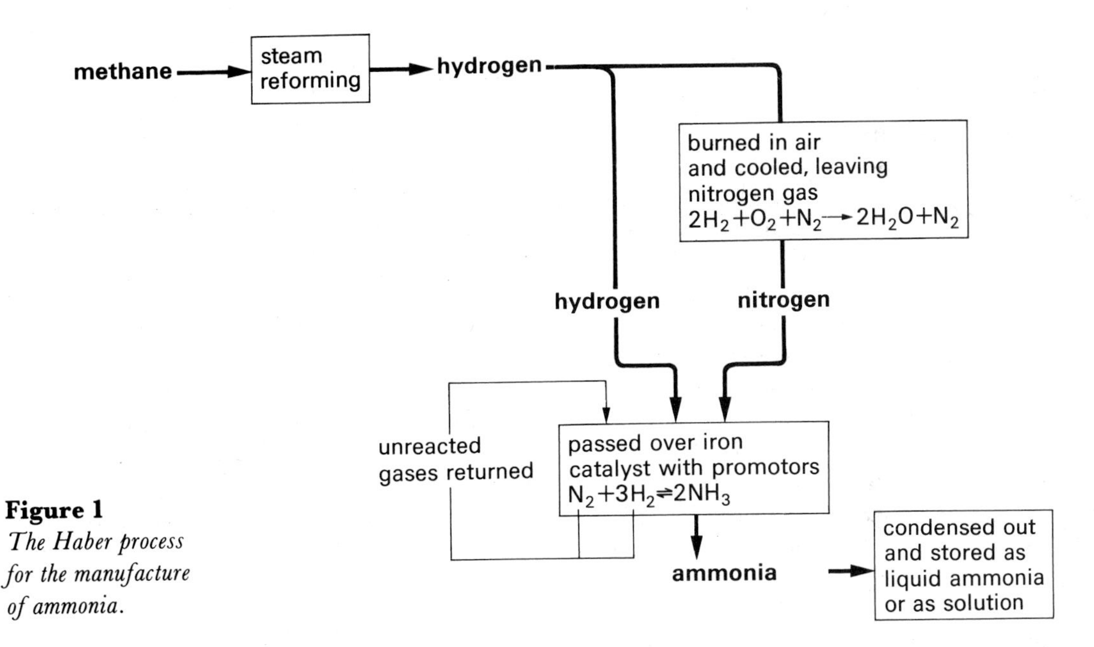


**Figure 1**
*The Haber process for the manufacture of ammonia.*

## 16.3 The laboratory preparation of ammonia

Ammonia is made whenever any ammonium salt reacts with any alkali:

$$\underset{\text{ammonium salt}}{NH_4^+} + \underset{\text{alkali}}{OH^-} \longrightarrow \underset{\text{ammonia}}{NH_3(g)} + H_2O(g).$$

This is in fact how ammonium salts are tested and identified. The ammonium salt and alkali most often used in the laboratory are ammonium chloride and calcium hydroxide:

ammonium chloride + calcium hydroxide ⟶ ammonia + calcium chloride + water.

$$2NH_4Cl(s) + Ca(OH)_2(s) \longrightarrow 2NH_3(g) + CaCl_2(s) + 2H_2O(g).$$

The apparatus for this reaction is shown in figure 2 on the next page.

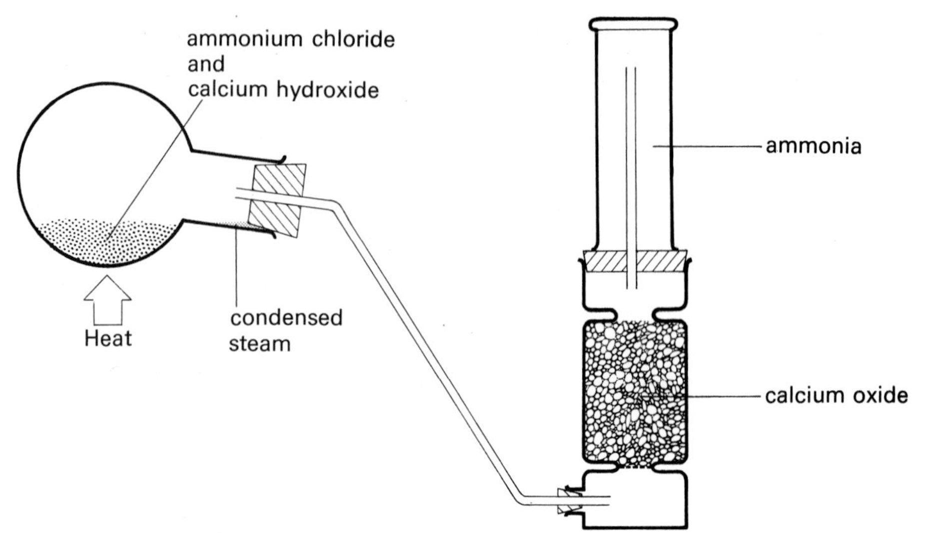


**Figure 2**
*The Preparation of ammonia.*

The reaction flask is sloped so that the steam produced will condense in the neck of the flask and not run back into the hot reaction mixture. The gas is allowed to rise through a tower of calcium oxide before it is collected upwards into a gas jar. The calcium oxide dries the gas. Normally concentrated sulphuric acid or calcium chloride would be used to dry a gas, but since ammonia is an alkali, it would react with the acid:

$$\underset{\text{alkali}}{2NH_3(g)} + \underset{\text{acid}}{H_2SO_4(l)} \longrightarrow \underset{\text{salt}}{(NH_4)_2SO_4(aq)}.$$

Calcium chloride also reacts with ammonia. So calcium oxide is the most suitable drying agent.

## 16.4 The physical properties of ammonia

**1** Ammonia is a colourless gas with a choking smell.

**2** It is lighter than air.

**3** It is alkaline and turns damp red litmus blue.

**4** It is extremely soluble in water and will perform the fountain experiment. (See chapter 14.)

## 16.5 The chemical reactions of ammonia

**Ammonia is a reducing agent.** It will reduce copper(II) oxide to copper:

$$\underset{\text{black}}{3CuO(s)} + 2NH_3(g) \longrightarrow \underset{\text{pink}}{3Cu(s)} + 3H_2O(g) + N_2(g).$$

It will also reduce lead oxide to lead:

$$\underset{\text{yellow}}{3PbO(s)} + 2NH_3(g) \longrightarrow \underset{\text{silver coloured}}{3Pb(s)} + 3H_2O(g) + N_2(g).$$

The apparatus for this reaction is shown in figure 3.

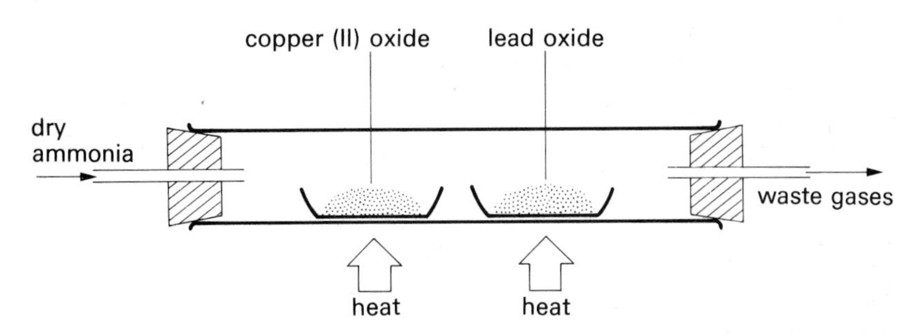


**Figure 3**
*Ammonia is a reducing agent.*

**Ammonia gas reacts with hydrogen chloride gas.** It forms a solid called ammonium chloride:

$$NH_3(g) + HCl(g) \rightleftharpoons NH_4Cl(s).$$

This happens when a gas jar of ammonia is opened over a gas jar of hydrogen chloride. (See figure 4.)

hydrochloric acid gas

white smoke of solid ammonium chloride

ammonia gas

**Figure 4**
*Ammonia reacts with hydrogen chloride.*

This is called sublimation, because the two gases react directly forming a solid. No liquid state is involved.

**Ammonia dissolves in water.** It forms an alkali called ammonium hydroxide:

$$NH_3(g) + H_2O(l) \rightleftharpoons NH_4OH(aq).$$

Ammonium hydroxide is a weak alkali because it is not completely ionised in solution:

$$NH_4OH(aq) \rightleftharpoons NH_4^+(aq) + OH^-(aq).$$

Nevertheless, it will perform the normal alkali reactions, as shown in the next section.

16.6
## The reactions of ammonium hydroxide

**'Undissolving' of ammonium hydroxide.** A concentrated solution of ammonium hydroxide smells strongly of ammonia especially in a warm room, because the ammonia gas is 'undissolving' from the solution. This also happens when ammonium hydroxide is warmed:

$$NH_4OH(aq) \longrightarrow NH_3(g) + H_2O(l).$$

This provides a convenient way of making small quantities of ammonia gas. (See figure 5.)

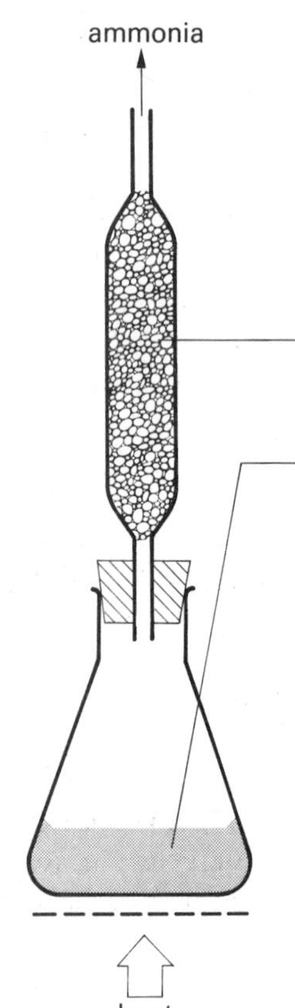


**Figure 5**
*Another way of making ammonia.*

**Ammonium hydroxide reacts with acids.** It forms salts and water. For example:

ammonium hydroxide + nitric acid ⟶ ammonium nitrate + water.

$$NH_4OH(aq) + HNO_3(aq) \longrightarrow NH_4NO_3(aq) + H_2O(l).$$

**Precipitation of insoluble metal hydroxides.** Ammonium hydroxide can be used to make precipitates of insoluble metal hydroxides. For example, when ammonium hydroxide is added to iron(II) chloride a dirty green-coloured precipitate is formed, which is iron(II) hydroxide:

ammonium hydroxide + iron(II) chloride ⟶ iron(II) hydroxide + ammonium chloride.

$$2NH_4OH(aq) + \underset{\text{pale green}}{FeCl_2(aq)} \longrightarrow \underset{\text{dirty green precipitate}}{Fe(OH)_2(s)} + 2NH_4Cl(aq).$$

Ammonium hydroxide can be used to precipitate zinc hydroxide:

ammonium hydroxide + zinc sulphate ⟶ zinc hydroxide + ammonium sulphate.

$$2NH_4OH(aq) + \underset{\text{colourless}}{ZnSO_4(aq)} \longrightarrow \underset{\text{white precipitate}}{Zn(OH)_2(s)} + (NH_4)_2SO_4(aq).$$

Reactions of this kind are called *ionic precipitation* reactions. The reaction of ammonium hydroxide with copper(II) ions is an interesting one. If ammonium hydroxide is added to blue copper sulphate solutions the reaction is at first the same as above. A blue gelatinous precipitate of copper(II) hydroxide is formed:

$$\underset{\text{}}{2NH_4OH(aq)} + \underset{\text{blue solution}}{CuSO_4(aq)} \longrightarrow \underset{\text{blue precipitate}}{Cu(OH)_2(s)} + (NH_4)_2SO_4(aq).$$

Then, when more ammonium hydroxide is added, the precipitate redissolves to form a dark blue solution. The extra ammonium hydroxide has formed a complex salt called tetrammine copper(II) sulphate:

$$CuSO_4 + \text{excess } NH_4OH(aq) \longrightarrow \underset{\text{dark blue solution}}{[Cu(NH_3)_4]SO_4(aq)}.$$

This is a test which shows the presence of copper(II) ions in solution.

## 16.7 Ammonium compounds

Ammonium compounds are formed when ammonium hydroxide reacts with an acid. Ammonium chloride is formed by the action of ammonium hydroxide on hydrochloric acid:

$$\text{ammonium hydroxide} + \text{hydrochloric acid (dilute)} \longrightarrow \text{ammonium chloride} + \text{water}.$$

$$NH_4OH(aq) + HCl(aq) \longrightarrow NH_4Cl(aq) + H_2O(l).$$

Just as ammonium hydroxide is unstable when heated, ammonium compounds decompose when they are heated. Ammonium chloride, for example, will decompose to give ammonia and hydrogen chloride:

$$\text{ammonium chloride} \xrightarrow{\text{heat}} \text{ammonia} + \text{hydrogen chloride}.$$

$$NH_4Cl(s) \xrightarrow{\text{heat}} NH_3(s) + HCl(g).$$

This is the reverse process of sublimation. It is called *thermal dissociation.*

Sometimes, when ammonium chloride is heated in a test tube, after thermal dissociation has occurred at the bottom of the tube, the hot gases will sublime back to ammonium chloride on the cold upper part of the test tube. (See figure 6.)

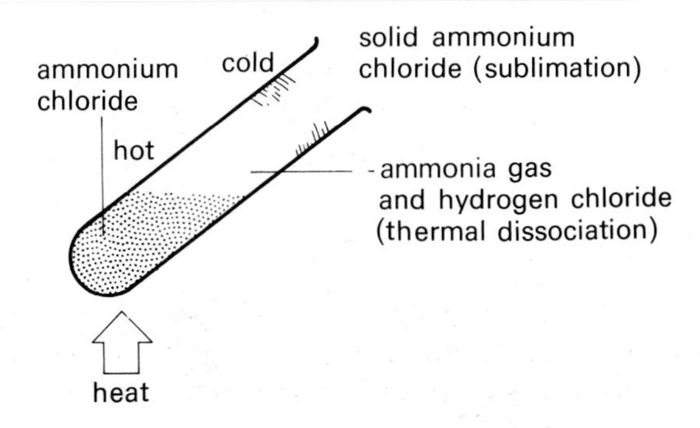


**Figure 6** *Sublimation and thermal dissociation can occur in the same tube.*

## 16.8 Nitric acid

Nitric acid is an extremely important chemical which is used in the manufacture of fertilizers such as ammonium nitrate ($NH_4NO_3$), and in the production of many explosives.

Names such as nitroglycerine and trinitrotoluene (TNT), are well known. Both of these need nitric acid for their manufacture. Gun powder was invented by the Chinese, thousands of years ago. It contains potassium nitrate, another compound of nitric acid.

Nitroglycerine is a very dangerous liquid to handle. It can explode simply as a consequence of vibration. About 100 years ago, a Swedish industrialist called Alfred Nobel invented a stable form of it by soaking the very unstable liquid into sticks of clay. This made it reasonably harmless since it could only be set off with a detonator. He called his invention dynamite.

Alfred Nobel made vast profits from the sale of dynamite, and when he died, he left much of his money in trust with the Swedish Government for the provision of prizes – large sums of money called Nobel Prizes, which are awarded to people who make great progress in the fields of science, literature or peacemaking.

*T.N.T. in use.*

## 16.9 The manufacture of nitric acid

Nitric acid is manufactured industrially by the oxidation of ammonia gas. This process is diagrammatically shown down the side of the page.

ammonia
air
cleaned mixed and compressed
900°C
platium and rhodium catalyst
cooled
nitrogen (IV) oxide
water
65% nitric acid

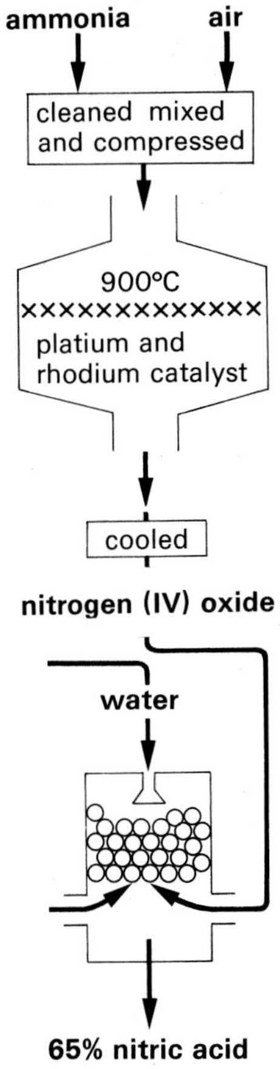

**Figure 7**
*The production of nitric acid as a flow scheme.*

Ammonia from the Haber process is mixed with an excess of air and the mixture is cleaned so as to be free of dust.

It is then compressed to several times atmospheric pressure.

It is then ready to be passed over a catalyst which consists of a gauze made of an alloy of platinum and rhodium. The gauze is heated to a temperature of 900 °C to initiate the reaction of the ammonia with the oxygen in the air. After it has started, the reaction is so strongly exothermic that it heats the incoming gases by itself, so the reaction stays at this temperature without any more heating.

The outcoming gases are now cooled, and nitrogen (IV) oxide is formed. This is the overall reaction:

$$4NH_3(g) + 7O_2(g) \longrightarrow \underset{\text{nitrogen(IV) oxide}}{4NO_2(g)} + 6H_2O(g).$$

It is then mixed with more air, and passed up an absorption tower, against a downward flow of water. The tower is packed with glass balls to provide a surface in the gases to react with the water:

$$4NO_2(g) + O_2(g) + 2H_2O(l) \longrightarrow 4HNO_3(aq).$$

The mixture coming out is 65% acid, and 35% water. Stronger acid may be made by very careful distillation of the solution.

## 16.10 The laboratory preparation of nitric acid

Nitric acid is made whenever a nitrate is heated with concentrated sulphuric acid:

$$\underset{\text{nitrate ion}}{NO_3^-(s)} + H_2SO_4(l) \longrightarrow \underset{\text{hydrogen sulphate ion}}{HSO_4^-(aq)} + \underset{\text{nitric acid vapour}}{HNO_3(g)}.$$

This method was used as the industrial preparation for a long time. The cheapest nitrate was sodium nitrate:

$$NaNO_3(s) + H_2SO_4(l) \longrightarrow \underset{\text{sodium hydrogen sulphate}}{NaHSO_4(aq)} + HNO_3(g).$$

Because nitric acid is such a corrosive liquid, an all-glass apparatus must be used. (See figure 8.)

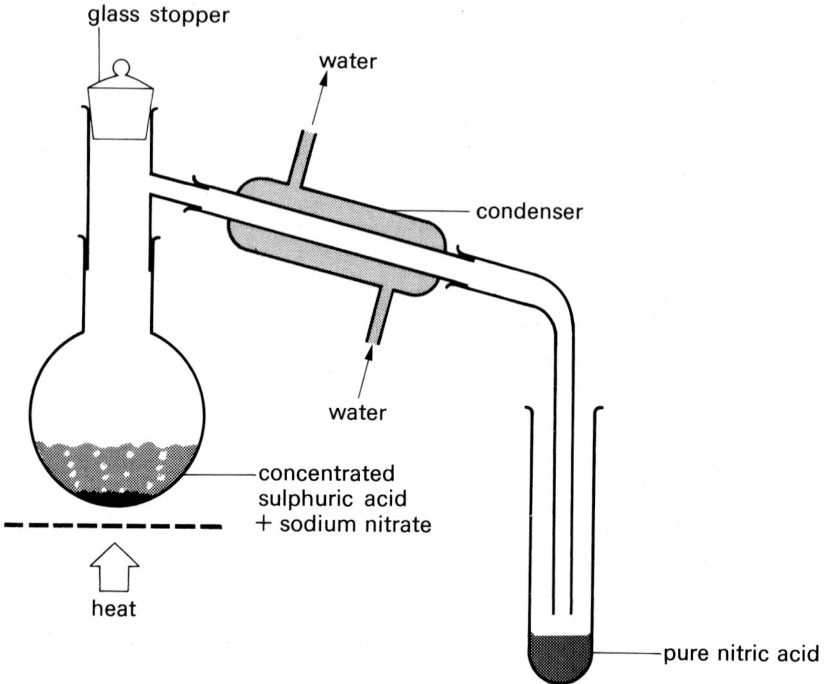


**Figure 8**
*The laboratory preparation of nitric acid.*

As the sodium nitrate is warmed with the acid, the nitric acid is formed as a vapour and must be condensed. If too high a temperature is used, the acid will decompose and become discoloured:

$$4HNO_3(l) \longrightarrow 4NO_2(g) + O_2(g) + 2H_2O(g).$$

It is the brown nitrogen(IV) oxide which causes the acid to become brown. Because of this, concentrated nitric acid is kept in brown bottles, to protect it from the light – another source of energy which will cause the acid to decompose.

The pure nitric acid formed by this method is a colourless, oily liquid. It can cause severe burns, turning the skin brown, and may also cause fires if allowed to come into contact with organic substances, such as sawdust or wood shavings.

## 16.11 The chemical reactions of nitric acid

**Concentrated nitric acid is a powerful oxidising agent.** As mentioned above, it will rapidly oxidize organic compounds. Dry sawdust provides a typical example:

dry saw-dust (a compound of carbon, hydrogen and oxygen) + nitric acid ⟶ carbon dioxide + water + nitrogen(IV) oxide.

This is a very exothermic reaction, and if the sawdust is very dry, it may catch fire.

**Concentrated nitric acid will oxidize metals.** It oxidizes them to salts called nitrates, and it is reduced to nitrogen(IV) oxide and water. For example:

copper + concentrated nitric acid ⟶ copper nitrate + nitrogen(IV) oxide + water.

$$Cu(s) + 4HNO_3(l) \longrightarrow Cu(NO_3)_2(aq) + 2NO_2(g) + 2H_2O(l).$$

**Dilute nitric acid shows typical acidic properties.** It reacts with bases to form salts plus water, and carbonates to form salts plus carbon dioxide plus water. However, it does not react with metals in the same way as other acids. It still acts as an oxidising agent, but instead of producing hydrogen when added to a metal, it produces nitrogen(II) oxide and water. In addition, dilute nitric acid reacts with copper, which has no reaction with other dilute acids such as hydrochloric acid.

copper + dilute nitric acid ⟶ copper nitrate + nitrogen (II) oxide + water.

$$3Cu(s) + 8HNO_3(aq) \longrightarrow 3Cu(NO_3)_2(aq) + 2NO(g) + 4H_2O(l).$$

Only when very dilute will nitric acid react with reactive metals such as magnesium and calcium, to form a salt plus hydrogen. For example:

calcium + very dilute nitric acid ⟶ calcium nitrate + hydrogen.

$$Ca(s) + 2HNO_3(aq) \longrightarrow Ca(NO_3)_2(aq) + H_2(g).$$

Because nitric acid is a monobasic acid it forms only normal salts, called *nitrates*.

## 16.12 Reactions of nitrates

**Warming with concentrated sulphuric acid makes nitric acid vapour.** Sodium nitrate is a typical example:

$$NaNO_3(s) + H_2SO_4(l) \longrightarrow HNO_3(g) + NaHSO_4(aq).$$

This reaction may be done in a test tube. As the nitrate and sulphuric acid are heated, small drops of nitric acid may be seen condensing at the cold top of the test tube. (See figure 9.)

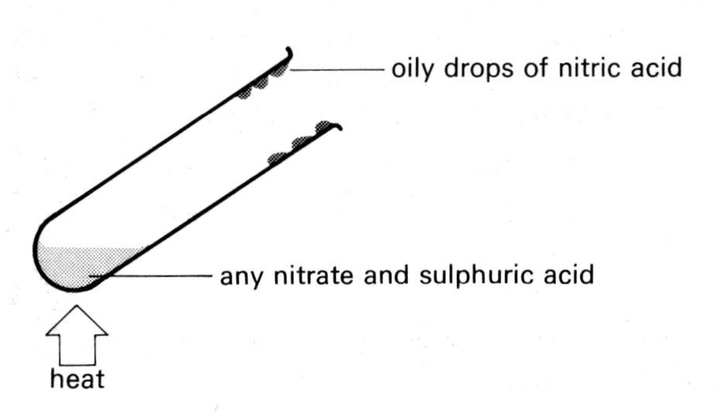


**Figure 9**
*The reaction of nitrates with concentrated sulphuric acid.*

**The Brown Ring test for nitrates.** This is a rather difficult test to perform, and it is best described by a series of diagrams. These are shown in figure 10.

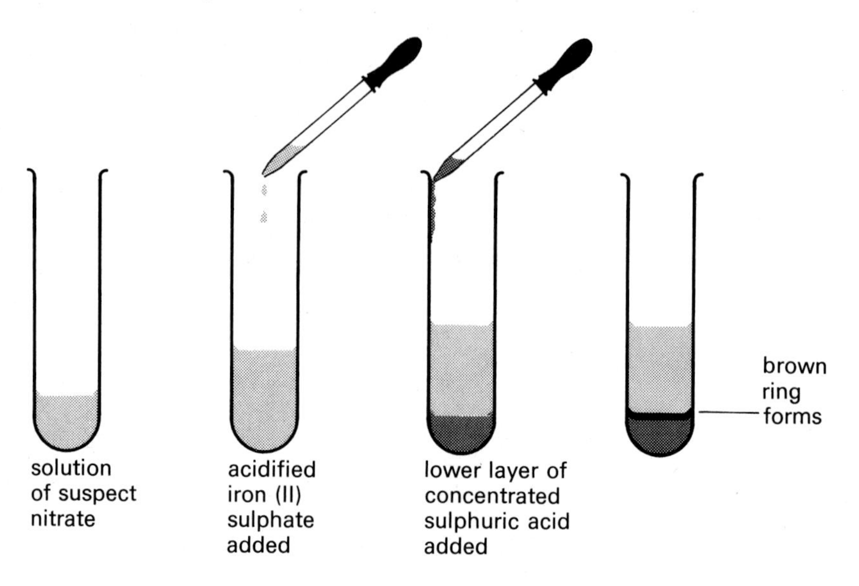


**Figure 10**
*The Brown Ring test for nitrates.*

The brown ring is formed because a small amount of nitric acid is formed when the sulphuric acid meets the nitrate solution at the interface between the two layers. The nitric acid reacts with the iron(II) sulphate to form a brown compound.

**Nitrates decompose when they are heated.** The way in which they do so depends upon how reactive the metal in the nitrate is. If it is a reactive metal, the nitrate will be quite stable and difficult to decompose. If the metal is less reactive, the nitrate will decompose more easily. But you can read more about this in chapter 18.

Potassium and sodium nitrates are quite stable when heated. Potassium nitrate, for example, will decompose, but not completely:

potassium nitrate ⟶ oxygen + potassium nitrite.

$$2KNO_3(s) \longrightarrow O_2(g) + 2KNO_2(s).$$

The majority of nitrates decompose more than this – in other words, right down to the oxide. Lead nitrate is typical:

lead nitrate ⟶ lead oxide + nitrogen(IV) oxide + oxygen.

$$2Pb(NO_3)_2(s) \longrightarrow 2PbO(s) + 4NO_2(g) + O_2(g).$$

colourless crystals ... yellow solid, brown gas

Lead nitrate crystals decompose explosively with small cracking noises. This is called *decrepitation*. In the case of copper nitrate:

copper(II) nitrate ⟶ copper(II) oxide + nitrogen(IV) oxide + oxygen.

$$2Cu(NO_3)_2(s) \longrightarrow 2CuO(s) + 4NO_2(g) + O_2(g).$$

blue crystals — black powder — brown gas

Nitrates of the noble metals, mercury, silver and gold, (i.e., the least reactive ones) decompose right down to the metal. For example:

mercury nitrate ⟶ mercury + nitrogen(IV) oxide + oxygen.

$$Hg(NO_3)_2(s) \longrightarrow Hg(l) + 2NO_2(g) + O_2(g).$$

**Summary**

At the end of this chapter you should be able to:

**1** Describe the Haber process for the production of ammonia.

**2** State the main uses of ammonia.

**3** Describe the physical properties of ammonia.

**4** Describe the reactions of ammonia as a reducing agent and as an alkali.

**5** Describe the industrial manufacture of nitric acid.

**6** Describe how nitric acid is made in the laboratory.

**7** Explain the chemical reactions of nitric acid as an oxidising agent, and as an acid.

**8** Describe the Brown Ring test for nitrates.

**9** Describe how nitrates behave when heated.

# Fertilizers

*It is only in the last hundred years or so that farmers have realized that their plants need help, if they are to grow so as to give him a good crop. Just what help do they need? Read on . . .*

If a plant is to grow healthily, it needs the right food – just as we do. It takes in carbon dioxide through its leaves, and with water which it takes in through its roots, turns them into starches and sugars by the process called photosynthesis. The energy for this reaction is supplied by the sun, and the catalyst is chlorophyll, a substance which all green plants contain. The water contains chemicals from the soil which the plant needs to grow. These chemicals contain three important elements: nitrogen, phosphorus and potassium.

**Nodules on the legumes . . .**

Nitrogen is usually present in the soil in the form of soluble nitrates. Without nitrogen, the plant cannot make proteins, and as the nitrogen also assists with the formation of chlorophyll, a lack of nitrogen would slow the process of photosynthesis as well. Plants which have a deficiency of nitrogen are stunted and yellow and are very susceptible to diseases. Some plants called *legumes* have special bacteria contained in little lumps on their roots, called *nodules*, which can convert nitrogen gas directly from the air into chemicals in the plant. Peas and beans are good examples of legumes.

Phosphorus occurs in the soil as phosphates. These are not very soluble so the phosphorus is only taken up by the plant very slowly. Phosphorus helps the roots of the plant to grow in its early life, and later on it quickens the formation of seeds and the ripening of fruits. Plants which lack phosphorus have stunted leaves and roots. Their blossom may fall off prematurely and fruit yields may be low.

Potassium compounds are contained in clay soil and humus. Their function is to help the plant resist disease and grow strongly. Plants lacking potassium are stunted and die easily.

**Nutrients removed by man**

In a situation where man is not intervening, plants release all the chemicals they use back to the soil when they die. But plants are eaten by animals and man, and the chemicals end up in their bodies as new proteins and salts, and only a little is returned to the land in the form of manure. Human manure is considered dangerous unless it has been treated at the sewage works, and even then its fertilizer value is rather small and uneconomic.

Two important words have been introduced. 'Economic' is important because a farmer wants to get his land into good growing

*Demonstrating the need for fertilizers.*

condition with the smallest cost and effort. 'Fertilizer' is the general name given to the range of chemicals that he adds to the soil to achieve this.

Artificial (because they are manufactured) fertilizers put back into the land the chemicals which have been removed by the plants which in turn have been removed and eaten by animals and people. When new fertilizers were tried in India, there was a 50% increase in the yield of rice, and in Indonesia, the yields of maize increased by 100% although some of these increases were due to improved techniques and pest control.

**84 million tonnes of it**

We make and use a lot of fertilizer. In 1973/74, the World production of fertilizer was 84 million tonnes, compared with one million tonnes at the start of the century. The developed countries of the world, such as the U.S.A., Europe, Russia and Japan, used the majority of this. The Third World countries used only 17%. Farmers in Great Britain consumed just less than 2 million tonnes.

There isn't just one kind of fertilizer. There are many different types. Different soils need different amounts of nitrogen, potassium or phosphorus. In addition to these primary chemicals, there are a number of secondary elements which need to be added to the soil in smaller amounts to ensure the healthy growth of plants. They are calcium, magnesium (for the chlorophyll), and sulphur. In smaller amounts still, certain 'trace elements' are needed as well. These include copper, boron, molybdenum, manganese, iron and zinc. These can be supplied in different types of fertilizer.

Nitrogenous fertilizers supply nitrogen to the soil. They are ammonium sulphate, ammonium nitrate, ammonium hydroxide and urea. They are made from ammonia which comes from the Haber process. They are very soluble compounds and their nitrate ions can easily be absorbed by the plants through their roots.

Phosphate fertilizers can be made from ground up bone, for example bone meal. Better still they are made from rock phosphate ores from Russia, the U.S.A. and North Africa. The ore is made into calcium phosphate, and this is sold as 'super phosphate', or 'triple super phosphate'. Basic slag, the waste product from blast furnaces is a good phosphate fertilizer because it is only sparingly soluble and releases its phosphorus to the soil at just the right speed.

Potassium fertilisers are usually made of potassium chloride, potassium sulphate, or other ores of potassium which occur naturally.

Most fertilizers are sold as a mixture of these primary elements.
NP fertilizers contain nitrates and phosphate;
PK fertilizers contain phosphates and potash;
NK fertilizers contain nitrates and potash.
You can guess the composition of NPK fertilizers. In addition, small amounts of secondary and trace elements can be included as required.

By law, the amount of potassium, phosphorus and nitrogen in a fertilizer must be printed on the bag.

## Questions

**1** In one of the stories about how ammonia got its name, it was suggested that it derives from a name for ammonium chloride, which was mined in Armenia. How do you think the Armenians got the ammonia out of the ammonium chloride? (There are two ways in which they might have done it.)

**2** Summarize the Haber Process by giving details of:

**a** The reactants, and where they come from,

**b** the reaction and its conditions,

**c** the products, and how they are separated and treated.

What is a catalyst? What does it do?

**3** Compost is good for the garden when it is dug into the ground. But you have to look after compost heaps. You can buy all sorts of things to tip over to make them rot quickly and form compost. One such substance is 'Sulphate of ammonia', and another is 'Quicklime'.

The index will help to find out the proper names and formulae for these substances. Bearing in mind the fact that compost heaps are wet and get hot, explain why these two substances should not be put onto the compost heap at the same time.

**4** Some hydrogen has become contaminated with ammonia in a factory. Explain how you would remove the ammonia to leave pure hydrogen. Sketch the apparatus which you would use. How would you test the hydrogen in the first place to see if it really was contaminated?

**5** You want a sample of copper, but you have only got copper oxide, ammonium chloride, and sodium hydroxide. Explain, with equations, how you would get the copper, and draw the apparatus you would use.

**6** An experiment was set up as shown in the diagram.

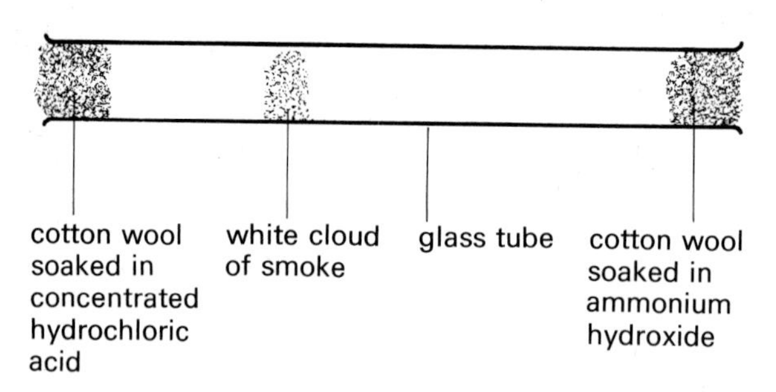


After a few minutes, a small cloud of white smoke appeared near one end of the glass tube.

**a** What was the white smoke?

**b** How was it formed?

**c** Why wasn't it formed immediately the plugs of cotton wool were put into place?

**d** Why wasn't the white cloud formed in the middle of the glass tube?

**7** Describe how you would prepare crystals of ammonium sulphate, starting from ammonium hydroxide and dilute sulphuric acid. The salt will need to be completely neutral, and not contaminated with excess acid or alkali.

Describe the apparatus you would use, and write the equation.

**8** Write equations for the ionic precipitation reactions which occur between ammonium hydroxide and solutions of:

**a** iron(III) sulphate
**b** lead nitrate, and
**c** magnesium chloride.

**9** Describe a way in which you could get a pure sample of ammonium chloride from a mixture of ammonium chloride and sodium chloride.

Draw a diagram of the apparatus you would use.

**10** Nitric acid is a dangerous chemical because it can cause fires. Explain how an accident like this might happen in a factory store room.

**11** Describe the reaction of dilute nitric acid with:

**a** sodium hydroxide solution
**b** calcium oxide
**c** calcium carbonate
**d** copper

**12** Some unlabelled chemical bottles have been found. One contains a liquid, which might be a solution of ammonium hydroxide solution; the other contains crystals which might be of potassium nitrate.

Describe one physical test, and two chemical tests to identify the solution, and two chemical tests to identify the solid. Give equations for all the reactions you mention.

# 17 Electrochemistry

## 17.1 Names and symbols

When we talk about electricity and chemistry, certain names and symbols have to be used.

A source of electricity is called a cell, and a group of cells is called a battery. The negative terminal of a battery is always drawn short, and the positive terminal long. In this book, the negative terminal is always drawn on the left hand side.

Switches are used in circuits to stop or start the flow of electricity. Bulbs will register whether or not a current is flowing. The symbols used for cells, batteries, switches and bulbs are shown in figure 1.

| electrical device | symbol |
|---|---|
| one cell (negative and positive terminals marked) | ▮\| |
| a battery of three cells | ▮\|▮\|▮\| |
| switch | —⁄— |
| bulb | —◯— |

**Figure 1**
*Cells, batteries, switches and bulbs, and their symbols.*

The things that carry the electric current into a liquid or a molten substance are called *electrodes*. They are usually made of inert substances such as graphite, platinum, silver and gold. Sometimes mercury is used.

The electrode which is connected to the positive side of the battery is called the *anode*. The electrode which is connected to the negative side of the battery is called the *cathode*. A battery is an 'electron pump'. It pushes electrons in at the cathode and sucks them out at the anode. Figure 2 shows the circuit diagram for the apparatus that would be used to pass electricity through a solution.

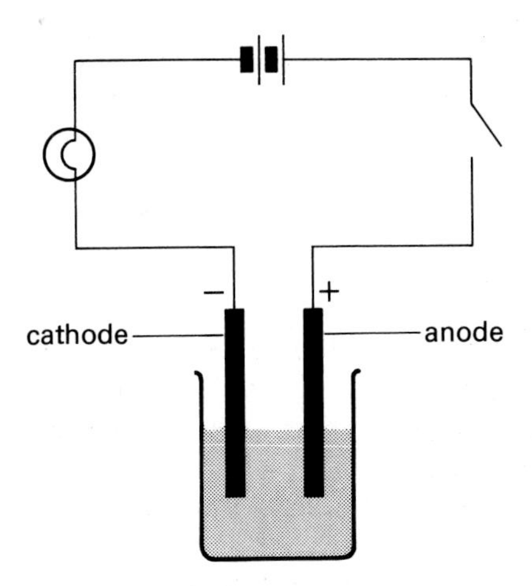


**Figure 2**
*A circuit diagram for passing electricity through a solution.*

## 17.2 Conductors and insulators

Solid substances which allow electricity (i.e. a flow of electrons) to pass through them are called *conductors*. Examples of conductors are metals, and graphite.

Solid substances which do not allow electricity to pass through them are called *non-conductors*, or insulators. Examples of non-conductors are: non-metals (except graphite), wood, glass ceramics, and plastics.

## 17.3 Electrolytes and non-electrolytes

Liquids which will allow electricity to pass through them are called *electrolytes*. Examples of electrolytes are:
solutions of acids in water,
solutions of alkalis in water,
solutions of salts in water,
and molten salts.

Liquids which do not allow electricity to pass through them are called *non-electrolytes*. Examples of non-electrolytes are: pure water, organic solvents such as ethanol and propanone, and molten covalent substances such as wax and naphthalene.

## 17.4 Electrolysis

The passage of an electric current through an electrolyte is called *electrolysis*.

Electrolytes are liquids which contain ions. The ions are free to move about. When two electrodes which have been connected in a circuit with a battery are put into the liquid electrolyte, the charged ions are attracted to the charged electrodes.

The positive ions are attracted to the negative electrode (the cathode) so they are called *cations*.

The negative ions are attracted towards the positive electrode (the anode) so they are called *anions*.

As the ions move they carry the electric charge through the electrolyte and a current flows around the circuit. As this happens, this means that electrolysis is taking place, and an *electrochemical reaction* is happening.

## 17.5 The electrolysis of molten potassium iodide

Potassium iodide has a melting point of 685 °C. When it is melted, the ions in the crystal lattice get enough energy to break apart, and move around independently. If electrodes are put into the molten salt, the ions are attracted towards them. Figure 3 shows the apparatus that could be used for this electrolysis.

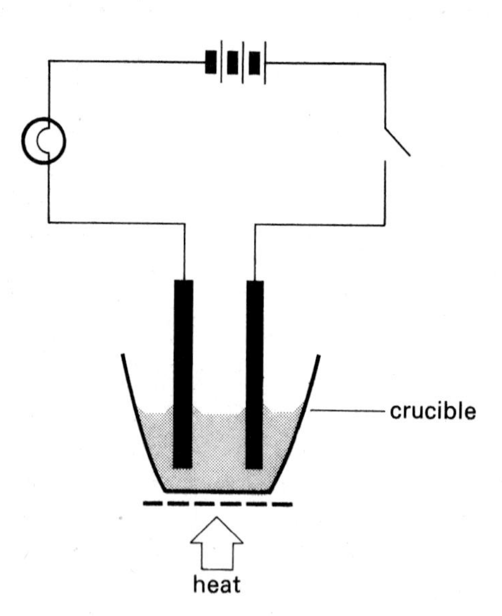


**Figure 3**
*The electrolysis of potassium iodide.*

The negative iodide ions are attracted towards the anode. They are called anions because of this. At the anode, they give up their electrons and turn into iodine molecules. The electrons go off round the circuit:

$$2I^- \longrightarrow 2I + 2e^-$$

$$2I \longrightarrow I_2(g)$$

The iodine is released as purple iodine vapour.

The positive potassium ions are attracted towards the cathode. They are called cations because of this. At the cathode, they take in the electrons which have come round the circuit, and form potassium atoms.

$$2K^+ + 2e^- \longrightarrow 2K(l).$$

At this temperature the potassium is formed as a liquid.

Note how the two electrode reactions balance in terms of electrons. For each electron released at the anode, one is taken in at the cathode. As the ions move towards their electrodes, electrons flow around the circuit. A current is flowing, so the bulb lights up. Because the ions are separating, electrolysis always means that the electrolyte is decomposing.

## 17.6 The electrolysis of molten lead bromide

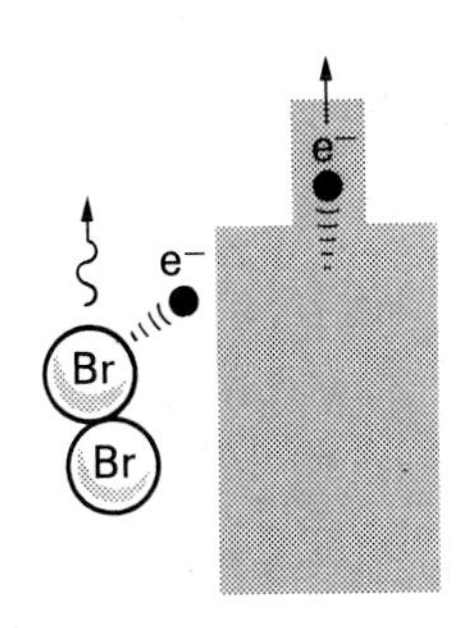


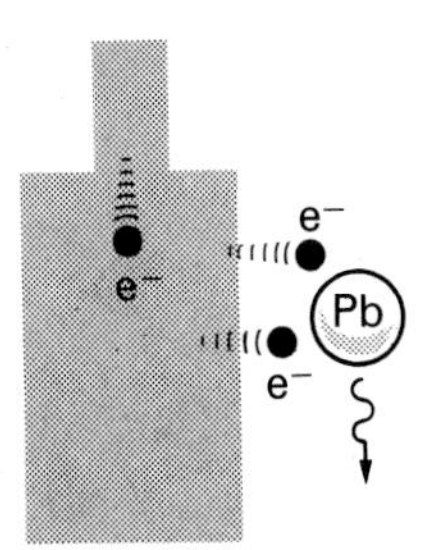


The melting point of lead bromide is 370 °C. Molten lead bromide contains lead ions and bromide ions. When two electrodes in a circuit are put into molten lead bromide, the ions are attracted towards the electrodes and a current flows in the circuit.

The bromide ions are attracted towards the anode. Because of this they are called anions. At the anode, they give up their electrons and turn into bromine molecules. At this temperature the bromine is given off as a brown vapour:

$$2Br^- \longrightarrow 2Br + 2e^-$$

$$2Br \longrightarrow Br_2(g).$$

The lead ions are attracted towards the cathode. Because of this they are called cations. At the cathode, they take in the electrons which have travelled round the circuit, and change into lead atoms. At this temperature the lead is formed as a molten blob of metal at the bottom of the crucible.

$$Pb^{2+} + 2e^- \longrightarrow Pb(l).$$

Now that you have seen the electrolysis of two molten salts, take another look at the electrode reactions. The ions reacting at the cathode always gain electrons. Gain of electrons is reduction. Reduction always happens at the cathode. The ions reacting at the anode always lose electrons. Loss of electrons is oxidation. Oxidation always happens at the anode.

If you want some more examples of the electrolysis of molten salts, look at Chapter 18. Sodium and aluminium are extracted from their ores by electrolysis, and copper is purified by electrolysis as well.

17.7
**The electrolysis of acidified water**

Pure water is a non-electrolyte because it is a covalent compound. But whenever an ionic substance is dissolved in it, the water ionises. Sulphuric acid could be used. There would then be three types of ions in solution:

from the water: $H_2O(l) \longrightarrow H^+(aq) + OH^-(aq)$;

from the acid: $H_2SO_4(aq) \longrightarrow 2H^+(aq) + SO_4^{2-}(aq)$.

When electrodes in a circuit are put into dilute sulphuric acid, the hydroxide ions and the sulphate ions are both attracted towards the anode. However, it is harder for the sulphate ions to give up their electrons than the hydroxide ions. So the hydroxide ions break up, forming oxygen and water:

$$4OH^- \longrightarrow 2H_2O + O_2 + 4e^-.$$

Bubbles of oxygen gas are formed at the anode. The electrons go round the circuit.

There are only hydrogen ions to be attracted towards the cathode, and they take in the electrons to form hydrogen molecules:

$$4H^+ + 4e^- \longrightarrow 2H_2(g).$$

Bubbles of hydrogen are formed at the cathode.

The apparatus that could be used for this electrolysis is shown in figure 4.

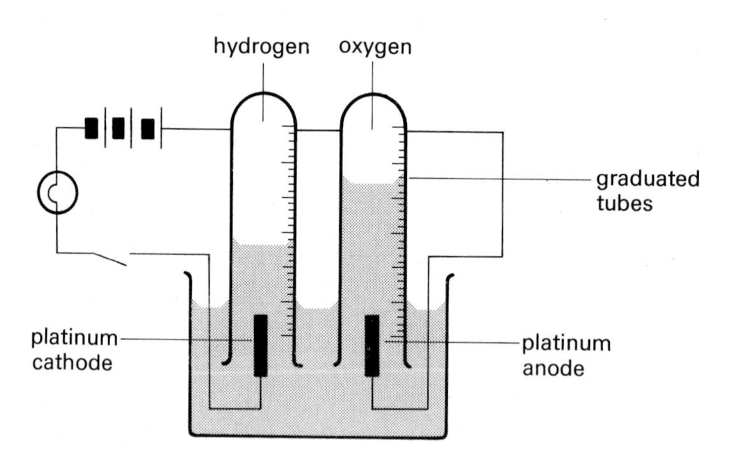


**Figure 4**
*The electrolysis of acidified water.*

Notice that the two electrode reactions are balanced, the same number of electrons are involved at each electrode. But you can see from the equations that one molecule of oxygen is produced for every two molecules of hydrogen. Each molecule of oxygen needs four electrons and each molecule of hydrogen needs 2 electrons. If the electrolysis is left running for a few minutes, it will be seen that the volume of hydrogen produced is twice that of oxygen.

## 17.8 The electrolysis of copper(II) sulphate

Electrolysis of copper(II) sulphate solution is different depending on whether platinum or copper electrodes are used.

**Platinum electrodes.** There are four sorts of ions in copper(II) sulphate solution.

from the water: $H_2O(l) \longrightarrow H^+(aq) + OH^-(aq)$;

from the copper sulphate:

$$CuSO_4(aq) \longrightarrow Cu^{2+}(aq) + SO_4^{2-}(aq).$$

When the platinum electrodes are put into the solution, both the sulphate ions and the hydroxide ions are attracted towards the anode, but as we have already seen, it is the hydroxide ions that break up to turn into water and oxygen:

$$4OH^- \longrightarrow 2H_2O + O_2 + 4e^-.$$

Oxygen gas is released at the anode.

Both the hydrogen ions and the copper ions are attracted towards the cathode. Copper ions will accept electrons more readily than hydrogen ions can. So it is copper that is formed:

$$2Cu^{2+} + 4e^- \longrightarrow 2Cu(s).$$

A brown coating of solid copper appears on the cathode.

Because copper ions and hydroxide ions are removed from the solution, hydrogen ions and sulphate ions are left behind after the electrolysis has gone on for some time. The solution will have become colourless sulphuric acid solution.

**Copper electrodes.** The reaction using copper instead of platinum electrodes is very different. All that happens at the anode is that the copper dissolves, and forms copper ions:

$$Cu(s) \longrightarrow Cu^{2+} + 2e^-.$$

At the cathode, copper ions turn into copper metal:

$$Cu^{2+} + 2e^- \longrightarrow Cu(s).$$

The copper anode gets thinner, and the copper cathode gets thicker. More details are given in the section on purification of copper. (Chapter 18.)

17.9

## Strong and weak electrolytes

Solutions and molten substances which contain a lot of ions are called strong electrolytes. Good examples are strong acids and alkalis, and salts – molten or in solution. They are strong electrolytes because they are fully ionised.

Sometimes, substances only ionise slightly when they are dissolved in water. The solution will then only contain a few ions, and it is only a weak electrolyte.

A good example of a weak electrolyte is ethanoic acid. In solution it is only partly ionised.

Strong electrolytes may be distinguished from weak electrolytes by testing how easily an electric current passes through them. (See figure 5.)

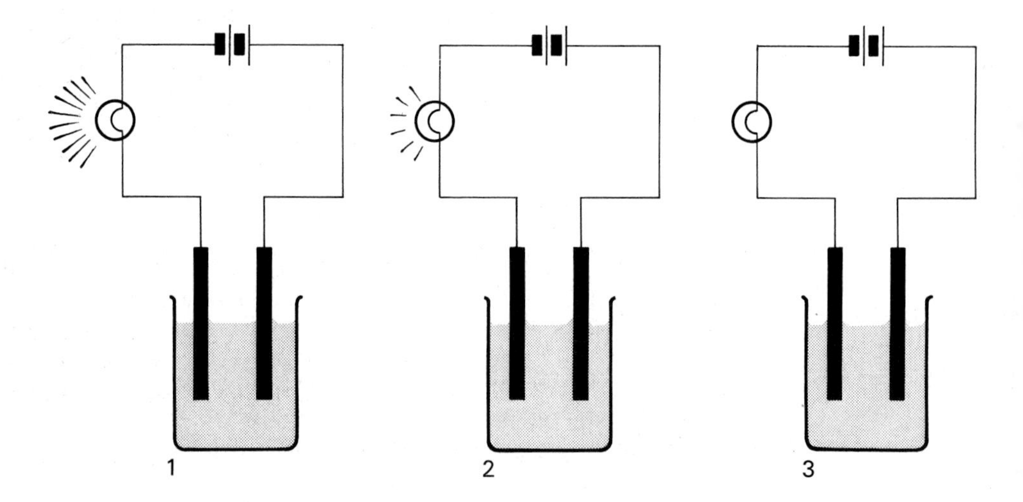


**Figure 5** *A comparison of strong, weak, and non-electrolytes.*

The solution in beaker 1 contains dilute hydrochloric acid which is a strong electrolyte. The bulb glows brightly.
The solution in beaker 2 contains ethanoic acid solution which is a weak electrolyte. The bulb glows dimly, because there are few ions to carry the electric current.
The solution in beaker 3 is sugar solution which is a non-electrolyte. The bulb does not glow at all because there are no ions to carry the electric current.

## 17.10 The industrial production of sodium hydroxide

Sodium hydroxide is produced by the electrolysis of brine in a device called a 'flowing mercury cathode cell'. It is shown in figure 6.

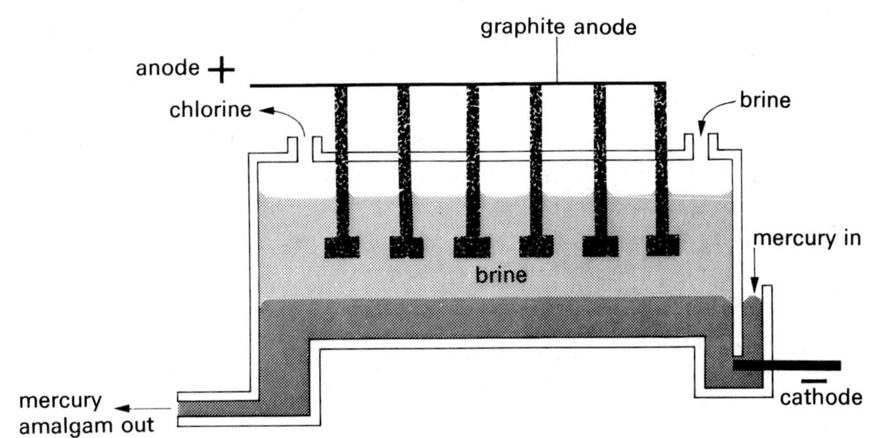


**Figure 6** *The flowing mercury cathode cell.*

The anode of this cell is made of graphite. The cathode is a continuously flowing stream of mercury. The electrolyte is brine, which is the common name for concentrated sodium chloride solution.

Because the sodium ions and chloride ions are in such high concentration, they are attracted to the electrodes in preference to $H^+$ and $OH^-$ ions.

At the anode, chloride ions give up electrons and form chlorine molecules:

$$2Cl^- \longrightarrow 2Cl + 2e^-$$

$$2Cl \longrightarrow Cl_2(g).$$

The chlorine is removed for other industrial uses.

At the cathode, sodium ions take in electrons and turn into sodium atoms. The sodium immediately dissolves in the liquid mercury to form a solution called an amalgam. The amalgam flows out of the cell:

$$2Na^+ + 2e^- \longrightarrow 2Na(s)$$

$$Na + Hg \longrightarrow Na/Hg.$$

The amalgam is then mixed with water. The sodium in the mercury reacts to form hydrogen and sodium hydroxide solution:

$$2Na(s) + 2H_2O(l) \longrightarrow 2NaOH(aq) + H_2(g).$$

The sodium hydroxide solution is carefully evaporated to give white, solid sodium hydroxide.

## 17.11 Electroplating

Electrodes may be caused to become coated with a layer of metal during electrolysis and this is called *electroplating*. Look at the apparatus in figure 7.

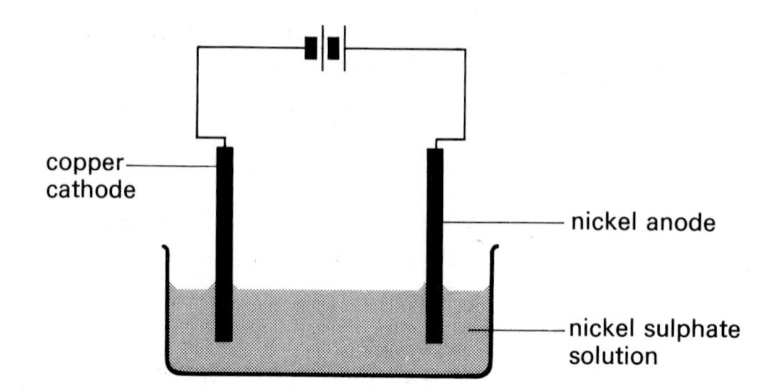


**Figure 7**
*Nickel plating.*

This cell contains a nickel anode and a copper cathode. The electrolyte is nickel sulphate solution. When the current is switched on the nickel anode dissolves to form nickel ions:

$$Ni(s) \longrightarrow Ni^{2+}(aq) + 2e^{-}.$$

At the cathode, the nickel ions in solution are turned into nickel atoms and are formed as a thin layer of metal on the electrode:

$$Ni^{2+}(aq) + 2e^{-} \longrightarrow Ni(s).$$

The copper cathode has been nickel plated. This process is called *electroplating.*

Any metal object which is to be electroplated must be cleaned so as to be free of dirt, corrosion, and grease. It is then made the cathode in the electrolysis cell. The anode is made of the metal which is to be used for the electroplating, and the electrolyte is chosen to have ions of the metal in it. The apparatus in figure 8 could be used to electroplate a nail with copper.

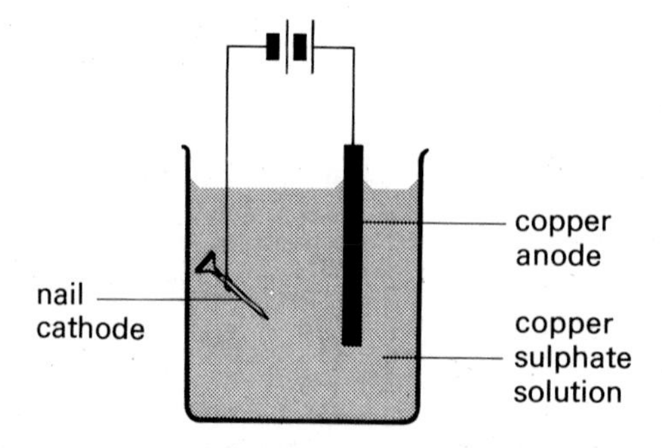


**Figure 8**
*Copper plating an iron nail.*

In industry a lot of care is taken to make sure that the layer of metal stays in place. If an iron object is to be chromium plated it has to go through four processes:

**1** It is thoroughly cleaned with sulphuric acid (this is called pickling), and washed with deionised water.

**2** It is then copper plated, because this sticks well.

**3** The copper layer is then nickel plated to prevent corrosion.

**4** Finally, the nickel layer is chromium plated, giving a bright, non-tarnishing finish to the iron.

## 17.12 Cells

Just as electricity can cause a chemical reaction to take place during electrolysis, so a chemical reaction can make an electric current flow around a circuit. A bulb may be made to glow dimly by connecting it to two electrodes( one of copper and one of zinc), and immersing them in dilute sulphuric acid. (See figure 9.)

The flow of electrons around the circuit means that there is an electric current.

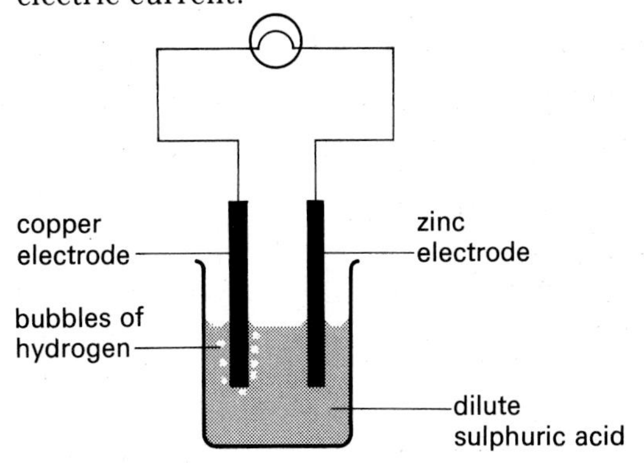


**Figure 9**
*A chemical cell.*

The zinc electrode dissolves in the acid and forms zinc ions:

$$Zn(s) \longrightarrow Zn^{2+}(aq) + 2e^-.$$

The electrons that are released flow round the circuit to the copper electrode. Here, hydrogen ions take in the electrons and form hydrogen gas:

$$2H^+(aq) + 2e^- \longrightarrow H_2(g).$$

The flow of electrons around the circuit means that there is an electric current, so the bulb lights. This is a *cell.* A chemical reaction is producing electricity.

This cell has two main disadvantages:

**a** Not all of the energy from the reaction is turned into electricity. A lot is made into heat – the cell gets hot.

**b** The copper electrode gets covered with a layer of hydrogen bubbles. These prevent the current from flowing. This is called *polarisation.* If the bubbles are cleared, the current flows again.

17.13
## The dry cell

This is the type of cell found in the batteries used in torches, radios etc. It is a little more complicated than the previous cell, but it does not get very hot (it does get warm), and it does not polarise too much. It is shown in figure 10.

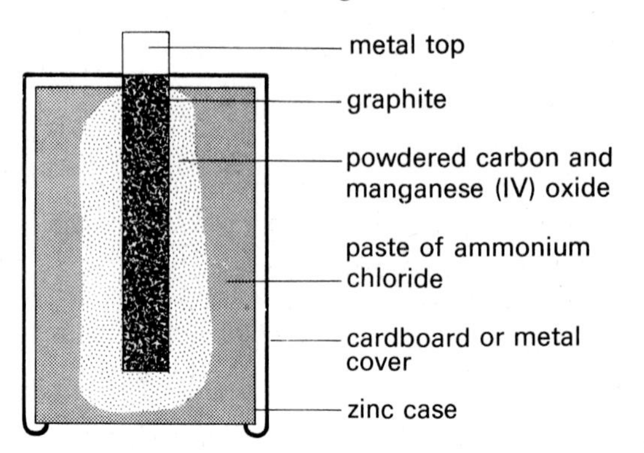


**Figure 10**
*The dry cell.*

The cell consists of a carbon rod which fits inside a porous bag of carbon powder mixed with manganese(IV) oxide. This bag is surrounded with a paste of ammonium chloride, and all this is contained in a zinc case. The top of the cell is sealed with pitch. A cardboard, or metal wrapper is added.

When the cell is connected into a circuit, the zinc slowly dissolves, releasing electrons. These go through the outside circuit, and so form the electric current. The terminals of the cell are on the top of the carbon rod and on the bottom of the zinc case.

When the chemicals have been used up, the battery is finished – it has gone flat. It cannot be used again, because the chemical reaction cannot be reversed.

This single cell has a voltage of 1·5 volts. Batteries with higher voltages (such as a 9 volt battery for a radio), contain a number of these cells connected end to end.

17.14
## A rechargeable cell – the accumulator

Dry cells have to be thrown away once they become flat. An accumulator (like the battery in a car) can be recharged when its chemical energy is used up. It consists of lead plates, which dip into dilute sulphuric acid. When the accumulator is charged, the positive plate has a layer of lead oxide built up on it. The negative plate stays as lead. As the accumulator is used, a chemical reaction takes place so that both of the plates become covered with lead sulphate. When this happens, the accumulator goes flat and must be recharged. When it has been recharged, the plates are restored to their previous condition, so the accumulator can be used to supply electricity again. In a motor car, the electric circuits are arranged so that after electricity has been drawn out of the battery, a dynamo charges it up. The worst problem is in winter, at night. Especially when it is raining, a lot of electricity is drawn from the battery. The dynamo may not be able to keep up with the demand, so the battery will go flat if these sort of driving conditions continue for a long time.

Usually, however, the battery in a car is kept charged by the dynamo. All that is needed is for the battery to be kept topped up with distilled water.

Each cell of an accumulator produces a voltage of 2 volts. Most car batteries consist of six of these cells together, and have a total voltage of 12 volts. (See figure 11.)

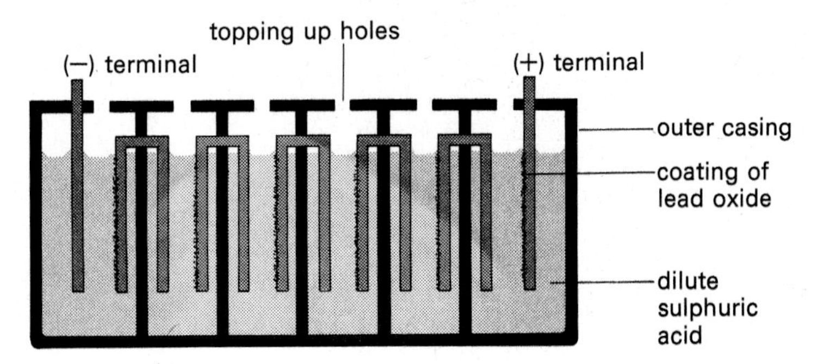


**Figure 11**
*A car battery.*

**Summary**

At the end of this chapter you should be able to:

**1** Draw the symbols for a cell, a battery, a bulb, a switch, and put them into a circuit diagram.

**2** Say which side of the battery the anode and cathode are connected to.

**3** Distinguish between conductors and non-conductors and give examples of each.

**4** Distinguish between electrolytes and non-electrolytes and give examples of each.

**5** Say what is meant by the term electrolysis.

**6** Describe the electrolysis of a molten salt.

**7** Describe the electrolysis of dilute sulphuric acid.

**8** Describe the electrolysis of copper(II) sulphate solution with platinum electrodes.

**9** Describe the electrolysis of copper(II) sulphate solution with copper electrodes.

**10** Distinguish between a strong and a weak electrolyte.

**11** Describe the industrial production of sodium hydroxide.

**12** Explain how an object may be electroplated and describe the sort of care that is taken before something is electroplated in a factory.

**13** Describe how electricity may be obtained from a chemical reaction.

**14** Describe the dry cell.

**15** Describe the accumulator.

# Hard sell for fuel cells

*Fuel cells powered the Apollo landing capsule.*

*There has been a lot of talk about 'the energy crisis' in the last few years. Some people believe that the crisis is one of the efficiency with which fuels are used, rather than the amount of fuel available. The following article tells you about one such group.*

Thirty-two American companies are trying to get the American government to give them money to help them develop fuel cells which they claim, will make the power stations which we need today completely unnecessary.

Fuel cells make electricity directly from fuels such as hydrogen or natural gas. Instead of burning the fuel to make steam, which is then used to drive generators, they convert the fuel directly to electricity. As a result, they can be up to three times as efficient as 'normal' systems. Until recently, however, there have been many technical problems which prevented them from being easy to make.

**Shocking discovery**

Fuel cells are not a new idea. Sir William Grove gave himself a nasty electric shock when he first discovered them, in 1839. But the idea did not really catch on – there was a gap of one hundred and twenty years before Francis Bacon in 1959, made a cell powerful enough to work a fork-lift truck. Ten years after that in 1969, they played a vital part in getting Apollo astronauts to the moon. By 1989, they may be providing heat and light in most of the homes in Britain . . .

**Redox reaction at heart of fuel cell**

When hydrogen burns, there is a redox reaction: the hydrogen is oxidized (it gives its electrons to the oxygen) and the oxygen is reduced (it accepts the electrons from the hydrogen). In order to do this in a fuel cell, the electrons are forced to travel via an outside circuit.

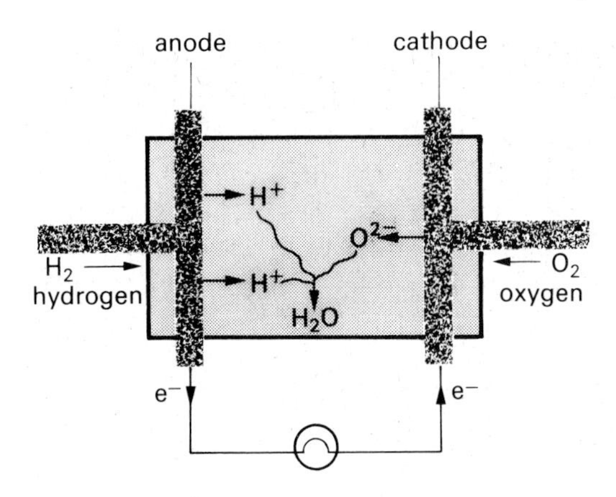


A simple fuel cell is shown in the diagram. The anode and cathode are made of a special porous material which conducts electricity, allows gas to go through it, and allows the electrolyte to penetrate it a little way. The anode and cathode are connected through an outside electrical circuit. Hydrogen gas is pumped into the anode, and oxygen into the cathode. When this is done, things start to happen . . .

**Ions one way, electrons the other**

At the anode, the hydrogen splits up to form electrons and $H^+$ ions. The ions drift out into the electrolyte, and the electrons go through the circuit, round to the cathode. At the cathode, the oxygen takes them in and forms $O^{2-}$ ions, which also drift out into the electrolyte. At the centre, the $H^+$ and $O^{2-}$ ions meet, to instantaneously form water, making space which more ions want to fill. There is a suction effect, which pulls more ions into the centre . . . they meet, form water, and so it goes on. All the time this happens, the electrons are going round the outside – in other words, there is an electric current, which can be used!

The first model that Bacon made was only powerful enough to work an ordinary light bulb. It took some years before he was able to make one with sufficient power to do useful work. His first working fuel cell was installed in a fork-lift truck, to demonstrate its power.

But it was not until the Apollo spacecraft programme, that research started in earnest. Fuel cells were designed to work off the hydrogen and oxygen that the spacecraft used for propulsion. Eventually cells were designed which made enough electricity to drive all the spacecraft systems, and enough 'waste' water to provide the crew with drinking water!

**Spare scientists get to work**

These cells were much too expensive for everyday use. When the Apollo programme finished, there were a lot of spare scientists without enough to do. So they set about trying to find everyday applications of fuel cell technology.

And on the whole, they failed. Fuel cells could be made, but they were always too expensive, and often contained highly explosive fuels. Their problems have not yet been solved – but there is hope. In 1969, the first fuel cell car was built. It had both fuel cells, and ordinary batteries built into it – it had a maximum speed of 80 k.p.h., and could travel 320 km before re-fuelling was necessary.

The possibility of fuel cells for domestic electricity supply is even more hopeful. A number of American companies are now working hard to develop a fuel cell for use *inside* the home which will run off natural gas, or coal. If these can be made safely, they will produce both electricity and heat for the home. With this very efficient system it may be possible to reduce the amount of fuel needed to one-third of that which is necessary today. Instead of needing huge, inefficient power stations and ugly electricity pylons, all that would be needed is a normal gas pipe connection to each home. The system would be non-polluting, and very quiet. Because they are so efficient, it is quite possible that they would be the cheapest way to make electricity – cheaper even than nuclear power. Is this fuel for thought?

*A fuel-cell powered car.*

## Questions

**1** Draw a labelled diagram of an electrolysis circuit containing a bulb, a switch, a four-celled battery, a carbon cathode and a copper anode, and a beaker of an electrolyte.

**2** Turn the following list, into groups of: conductors, non-conductors, electrolytes, and non-electrolytes. Wax, ethanol, sodium chloride solution, sugar solution, steel, plastic, brass, dilute hydrochloric acid, wood, sodium, naphthalene solution, molten lead chloride.

**3** Three electrolysis cells were set up like this:

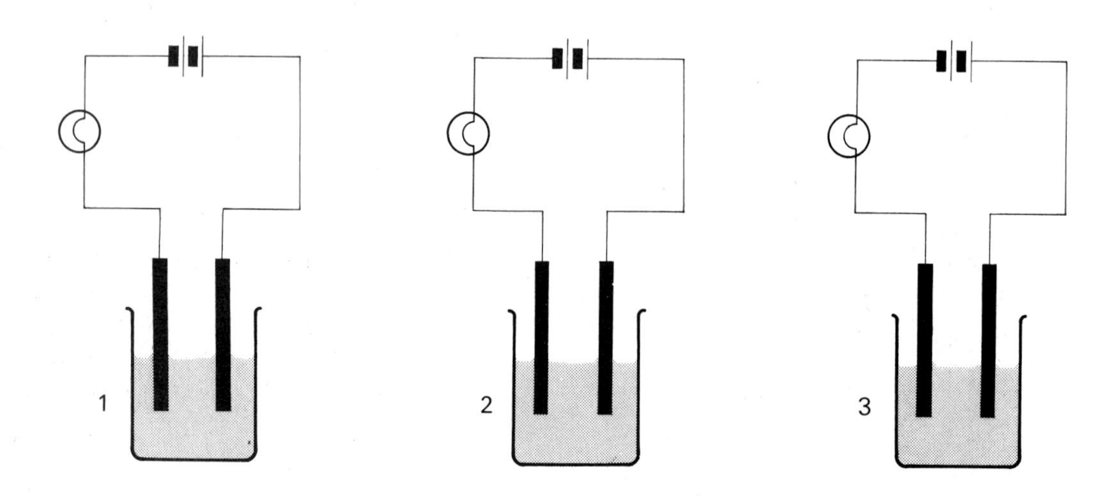


The bulb connected to cell 1 glowed dimly, and a few bubbles of gas appeared on both electrodes.

The bulb connected to cell 2 did not glow at all, and no bubbles of gas were seen.

The bulb connected to cell 3 glowed brightly, and bubbles of gas appeared on the anode. The cathode was seen to be coated with a brown solid.

Suggest what liquids might have been in each of the cells, and explain your reasons.

**4** What does the term electrolysis mean?

Draw the apparatus you would use to electrolyse molten magnesium chloride. What would be the products of this electrolysis? Write down the two electrode reactions and say which of them is an oxidation and which of them is a reduction.

What would the electrodes be made of, and why?. Why couldn't the anode be made of aluminium?

**5** Draw the apparatus that you would use if you wanted to gold plate a steel pen nib. What electrodes would you use? What electrolytes would you put in the electrolysis cell, and what would you do to the steel nib before starting the plating?

**6** Look at these diagrams:

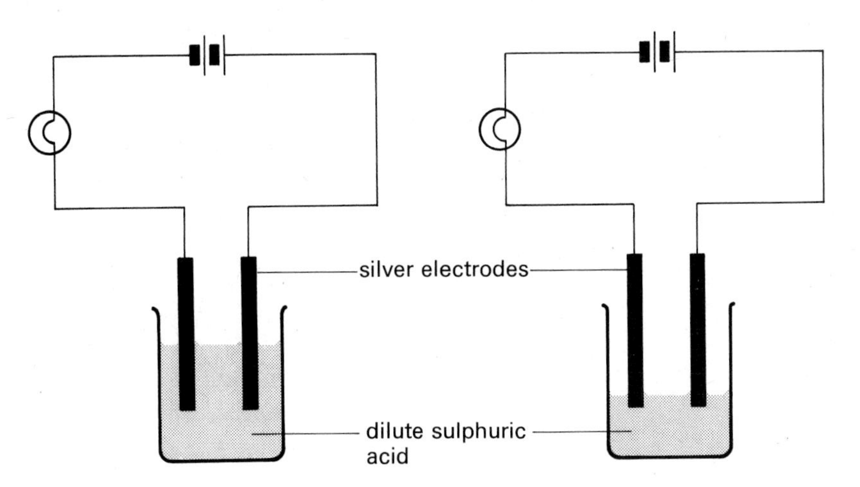


Which bulb will burn brighter? Why?

**7** A cardboard wrapped battery (not the metal covered sort), will ooze a white paste if it is left for a long time after it has gone flat. If the battery is in something made from iron, the paste will quickly make it rust. What is the paste, and why has it leaked out? Why does the metal container rust? (You may need to refer to other chapters.)

**8** Describe the advantage which an accumulator cell has over an ordinary dry cell.

An accumulator battery is used in a car. Under normal conditions the battery does not go flat, even over a period of several years. Why? What sort of conditions will make it go flat quickly? Explain.

The electrolyte in an accumulator cell is dilute sulphuric acid, but when it is topped up, only distilled water is used. Why does only water have to be added? Can you think why the water must be distilled?

**9** A scientist called Hoffmann designed the apparatus shown here for the electrolysis of liquids. Describe what you think will happen when the electrodes are connected to the battery. Explain what would be going on inside the dilute sulphuric acid, and at the electrodes.

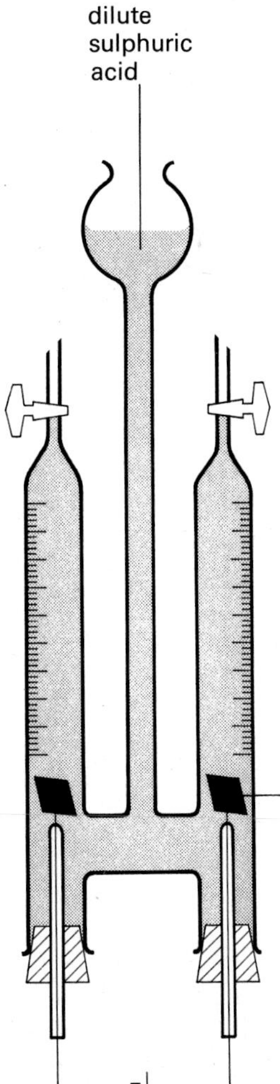

# 18 Metals and non-metals

## 18.1 It's that periodic table again

Elements are either metals or non-metals, and they occur in two halves of the periodic table. Figure 1 shows where the border line is.

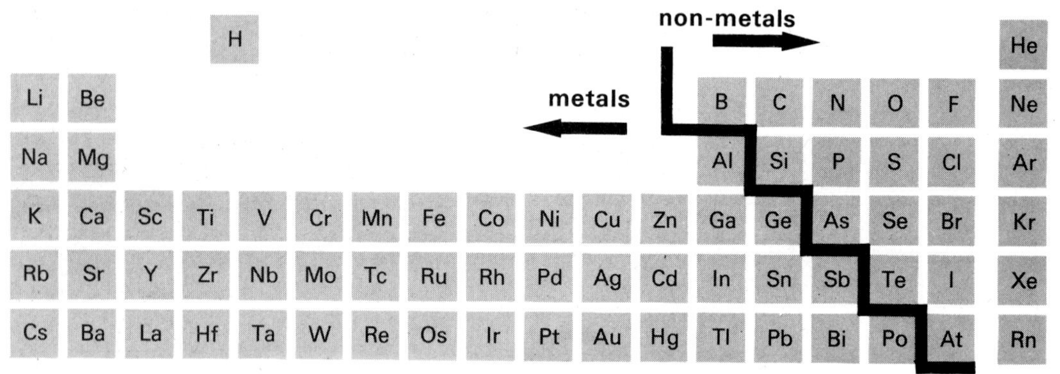


**Figure 1** *Metals and non-metals.*

You can see that the majority of the elements are metals, and they are all solids at room temperature, except mercury. The non-metals are all gases or solids at room temperature – with one exception. There is one liquid non-metal – bromine. This means that there are only two liquid elements in all.

The periodic table can be written in another way to show whether or not the best known elements form ions. (See figure 2.)

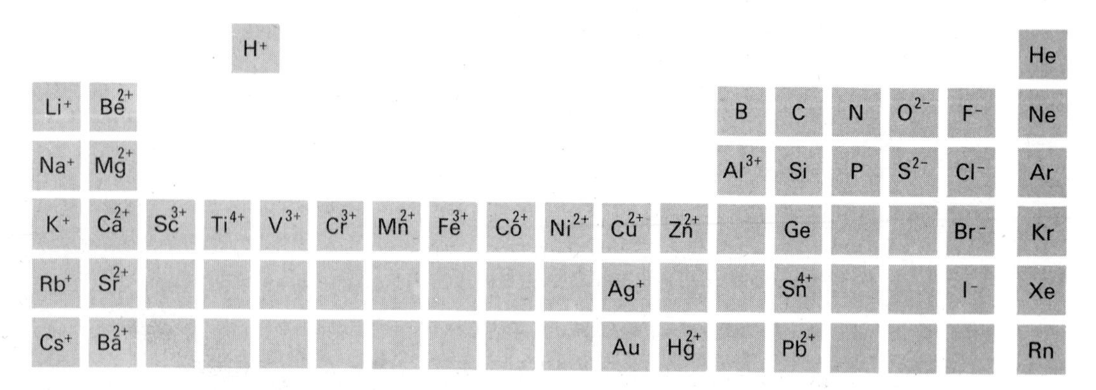


**Figure 2** *Some elements and their ions.*

Notice that the metals form positive ions, and the non-metals (if they form ions) form negative ones. The majority of metallic compounds are ionic. Some non-metallic compounds are ionic, and some are covalent.

## 18.2 What are metals?

If you were to ask a non-chemist what a metal was, he would probably tell you that it was a strong, heavy, shiny solid which was hard, and that it would be a good conductor of heat and electricity. He might add that if you were to hit it, it would clang, and that it was strong enough to be cut and moulded into all sorts of shapes and sizes.

That is certainly true of many metals. They are generally shiny when they are polished, although they may become corroded. Many of them have a high density, and they make a noise when they are hit (they are said to be *sonorous*). Many metals can be hammered out flat (they are said to be *malleable*) and may be drawn into wires (they are said to be *ductile*). But this isn't true of all metals. Sodium is lighter than water, and can be cut with a knife. Potassium and lithium are very similar to sodium; magnesium and calcium are also lightweight metals.

To get a true picture of what a metal is, we must add some chemical properties to the physical ones already described.

**Metals react with air or oxygen.** They form *basic oxides*. For example:

copper + oxygen ⟶ copper(II) oxide.

$$2Cu(s) + O_2(g) \longrightarrow 2CuO(s).$$

**Metals react with acids.** They form a salt plus hydrogen. For example:

magnesium + hydrochloric acid (dilute) ⟶ magnesium chloride + hydrogen.

$$Mg(s) + 2HCl(aq) \longrightarrow MgCl_2(aq) + H_2(g).$$

**Metals are reducing agents.** In any reaction of the above two types, the metal forms a positive ion and gives out electrons to the substance with which it is reacting:

$$Mg \longrightarrow Mg^{2+} + 2e^-.$$

Because of this we say that metals are reducing agents.

## 18.3 Metallic structure

Two properties that have not yet been discussed are the electrical and thermal conductivity of metals. They are good conductors of heat and electricity. We can understand this a little more easily if we look at the structure of metals.

The atoms in a piece of metal are closely packed together in a regular way. This is shown in figure 3**a**.

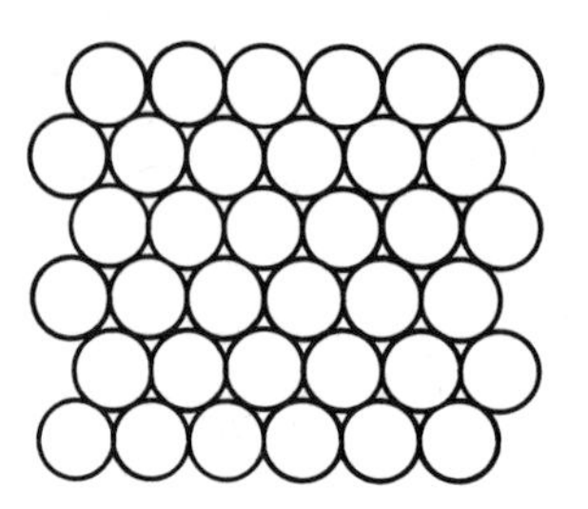

**Figure 3a**
*The arrangement of atoms in a metal.*

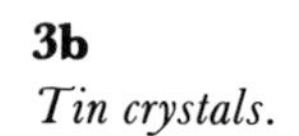

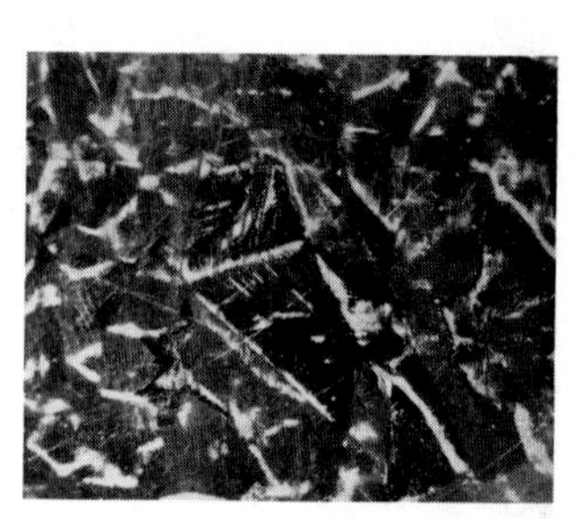

**3b**
*Tin crystals.*

In other words, metals contain a lattice of atoms – so it is not surprising that metals form crystals. You can sometimes see these crystals on the inside of a tin can. The metal there is tin, which has been thinly coated onto the steel can. (See figure 3**b**.)

When the metal atoms are packed tightly together in this way, the outside shell electrons from each atom become detached and form a 'sea' of electrons on the surface of the metal. Since electricity is just a flow of electrons along a wire, this 'sea' of electrons enables the metals to conduct electricity well.

The atoms of a metal vibrate. This means that heat energy can be passed through a piece of metal as the vibration energy passes from atom to atom. Metals conduct heat well.

## 18.4 What are non-metals?

Physically, the non-metals are the opposites of metals.

Solid non-metal elements do not clang when they are hit. If they are stretched or hammered, they break or shatter.

For example, sulphur is a weak, brittle solid, and carbon (in the form of graphite) is very soft.

Non-metals are not sonorous, malleable or ductile; they are weak and they do not conduct heat or electricity at all well. There is one exception to this last property, however – graphite is a good conductor of electricity. In addition, non-metals are the opposites of metals chemically.

**Non-metals burn in air or oxygen.** They form acidic oxides. For example:

sulphur + oxygen ⟶ sulphur dioxide.

$S(s) + O_2(g) \longrightarrow SO_2(g).$

**Non-metals are oxidising agents.** Whenever non-metals do react to form ions, they form negative ones. They take in electrons from the substance they are reacting with. For example:

$S + 2e^- \longrightarrow S^{2-}$.

Non-metals are said to be oxidising agents because of this.

**Non-metals have no reaction with dilute acids.** Carbon or sulphur, for example, will not react with any dilute acid.

## 18.5 Non-metallic structure

Non-metals can be either gases, liquids or solids. Typical gases are oxygen ($O_2$), nitrogen ($N_2$), chlorine ($Cl_2$) and hydrogen ($H_2$). They are all in pairs of atoms – they are diatomic molecules.

Sulphur ($S_8$), phosphorus ($P_4$) and bromine ($Br_2$) are amongst the solids and liquids made up from molecules of various sizes. Graphite and diamond are both solids which form giant structures. Theoretically, there is no limit on their size.

Non metals except for graphite do not conduct electricity. There is no 'sea' of electrons to carry the current since they are tightly bonded.

## 18.6 Metals and non-metals: a summary

You can see that the properties of metals are very different to those of non-metals. Figure 4 compares some of the physical and chemical properties of metals and non-metals.

| property | metals | non-metals |
|---|---|---|
| bendability | ductile and malleable | the solids are brittle |
| sonority | 'clang' | no 'clang' |
| electrical and thermal conductivity | high | low |
| corrodability | frequently corroded | do not corrode |
| reaction with oxygen | form basic oxides | form acidic oxides |
| reaction with acids | form salts + hydrogen | no reaction |
| ions formed | positive ions | negative ions or no ions at all |
| oxidising agent or reducing agent | reducing agent | oxidising agent |

**Figure 4** *The properties of metals and non-metals: a summary.*

## 18.7 The Activity Series

In simple terms, this is a list of common metals in the order of their activity with such things as air, water and dilute acids. More correctly, it is a list of the metals in the order of ease with which they form ions. The Activity Series for the common metals is shown in figure 5.

| metal | tendency to ionize |
| --- | --- |
| potassium<br>calcium<br>sodium<br>magnesium<br>aluminium<br>zinc<br>iron<br>tin<br>lead | Metals at the top of the series form positive ions very readily. |
| hydrogen ———— | Hydrogen is put in here as a reference point, for the reaction of metals with acids and water. |
| copper<br>mercury<br>silver<br>gold | Metals at the bottom are reluctant to form ions. |

**Figure 5**
*The Activity Series.*

## 18.8 The reactions of metals with oxygen

Many different chemical reactions can be explained by referring to this series. The vigour of the reaction with oxygen depends upon the position of the metal in the Activity Series. The metals at the top react most vigorously because they form ions most readily. Figure 6 gives a summary of these reactions. You will find a detailed account of how metals (and non-metals) react with oxygen in chapter 10.

| metal | vigour of reaction | nature of the oxide and solubility in water |
| --- | --- | --- |
| potassium<br>calcium<br>sodium | These metals burn very brightly after melting. | Oxides are soluble and their solutions are alkaline. |
| magnesium | | Oxide slightly soluble. |
| aluminium<br>zinc<br>iron | These metals will burn when powdered, the flames getting less bright down the Series. | These oxides are not soluble. |
| lead<br>copper<br>mercury | These metals do not catch fire, but become coated with layers of oxide. | |
| silver<br>gold | These metals do not react. | |

**Figure 6**
*The Activity Series and reactivity of the common metals.*

Remember that metal oxides are basic oxides.
Basic oxides react with acids to form a salt plus hydrogen.
Soluble basic oxides are called alkalis.

## 18.9 Reactions of metals with water

Once again, the vigour of this reaction depends upon the position of the metal in the Activity Series. In other words, upon how easily the metal forms ions.

Figure 7 on this page and the next summarises the reactions of metals with water.

| metal | description of the reaction | equation |
|---|---|---|
| **potassium** | As soon as a piece of potassium is put on water it melts, rushes around on the surface, getting smaller as it reacts. The heat of the reaction sets fire to the hydrogen which is evolved. It burns with a lilac flame. After a few seconds, the potassium has gone, and an alkaline solution of potassium hydroxide is left. | $2K(s) + 2H_2O(l) \longrightarrow H_2(g) + 2KOH(aq)$. |
| **sodium** | Sodium reacts in a similar way to potassium. It melts and rushes around on the surface of the water but the reaction is not hot enough to ignite the hydrogen. The sodium reacts with the water, leaving a solution of sodium hydroxide. | $2Na(s) + 2H_2O(l) \longrightarrow H_2(g) + 2NaOH(aq)$. |
| **calcium** | Although calcium comes higher in the Series than sodium, its reaction with water is less vigorous, because the calcium hydroxide that is formed is not very soluble. It forms a protective coating on the calcium which slows the reaction down. The calcium sinks, bubbles, but does not melt. The solution goes cloudy because of the insoluble alkali. | $Ca(s) + 2H_2O(l) \longrightarrow H_2(g) + Ca(OH)_2(aq)$. |

*continued overleaf*

| metal | description of the reaction | equation |
|---|---|---|
| **magnesium** | Magnesium ribbon hardly reacts with cold water at all. One or two bubbles may form on the surface, but you would have to wait several days to collect a test tube of hydrogen using the apparatus shown in figure 8**a**. The reaction is a little quicker with hot water, but becomes very vigorous if steam is used. Figure 8**b** shows the apparatus that could be used. The magnesium glows very brightly as it reacts with the steam, and white magnesium oxide is formed. The hydrogen which is formed is often ignited by the heat of the reaction. | $Mg(s) + 2H_2O(l) \longrightarrow Mg(OH)_2(s) + H_2(g)$. $Mg(s) + H_2O(g) \longrightarrow MgO(s) + H_2(g)$. |
| **aluminium** | Aluminium will react with steam in the apparatus described in figure 8**b**, but the reaction is much less vigorous than that with magnesium. | $2Al(s) + 3H_2O(g) \longrightarrow Al_2O_3(s) + 3H_2(g)$. |
| **zinc** | Zinc reacts with steam in the apparatus shown in figure 8**b**, but the reaction is even less vigorous than that with aluminium. | $Zn(s) + H_2O(g) \longrightarrow ZnO(s) + H_2(g)$. |
| **iron** | Although iron rusts, its reaction with water alone is very slow. It will just react with steam, but a very high temperature is needed. | $3Fe(s) + 4H_2O(g) \longrightarrow Fe_3O_4(s) + 4H_2(g)$. |
| **tin** **lead** | The reaction of these elements with steam is very, very slow, but they eventually form a coating of the metal oxide. | |
| **hydrogen** | At this point, hydrogen is put into the Series as a reference. Metals below hydrogen do not react with water or steam at all. | |

**Figure 7** *Reactions of metals with water.*

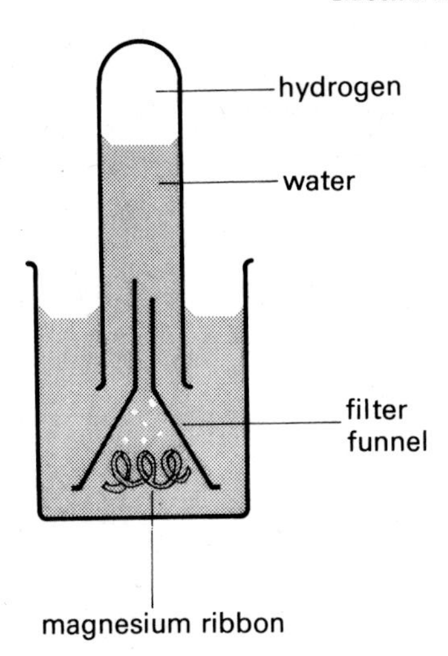


**Figure 8a**
*Reacting magnesium with cold water.*

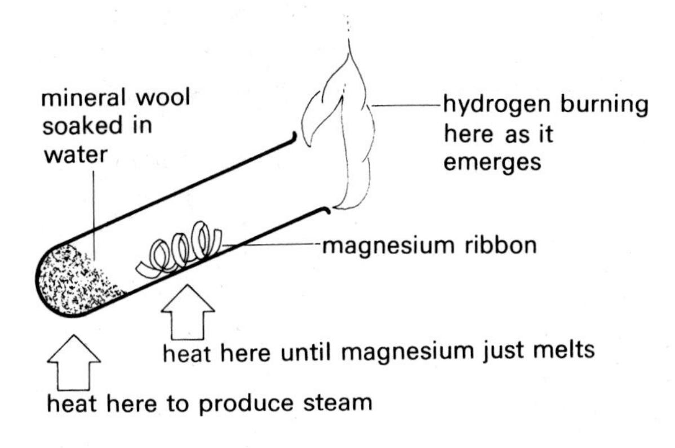


**8b**
*Reacting magnesium with steam.*

## 18.10 Reduction reactions

The Activity Series can also be used to predict how well the oxide of a metal can be reduced, using hydrogen, carbon and carbon monoxide as reducing agents. Figure 9 shows the details. You will see that hydrogen is not as good as the other two. In all cases, the lower the metal, the easier the reduction.

| metal | hydrogen | carbon and carbon monoxide |
|---|---|---|
| **potassium** | | |
| **calcium** | | |
| **sodium** | | |
| **magnesium** | | |
| **aluminium** | | |
| **zinc** | | Carbon and carbon monoxide will reduce the oxides of metals below this line. For example: $2ZnO(s) + C(s) \longrightarrow 2Zn(s) + CO_2(g)$. This needs a higher temperature than: $2PbO(s) + C(s) \longrightarrow 2Pb(s) + CO_2(g)$. As an example of carbon monoxide as a reducing agent: $ZnO(s) + CO(g) \longrightarrow Zn(l) + CO_2(g)$. |
| **iron** | Hydrogen will reduce the oxides of metals below this line. For example: $Fe_2O_3(s) + 3H_2(g) \longrightarrow 2Fe(s) + 3H_2O(g)$. This needs a high temperature. $CuO(s) + H_2(g) \longrightarrow Cu(s) + H_2O(g)$. This needs a lower temperature. | |
| **lead** | | |
| **copper** | | |
| **silver** | | |
| **gold** | | |

**Figure 9** *Reduction of metals.*

18.11
## Reduction of one metal oxide by another metal

One metal will reduce the oxide of any metal which is *lower* than itself in the Activity Series. For example: if iron powder is mixed with copper(II) oxide powder, and the mixture is heated, the iron reduces the copper(II) oxide to copper, and iron(III) oxide is formed. It is an exothermic reaction, and a glow spreads through the mixture once it has started:

$$2Fe(s) + 3CuO(s) \longrightarrow Fe_2O_3(s) + 3Cu(s).$$

The reaction takes place because iron is higher in the Activity Series, and that means that it will form ions more readily than copper. Consequently, the iron forms ions, and the copper ions turn into copper atoms.

This is an example of a redox reaction. Splitting the equation into two halves we have:

$$2Fe \longrightarrow 2Fe^{3+} + 6e^- \quad \text{(loss of electrons = oxidation)}$$

$$3Cu^{2+} + 6e^- \longrightarrow 3Cu \quad \text{(gain of electrons = reduction).}$$

If the reducing metal is high in the Activity Series, then the reaction is very exothermic.

For example, if magnesium powder is heated with copper(II) oxide powder, the reaction goes off with an explosive flash:

$$Mg(s) + CuO(s) \longrightarrow MgO(s) + Cu(s).$$

Reactions of this sort have been used to extract the metals chromium and manganese, by heating their ores with aluminium powder. The method was known as the Thermit process. In the days of trams, and before efficient welding apparatus, tram lines were welded together by a similar process. Aluminium powder was mixed with iron(III) oxide, and a pile of the mixture was made over the gap in the lines. The reaction was started, and it was so exothermic that after the sparks had cleared, a solid lump of iron was left on the gap:

$$2Al(s) + Fe_2O_3(s) \longrightarrow Al_2O_3(s) + 2Fe(s).$$

Figure 10 shows the apparatus that was used.

**Figure 10** *The Thermit process.*

## 18.12 Reactions of metals with dilute acids

Once again, the vigour of the reaction of metals with all dilute acids except nitric acid depends upon the position of the metal in the Activity Series. Figure 11 sums up the information, using hydrochloric acid as an example. But remember that nitric acid does not react the same way as other acids, and that potassium, calcium and sodium are dangerously reactive with all acids.

| **metal** | **description of reaction** | **equation** |
|---|---|---|
| potassium | All these metals react with acid | $2K + 2HCl \longrightarrow 2KCl + H_2$ |
| calcium | to form a salt plus hydrogen. | $Ca + 2HCl \longrightarrow CaCl_2 + H_2$ |
| sodium | The ones at the top react | $2Na + 2HCl \longrightarrow 2NaCl + H_2$ |
| magnesium | violently but the vigour of the | $Mg + 2HCl \longrightarrow MgCl_2 + H_2$ |
| aluminium | reactions decreases as you go | $2Al + 6HCl \longrightarrow 2AlCl_3 + 3H_2$ |
| zinc | down the Series. | $Zn + 2HCl \longrightarrow ZnCl_2 + H_2$ |
| iron | The reaction with lead is very | $Fe + 2HCl \longrightarrow FeCl_2 + H_2$ |
| lead | slow. | $Pb + 2HCl \longrightarrow PbCl_2 + H_2$ |
| **hydrogen** | **Metals below this point have no reaction at all with dilute acids.** | |

**Figure 11** *Reactions of metals with dilute hydrochloric acid.*

## 18.13 Displacement reactions

Another type of redox reaction is the displacement reaction which has already been covered in chapter 7.

Any metal will displace a metal lower in the Activity Series from a solution of one of the lower metal's salts. For example: if a piece of magnesium ribbon is placed in a solution of copper(II) sulphate, a black coating of copper forms on the surface of the magnesium and the blue colour of the solution fades, as magnesium sulphate is formed:

$$Mg(s) + CuSO_4(aq) \longrightarrow MgSO_4(aq) + Cu(s).$$

Magnesium is higher in the Series than copper, so it forms the ions, and the copper ions turn into atoms.

It is a redox reaction:

$$Mg \longrightarrow Mg^{2+} + 2e^- \quad \text{(loss of electrons − oxidation)}$$

$$Cu^{2+} + 2e^- \longrightarrow Cu \quad \text{(gain of electrons − reduction)}$$

If a coil of copper wire is placed in a solution of silver nitrate, silver crystals of silver metal soon begin to grow on the copper coil:

$$Cu(s) + 2AgNO_3(aq) \longrightarrow 2Ag(s) + Cu(NO_3)_2(aq).$$

## 18.14 The stability of metal salts to heat

The position of a metal in the Activity Series can tell you how stable some of its salts will be when they are heated. Figure 12 shows how this applies to carbonates and nitrates.

| metal | carbonate | nitrate |
|---|---|---|
| **potassium**<br>**sodium** | These carbonates do not decompose when heated. | These nitrates form oxygen + a nitrite:<br>$2KNO_3 \longrightarrow 2KNO_2 + 3O_2$. |
| **magnesium**<br>**aluminium**<br>**lead**<br>**copper** | These carbonates decompose. The ease of decomposition increases as you go down the Series.<br>carbonate $\longrightarrow$ oxide + carbon dioxide | These nitrates give: oxide + oxygen + nitrogen(IV) oxide.<br>e.g.:<br>$2Pb(NO_3)_2 \longrightarrow 2PbO + 4NO_2 + O_2$. |
| **mercury**<br>**gold** | | These nitrates give metal + nitrogen(IV) oxide + oxygen<br>e.g.:<br>$Hg(NO_3)_2 \longrightarrow Hg + 2NO_2 + O_2$. |

**Figure 12** *Stability of salts for metals in the Activity Series.*

## 18.15 Extraction of metals from their ores

Once again, the way in which a metal is extracted from its ore depends upon the position of the metal in the Activity Series.

Very reactive metals like potassium, calcium, sodium, magnesium and aluminium have to be extracted from their molten ore by electrolysis.

Less reactive metals like zinc and iron can be extracted from their ores by reduction with carbon monoxide or carbon.

The least reactive metals can be extracted from their ores by just roasting the ore in air at a high temperature.

In each case, the extraction of the metal is a reduction, because metal ions in the ore have to be changed into metal atoms:

$$M^{n+} + ne^- \longrightarrow M.$$

For example: $Zn^{2+} + 2e^- \longrightarrow Zn$.
This is gain of electrons – reduction.

Look at the examples that follow. You will see that the most easily extracted metals are considered first. The most difficult metals to extract come last.

## 18.16 Copper

The main ore of copper is a sulphide which has the formula $CuFeS_2$. It is mined mainly in Zambia and Canada.

First the ore is crushed and the powder is added to a vat of oily, frothy water which is being constantly stirred. The ore floats on the froth while the impurities of stones and sand sink to the bottom. This technique is known as froth flotation.

The purified ore is dried and roasted in air in a furnace. Limestone is added. Several complicated reactions take place, but overall, the iron part of the ore reacts with silica in the brick lining of the furnace and forms a slag:

$$FeO(s) + SiO_2(s) \longrightarrow FeSiO_3(l).$$

The copper sulphide part of the ore is burnt to form sulphur dioxide and copper:

$$CuS(s) + O_2(g) \longrightarrow SO_2(g) + Cu(l).$$

The impure, molten copper is run off into moulds and is purified by electrolysis.

Plates of impure copper are put into tanks of acidified copper(II) sulphate solution alongside other plates of pure copper. The impure copper is connected to the positive electricity supply, and the pure copper is connected to the negative electricity supply. (See figure 13.)

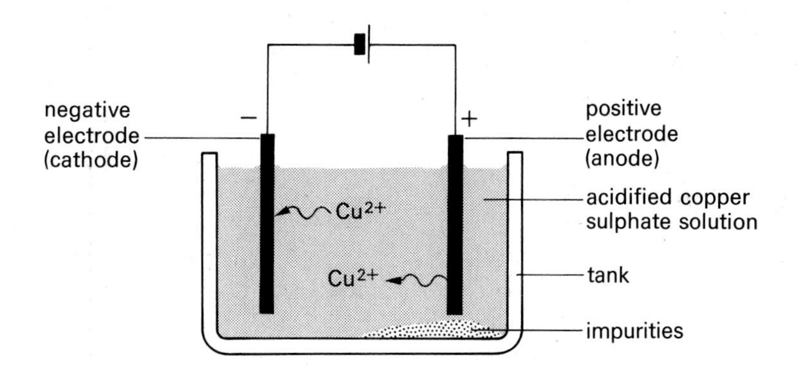


**Figure 13**
*The purification of copper by electrolysis.*

When the electricity is switched on, the impure copper dissolves:

$$Cu(s) \longrightarrow Cu^{2+}(aq) + 2e^-.$$

The impurities from the copper form a sludge at the bottom of the tank. This is very valuable because it contains small quantities of silver and gold.

At the cathode, copper ions are deposited as pure copper, making it thicker and thicker, as the anode gets thinner and thinner.

Copper is becoming an increasingly expensive metal. Because it is such a good conductor of electricity it is used for wires in electrical cables, but its use in water piping is becoming less common.

## 18.17 Iron and steel

The main ores of iron are:

magnetite $Fe_3O_4$
haematite $Fe_2O_3$
and siderite $FeCO_3$

Although this country has some iron ore, it is of rather poor quality, so we import a lot of high quality ore from Sweden.

The iron ore is reduced in a blast furnace. (See figure 14.) The blast furnace is about 30 metres high and is made of steel. It is lined with fireproof bricks on the inside.

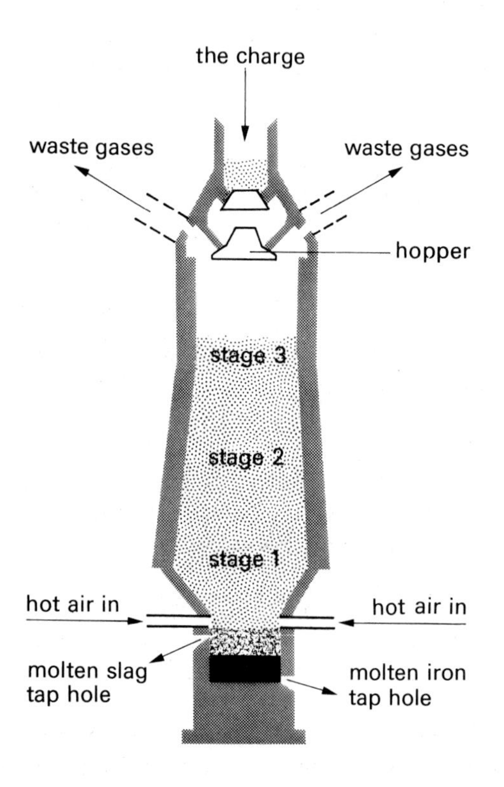

**Figure 14**
*A blast furnace.*

A mixture of the ore haematite, limestone (calcium carbonate) and coke (specially prepared from high quality coal) is put in at the top of the furnace by means of a hopper. This mixture is called the charge.

Hot air is blown in through the base of the furnace through pipes. In the lower part of the furnace, at Stage 1, the coke burns in the air to form carbon dioxide:

$$C(s) + O_2(g) \longrightarrow CO_2(g).$$

This is an exothermic reaction and the temperature at this part of the furnace can be as high as 1 900 °C.

As the carbon dioxide moves up through the furnace, more carbon reduces it to carbon monoxide:

$$CO_2(g) + C(s) \longrightarrow 2CO(g).$$

This reaction is endothermic, so at Stage 2, the temperature has fallen to about 1100 °C.

At Stage 3, the carbon monoxide reduces the iron ore to form iron and carbon dioxide:

$$Fe_2O_3(s) + 3CO(g) \longrightarrow 2Fe(l) + 3CO_2(g).$$

The temperature of this reaction is about 600 °C. The waste gases continue to rise through the furnace. They contain nitrogen (from the air), hydrogen (from reactions with any water), carbon monoxide, and carbon dioxide. They are very useful because after cleaning, there is enough hydrogen and carbon monoxide in the gas for it to be burned and used to heat the air which is pumped in through the pipes at the bottom of the furnace. As little as possible of the heat is wasted. The molten iron formed in the reduction of the iron ore trickles down to the bottom of the furnace. However, it is very impure, and contains quite large quantities of carbon and silica (sand). This is where the limestone is used. It has already decomposed in the hot part of the furnace to form calcium oxide and carbon dioxide:

$$CaCO_3(s) \longrightarrow CaO(s) + CO_2(g).$$

The calcium oxide (a basic oxide) reacts with the silica (an acidic oxide) to form a salt called calcium silicate:

$$CaO(s) + \underset{\text{silica}}{SiO_2(s)} \longrightarrow CaSiO_3(l).$$

This salt is formed in a molten state and is called *slag*. The slag runs down to the bottom of the furnace where it floats on the surface of the iron.

When sufficient iron has been formed, both the slag and the iron are tapped off through separate holes. The slag, when solidified, is broken up and used for road making. The iron is run off into moulds to form ingots of *cast iron*.

Cast iron is a very impure form of iron because it still contains up to 4% of carbon. This carbon makes the iron very brittle and hard, so that it will snap when bent or stretched. It is used to make objects which do not have to take great loads, but which need to be cheaply cast into exact shapes. Two examples of the use of cast iron is for the manufacture of bunsen burner bases, and for car engine blocks.

Some cast iron is made into wrought iron, in a process called 'puddling'. The cast iron is partly melted, and then is stirred, so that some of the carbon is oxidised out of the metal. Then the hot iron is squeezed, hammered, and thumped between giant rollers and hammers so that the remaining carbon is literally squeezed to the surface and ejected as scale. Wrought iron is much softer and can be worked and bent without danger of breaking. It is used to make chains, bolts and ornamental metalwork, such as garden gates.

The majority of cast iron is turned into steel. To make steel, all the impurities have to be removed from the iron first.

In the *Linz-Donawitz* process, up to 300 tonnes of iron at a time are melted in big brick lined vats called converters. Oxygen under high pressure is then blown onto the surface of the iron and calcium oxide is added along with a large quantity of scrap iron.

The converter rotates as the oxygen is blown in, for about an hour. During this time, the impurities of carbon and phosphorus which are still in the iron are blown out as gaseous oxides. The other impurity, silica, reacts with the calcium oxide and forms slag, which is skimmed off. Finally, the process is carefully controlled so that only a small but exact amount of carbon is left. It is now *steel*.

Very hard steel, such as that used for making tools like files and drills, contains between 0·7 and 1·5% of carbon.

Other steels, which are softer, contain smaller amounts of carbon down to about 0·2%.

In addition, other metals can be put into the steel while it is in the converter to make different types of steel:
Armour plated steel contains manganese.
Tool steel contains tungsten.
Ball bearing steel contains chromium.
Stainless steel contains chromium and nickel.
Magnet steel contains cobalt.

## 18.18 Aluminium

Like sodium, aluminium is extracted from its ore by electrolysis. The ore is bauxite ($Al_2O_3 . 2H_2O$). Bauxite has a very high melting point, so it is first dissolved in molten cryolite, ($Na_3AlF_6$), another ore of aluminium. The cryolite acts as a solvent so that the ions in the bauxite may be set free. The electrolysis cell is shown in figure 15.

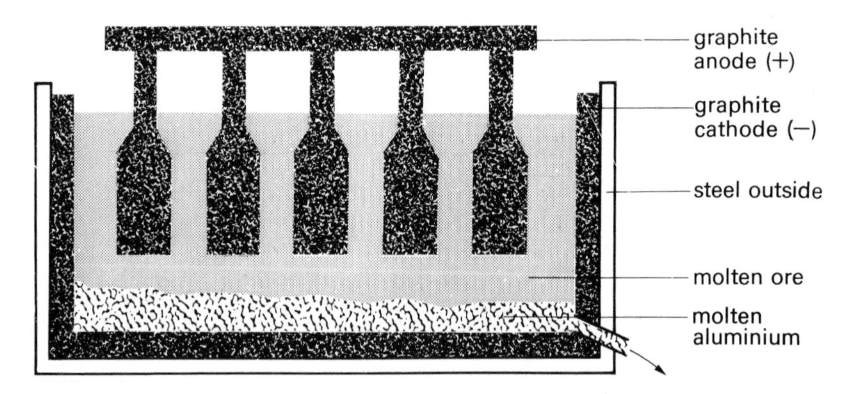


**Figure 15** *The cell for the production of aluminium.*

The cell is lined with graphite, which acts as the cathode. The anode dips down into the molten mixture of ores. This mixture contains two important ions from the bauxite. They are $Al^{3+}$ and $O^{2-}$.

When the electricity is switched on, the positive ions are attracted towards the cathode. Here they are turned into aluminium:

$$Al^{3+} + 3e^- \longrightarrow Al(l).$$

The molten aluminium collects at the bottom of the cell and is tapped off. The negative oxygen ions are attracted towards the positive anode where they turn into oxygen molecules:

$$2O^{2-} \longrightarrow O_2 + 4e^-.$$

Aluminium is a very important metal. It is used in the manufacture of aeroplanes, trains and buses, electric cables, cooking foil – any application where a lightweight metal is needed.

## 18.19 Sodium

Sodium is extracted by the electrolysis of molten sodium chloride. The cell that is used is about 3 metres high and is made of steel. Figure 16 shows the cell cut down the middle.

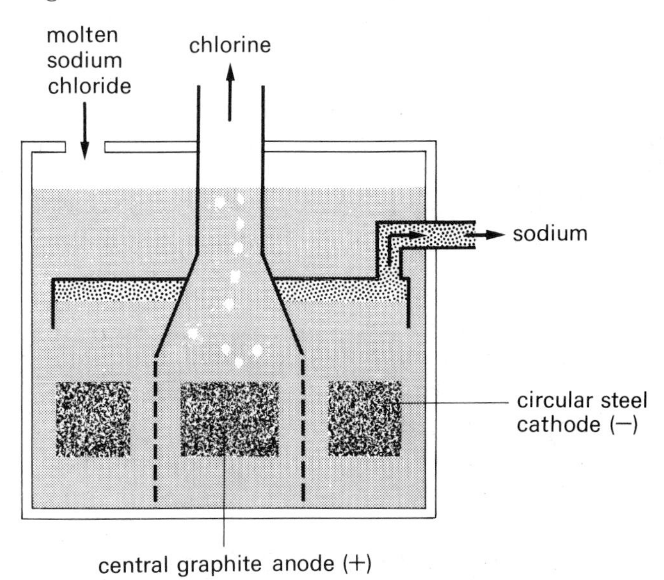


**Figure 16**
*The cell for the production of sodium.*

Inside is a graphite anode, surrounded by a circular steel cathode. In between the electrodes is molten sodium chloride which is made of sodium ions and chloride ions:

$$NaCl \longrightarrow Na^+ + Cl^-.$$

The positive sodium ions are attracted to the negative steel cathode, where they turn into molten sodium metal. This is sucked out of the cell and cooled to a solid.

$Na^+ + e^- \longrightarrow Na.$ This is reduction.

The negative chloride ions are attracted to the positive graphite anode. The extra electrons which made them be ions are taken away, so that chlorine gas is made. This is withdrawn from the cell and used for other industrial purposes.

$2Cl^- \longrightarrow 2Cl + 2e^-.$ This is oxidation.

$$2Cl \longrightarrow Cl_2(g).$$

## 18.20 Alloys

Alloys are mixtures of metals. Pure metals are often soft and easily bent or distorted. This is because the regular arrangement of atoms in a metal can allow the layers of atoms to slide over each other. (See figure 17.)

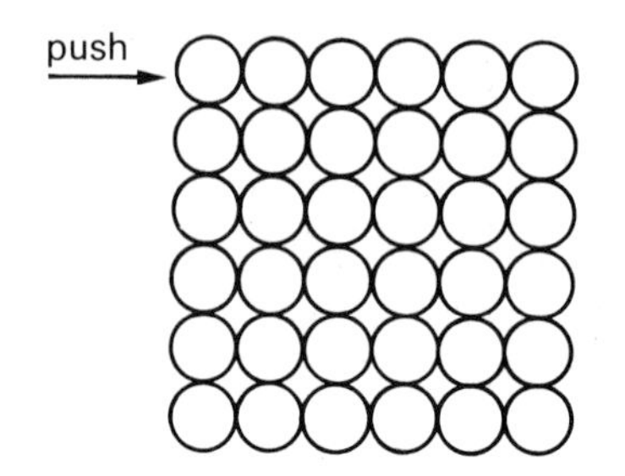


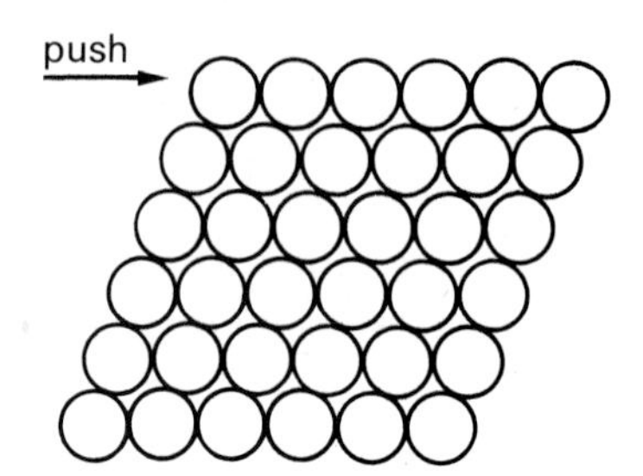


**Figure 17**
*Layers of atoms can slide over each other.*

Sometimes this is a useful property. It enables gold to be beaten into thin foil, and copper to be drawn into wires. However, it would be a great disadvantage if you wanted to build a bridge or an aeroplane.

In an alloy, a carefully calculated amount of another element, usually a metal, is added so that these new atoms form a sort of wedge for the pure metals' structure, making it stronger and less easily distorted. Too much can destroy the crystal structure of the metal and make it even weaker.

There are many examples of alloys in everyday use. (See figure 18.)

| alloy | use | composition |
|---|---|---|
| **steel** | girders, bridges, etc. | iron and about 1% carbon |
| **solder** | used for 'welding' electrical wires together. | equal amounts of lead and tin. |
| **brass** | electrical connections and machine bearings. | 60% copper and 40% zinc. |
| **bronze** | machine parts. | 90% copper and 10% tin. |
| **cupronickel** | for 'silver' coins. | 75% copper and 25% nickel. |
| **constantan** | for electrical wires in thermocouples. | copper and nickel. |
| **duralumin** | a light alloy for aircraft parts. | 95% aluminium, 4% copper and some magnesium, manganese, iron and silica. |

**Figure 18** *Some common alloys and their uses and composition.*

Sometimes an alloy will have a much lower melting point than any of its components. One example is solder. Another amazing example is Woods metal. This is made of bismuth, lead, tin and cadmium. It has a melting point of 71 °C, so it will melt in hot water! It was used in the newspaper industry as an easily melted type metal.

New alloys are often made for specific purposes. When Concorde was first being designed, a new alloy had to be made that was light, but which would withstand the very high temperatures that the outside of the aircraft would reach when flying supersonically. Recently, in 1976, a Russian pilot landed his Mig fighter in Japan. This was a new aeroplane that had never been closely studied by the Western countries and almost certainly the first thing that Western scientists did was to examine the alloys from which it was made.

## 18.21 Corrosion

Metals and alloys corrode, (although one property of an alloy, like brass, may be that it is less easily corroded than a pure metal). Only the noble metals such as silver and gold, will *not* corrode in air and water.

Lithium, sodium and potassium have to be kept under oil to protect them from air and moisture. Nevertheless, they still become corroded very soon.

Magnesium and calcium are usually covered with a thin coating of oxide, and pink, new copper soon becomes brown as copper oxide forms. Think of newly minted pennies. In the open air, copper covered roofs turn green because of the formation of *verdigris*, a mixture of copper carbonate and copper sulphate.

Sometimes corrosion on metals is an advantage, because it will protect the metal underneath once it has been formed.

However, iron's form of corrosion – rust – is never thought of as beneficial.

**Rust.** Rust forms whenever air and water attack iron together. The rust eats into the metal and unless it is completely removed from the iron it will go on developing, even below a layer of paint. Figure 19 shows an experiment that may be performed to demonstrate that both water and air are needed for rusting.

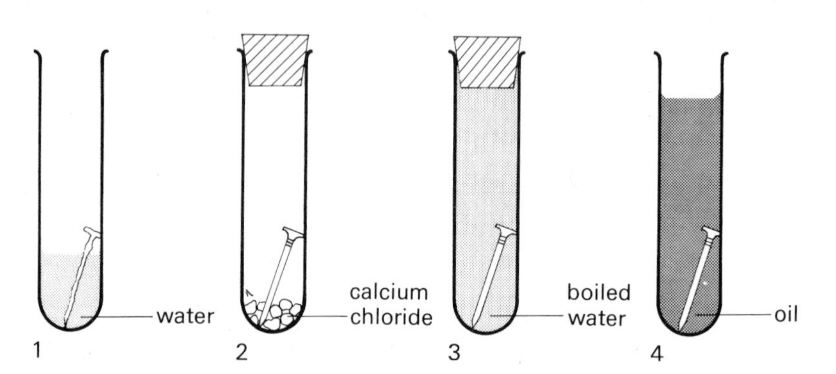


**Figure 19**
*A rusting experiment.*

Test tube 1 contains an iron nail, open to the air and immersed in water. The nail will soon go rusty.

Test tube 2 contains an iron nail in dry air. The calcium chloride removes any moisture. The nail does not go rusty.

Test tube 3 contains air free, boiled water. The nail does not go rusty.

Test tube 4 contains an iron nail completely immersed in dry oil. The nail does not go rusty.

Both air and water are needed to make iron rust. If either one of these is excluded, the metal is protected.

## 18.22 Ways of protecting iron from rust

**Paint the iron, or keep it coated in oil.** Bridges and iron railings have to be painted if they are to last. Machine parts cannot be painted, so they must be kept oily.

Similarly iron or steel parts in instruments such as cameras or typewriters have to be protected from damp air. They cannot be painted or oiled, so very often, a small bag containing silica gel, a water absorbing material, is put inside the instrument case.

**Galvanising.** Articles made of iron may be dipped into a bath of molten zinc after they are made. This leaves a thin coating of zinc on the surface of the iron. This process is called *galvanising*. The zinc provides a protective coating. It will of course corrode itself, but after an initial layer of zinc oxide has been formed, corrosion will stop. Paints containing zinc can be bought as undercoats to put on iron and steel before they are painted with ordinary paint.

However, the zinc layers may be scratched, and the iron could be exposed. (See figure 20**a**.)

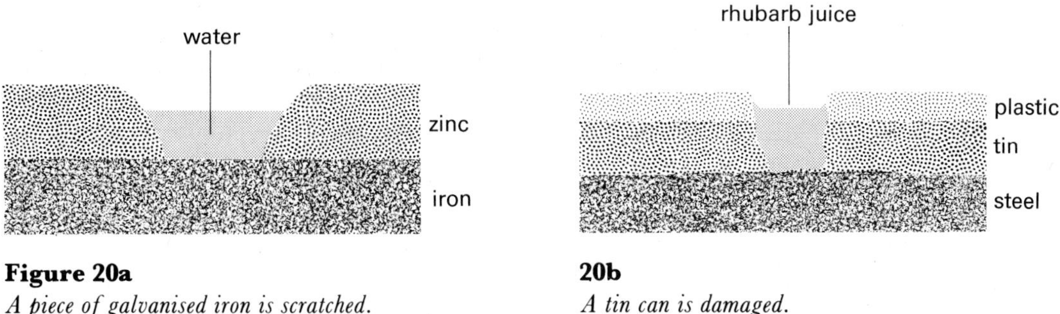


**Figure 20a**
*A piece of galvanised iron is scratched.*

**20b**
*A tin can is damaged.*

Zinc is higher in the Activity Series than iron, so it forms ions more readily. Also, the zinc and iron are in contact with water, so an electrical cell is set up. The zinc reacts in preference to the iron, to form zinc ions. The iron does not react, thus remaining intact.

**Tinning.** Tin cans are really made of steel. Only the inside of the can is coated with a thin layer of tin. This in turn, is often covered by a plastic material. The plastic should prevent any corrosion at all. (Remember that the things that are put into cans are often quite acidic – think of rhubarb.) If the plastic does become scratched, the tin comes into contact with the contents, so it corrodes. Tin is not a very reactive metal, and so it reacts very slowly. If the tin coating is damaged as well, the iron becomes exposed. (See figure 20**b**.)

This time, the iron in the steel is higher in the Activity Series than the other metal, so the iron reacts with the rhubarb juice. This may cause a dangerous build up of gas, and poisonous substances may get into the food. Beware dented tins.

**Gas and water anodes.** How often have you seen a small stone slab at the side of the road, or on a grass verge, with the inscription 'gas anode' or 'water anode'? Never? Have a good look around. This is the Gas or Water Board's way of protecting their pipes from rusting. The pipes, made of steel, are painted, but after a time, corrosion is bound to set in. It costs money to dig up pipes and replace them, so, under the 'anode', they put a small bag or rod of magnesium. This is connected to the pipe by a wire, so that a small electrical cell is set up. Magnesium is higher in the Activity Series than iron, so that it corrodes and not the iron pipe. The magnesium can be renewed easily. (See figure 21.)

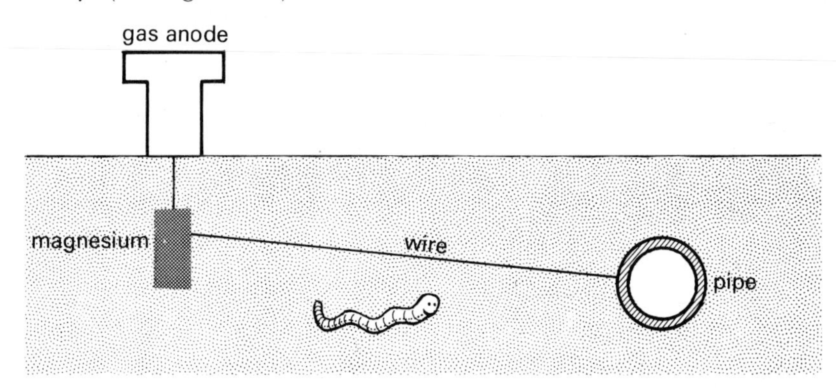


**Figure 21**
*Gas and water anodes protect pipes.*

## 18.23 Groups of elements: the alkali metals of Group I

In a much earlier chapter, you read that elements in the same group of the periodic table have the same number of outside shell electrons. This gives them similar physical and chemical properties. This is certainly true of Group VII, the Halogens, as you saw in chapter 14. Another good example is the metals of Group I. They all have one outside shell electron:

| | | | | | |
|---|---|---|---|---|---|
| Li | lithium | 2. | 1. | | |
| Na | sodium | 2. | 8. | 1. | |
| K | potassium | 2. | 8. | 8. | 1. |

They are very much alike physically:

**a** They are all soft enough to be cut with a knife, although their softness increases as you go down the Group.

**b** They are all kept under oil to protect them from air and moisture. Nevertheless, they all corrode. When cut they are silver on the inside, but this tarnishes in a matter of seconds.

**c** They are all lighter than water – they all float.

**d** They all have low melting points:

Li = 186 °C
Na = 98 °C
K = 63 °C.

**e** They are all good conductors of electricity.
We have seen many of their chemical reactions already. Let's summarise them:

**f** They all burn in air with coloured flames to form basic oxides. Lithium burns with a crimson flame:

$$4Li(s) + O_2(g) \longrightarrow 2Li_2O.$$

Sodium burns with a bright yellow flame:

$$2Na(s) + O_2(g) \longrightarrow Na_2O_2(s).$$

Potassium burns with a lilac flame:

$$K(s) + O_2(g) \longrightarrow KO_2(s).$$

**g** They all react violently with water to form hydrogen and a solution of an alkali. For example:

Lithium fizzes and gets smaller as it gives off hydrogen. The solution that is left is alkaline lithium hydroxide:

$$2Li(s) + 2H_2O(l) \longrightarrow 2LiOH(aq) + H_2(g).$$

Sodium melts, and rushes around as it fizzes violently, evolving hydrogen. Alkaline sodium hydroxide is left in solution:

$$2Na(s) + 2H_2O(l) \longrightarrow 2NaOH(aq) + H_2(g).$$

Potassium melts, and reacts so violently that the hydrogen which is evolved catches fire, and burns with a lilac flame. The solution that is left is alkaline potassium hydroxide:

$$2K(s) + 2H_2O(l) \longrightarrow 2KOH(aq) + H_2(g).$$

**Summary**

At the end of this chapter you should be able to:

**1** Distinguish between metals and non-metals in the periodic table.

**2** Describe the physical and chemical properties of metals and compare them with those of non-metals.

**3** Explain why metals are good conductors of heat and electricity.

**4** Write down the common metals in the order in which they appear in the Activity Series.

**5** Explain how the reactions of metals with air, water and dilute acids follow the order of the Activity Series.

**6** Compare the reducing powers of hydrogen, carbon, carbon monoxide, and other metals on metal oxides.

**7** Describe what is meant by a displacement reaction in terms of redox and the Activity Series.

**8** Show that the extraction of metals from their ores is a reduction.

**9** Describe the production of copper from its ore.

**10** Describe the production of iron and steel.

**11** Describe the extraction of aluminium from bauxite.

**12** Describe the extraction of sodium from sodium chloride.

**13** Describe the advantages of alloys over pure metals and give examples of some common alloys.

**14** Explain the cause of rusting in iron and describe the ways in which it may be prevented.

**15** Show that the members of Group I of the periodic table are very similar in physical and chemical properties.

# Concorde– and Samurai!

*Concorde, and the ancient Japanese Samurai swordsmen, have at least one thing in common. Concorde has problems slicing through air. The Samurai had problems slicing through something else . . .*

'Nose down'. The co-pilot lifts the nose control and moves it to the fully lowered position. In front, the nose slides down below the windscreen lower edge, improving the already good visibility.
'1000 feet.' The Engineer Officer calls out the altimeter readings as we close in. The approach continues.
'100 feet'
'50'
'40'. A touch of the button on numbers 1 and 4 throttles and the auto-throttle disconnects.
'30'
'20'. A slight ease back on the control column stops the nose lowering too quickly.
'15'
'Close the throttles, not slowly, but steadily', and Concorde sits down nicely on the runway. Straight into idle reverse, lower the nose wheel gently, and then turn the engines to full reverse thrust.
'110 knots'. The outer engines idle.
'75 knots'. All engines idle.
'50 knots'. Concorde hardly appears to be moving, but the Horizontal Situation Indicator makes sure we do not taxi too fast. We nose into Bay 7, and one more successful Concorde flight is completed.

**Best of its kind in the world**
Concorde is the result of British and French aeronautical technology, and it is the best of its kind in the World. It cost a great deal of money to design and build – millions and millions of pounds, and lots of problems had to be solved.

For example, most jet passenger planes fly at five or six hundred kilometres per hour. That's pretty fast, but at top speed, Concorde reaches twice the speed of sound – about 2400 km per hour. At this speed, the outside of the fuselage reaches a temperature of 150 °C. You could cook your dinner on it. This high temperature is caused by resistance between the molecules of oxygen and nitrogen in the air, and the metal of the plane as it rushes along. When meteorites enter the Earth's atmosphere at high speed, they catch fire. It would be disastrous if Concorde did this, so it cannot be made of thin steel plate or aluminium like any ordinary aircraft.

**New alloys designed**
Instead, a new alloy of steel and titanium was designed to withstand the high temperatures that Concorde would experience. That meant new tools to cut and form the new metal, and rigorous tests to see how it would behave when it was made to fly. Was it light enough? Was it strong enough?

The wings and tail plane of Concorde are very big and titanium steel would have been too heavy for them. Instead, a new aluminium alloy was made that was light enough, but which would still withstand the cruel changes of temperature from near freezing to 150 °C that the aeroplane would have to go through. In addition, Concorde's wings droop when it is on the ground, and are lifted up by the pressure of the air rushing over them when it is flying. The alloy had to be strong enough to withstand this flexing.

**Planes of the future**

Research into new alloys for aeroplanes has not stopped there. An American aircraft company is making experimental aeroplanes which are constructed from aluminium and graphite, steel, tungsten and fibre glass. These substances are made into mixtures and alloys and they are much lighter and cheaper than conventional alloys. They can be shaped more easily into the aeroplane parts.

The wings and engine mountings are made of graphite and fibre glass and the main body of the aeroplane is manufactured from titanium and aluminium alloys. These experimental test planes will be carried up to 15 000 metres under B52 bombers and then released. They will fly at the speed of sound and when they turn in the air, forces eight times as strong as that of gravity will build up in the body of the aeroplane. These will be the fighter planes of the future.

**Samurai!**

The process of making a Samurai sword reflects the delicate control of carbon and of heat treatment by which a steel object is made to fit its function perfectly. Even the piece of steel is not simple, because a sword must combine two different and incompatible properties of metals. It must be flexible, and yet must be hard. Those are not properties which can be built into the same material unless it consists of layers. In order to achieve this, the piece of steel is heated and cut, and then doubled over. It is hammered and folded again and again – fifteen times, so that within the steel, more than thirty thousand layers have been created. Each layer must be bound to the next, which has a, different property. It is as if the craftsman were trying to combine the flexibility of rubber with the hardness of glass, and the sword, essentially, is an immense sandwich of these two properties.

At the last stage, the sword is covered with clay to different thicknesses, so that when it is heated and plunged into water it will cool at different rates. The temperature of the steel for this final moment must be judged precisely, and it was the practice of the Japanese Samurai craftsmen to watch the sword being heated 'until it glows to the colour of the morning sun.'

The climax, not so much of drama as of Chemistry, is the quenching, which hardens the sword and fixes the different properties within it. Different crystal shapes and sizes are produced by the different rates of cooling: large smooth crystal at the flexible core of the sword, and small jagged crystals at the cutting edge.

But the test of the sword is 'Does it work?' 'Can it cut the human body in the way that Samurai ritual lays down? The body is replaced by a bale of straw nowadays, but in the past, a new sword was tested more literally, by using it to execute a prisoner.

## Questions

**1** Think of as many reasons why:

**a** You couldn't make electric wires out of phosphorus.

**b** You can't build bridges out of graphite. (Or even diamond for that matter.)

**c** You can't make bells out of sulphur.

**2** Although you may never have seen them, or read about them before, what sort of reactions would you expect between:

**a** Gallium and dilute hydrochloric acid.

**b** Caesium and oxygen.

**c** Selenium and oxygen.

**d** Chlorine oxide and water.

Look at the positions of these elements in the periodic table in figure 1. This will help you to decide whether they are metals or non-metals. Then you may have to do some revision in other chapters. Don't forget the index.

In the case of reactions **a**, **b**, and **c**, write word equations for what is happening.

**3** Explain, with diagrams, why a piece of iron will conduct electricity, but a piece of sulphur will not. Why don't gases conduct electricity? How does their structure prevent them?

**4** Without looking back in the chapter, put these elements into the correct order in the Activity Series:

copper potassium zinc magnesium.

Describe the reactions of these metals with **a** air, **b** water and **c** dilute hydrochloric acid. Does this fit in with the order you gave?

**5** Iron will react with steam, but only at a very high temperature. Write the equation for this reaction.

The apparatus in figure 8**b** is no good for the reaction, because the glass will melt before the required temperature is reached.

Design a better apparatus for the experiment. You will need: something in which to produce the steam, something in which to heat the iron, something in which to collect the hydrogen after it has been produced.

Draw a diagram of the assembled apparatus and describe how it would work.

**6** Below is a list of possible reactions, but you are only given the left hand side of the equation. In each case:

say if the reaction will work,
say whether heat is needed,
write the equation using symbols and formulae.

**a** calcium + dilute hydrochloric acid $\longrightarrow$

**b** copper + dilute sulphuric acid $\longrightarrow$

**c** gold + oxygen $\longrightarrow$

**d** sodium + oxygen $\longrightarrow$

**e** copper + oxygen $\longrightarrow$

**f** carbon + magnesium oxide $\longrightarrow$

**g** carbon monoxide + silver oxide ⟶
**h** carbon + copper oxide ⟶
**i** hydrogen + aluminium oxide ⟶
**j** carbon monoxide + potassium oxide ⟶
**k** magnesium + lead oxide ⟶
**l** hydrogen + mercury oxide ⟶
**m** copper + zinc oxide ⟶
**n** magnesium + lead nitrate solution ⟶
**o** silver + zinc sulphate solution ⟶

From these reactions, choose: an exothermic reaction, a redox reaction, a synthesis reaction, a displacement reaction.

**7** How easy do you think it would be to extract sodium from molten sodium chloride in the laboratory?

If you were given a crucible, a large battery, some wires, two graphite rods and a high temperature bunsen burner, draw the apparatus that you would use. Explain what would happen when you started the experiment.

The melting point of sodium chloride is 801 °C and the melting point of sodium is 97 °C. What state will the sodium be produced in? What will happen when the sodium floats to the surface of the molten sodium chloride during the reaction?

**8** Explain:
**a** how cast iron is made from iron ore,
**b** how wrought iron is made,
**c** how steel is made.

Why are there so many different types of steel? Give one example and explain how it is made and what it is used for.

**9** Why is rust different from other types of corrosion? Write short notes on how rust is prevented by:
**a** galvanising,
**b** tin-plating,
**c** excluding air and water,
**d** burying bags of magnesium.

**10** What is an alloy? Why are alloys used instead of pure metals? Give an example of a commonly used alloy. What does it contain, and what is it used for?
See how many different alloys you can find in the laboratory, and at home. You could start with your bike, or Dad's car. You can tell them by different appearance, density, and tendency to rust.

**11** Describe the experiment you would perform to show that air and water are both needed to make iron rust. Say what apparatus you would use, and explain the reasons for the things you would do.

**12** Why should you beware of dented tins of rhubarb?

**13** Elements which have the same number of outside shell electrons are said to have similar physical and chemical properties.
Is this true for **a** the Halogens **b** the Alkali metals?
Give lots of reasons for your answers.

**14** Here are some clues about some mystery elements. Can you say whether each one is a metal or a non-metal?

**a** is a gas used to fill light bulbs,
**b** is a solid mixed with lead to make solder,
**c** has a melting point of 1240 °C,
**d** burns in oxygen to form an oxide which turns damp litmus red,
**e** clangs when you hit it,
**f** will snap if you try to bend it,
**g** will not conduct electricity,
**h** is made into wires,
**i** is a very poor conductor of heat,
**j** reacts with an acid to give a salt plus hydrogen.

**15** Metal A does not react with water, dilute hydrochloric acid, or steam.
Metal B reacts violently with water, and forms an alkaline solution.
Metal C corrodes slowly in the air but only reacts with steam.

**a** Put the three metals in the order in which they would occur in the Activity Series.
**b** Put hydrogen into the list as a reference point.
**c** Which of these metals would oxidise most easily?
**d** Which one would not be reduced by carbon?
**e** Will B reduce the oxide of A?
**f** Will C displace A from a solution of the chloride of A in water?
**g** The nitrate of which of the metals is the most difficult to decompose by heating?

# 19 Fuels

## 19.1 What is a fuel?

Fuels are substances which may be made to do work. When they burn, energy is released. This energy can be in the form of heat, light, electricity or movement. Figure 1 shows one example of this. Energy stored in oil can be changed into heat, light, electric or movement energy:

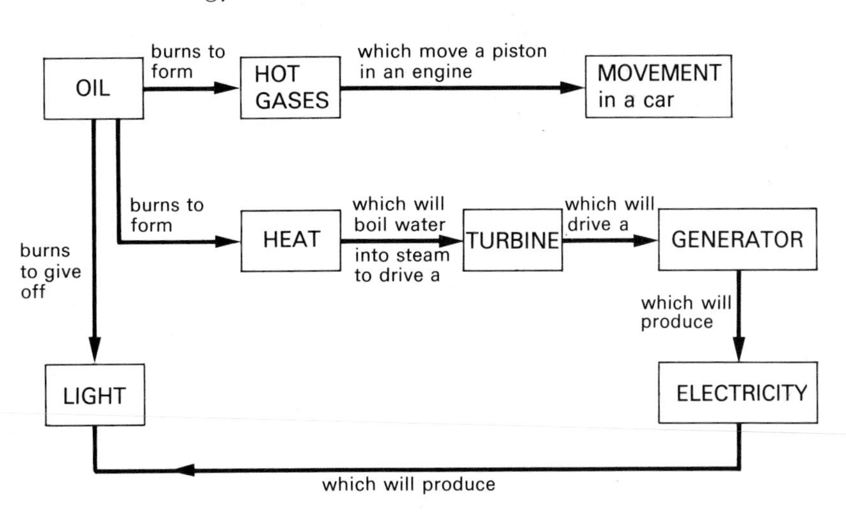


**Figure 1**
*Fuels and energy.*

You can probably think of lots more examples of energy changes, but all of them have the same sort of equation:

$$\text{fuel} + \text{oxygen} \longrightarrow CO_2 + H_2O + \text{energy in the form of} \begin{cases} \text{heat} \\ \text{light} \\ \text{sound} \\ \text{movement} \\ \text{electricity.} \end{cases}$$

## 19.2 What makes a good fuel?

If you were to think of some common fuels you could easily make a list of advantages and disadvantages for each one.
Here are some 'rules' for the ideal fuel:

**1** It must be easily and cheaply mined or manufactured.

**2** It must be easily transported from the place where it is found to the place where it is to be used.

**3** It must burn easily and produce a lot of energy.

**4** It must not give off dangerous fumes.

**5** It must not have a lot of waste products such as ash and smoke.

How many of our fuels satisfy all of these rules? None.

The amount of heat energy that a fuel can produce may be measured in energy units called joules. The larger the number of joules, the better the fuel. Some common fuels are compared in figure 2.

| fuel | heat energy value in joules per kilogram |
|---|---|
| wood | 21 000 |
| coal | 34 000 |
| petrol | 42 000 |
| natural gas | 56 000 |
| bread | 10 600 |
| sugar | 16 510 |
| butter | 31 220 |
| yoghurt | 2 390 |

**Figure 2**
*Some heat energy values for different fuels.*

From these figures you can see that gas will make a hotter fuel than wood, because it can form more heat energy.

Foods are also fuels, for the body. They provide energy and heat and you can see from the figures, that sugar will give you more energy than yoghurt. If you don't use the energy up, you get fat instead.

## 19.3 Coal

Coal is a fossil fuel. It is made from the remains of trees and plants that lived millions of years ago.

Long before man, trees grew, died, and decayed all in the same place. Their remains built up in layers. Over millions of years, this dead organic matter was covered by layers of silt and mud from rivers and rain. Changes took place in the structure of the Earth, and the silt and mud turned into sandstone and shale, compressing the organic matter, slowly changing it into the substance we recognize today as coal. We still have lots of coal in this country, even though it has been mined for many years. Different areas produce coals of different types: old, hard coal is black and bituminous. Younger coal called lignite, is softer and brown. Figure 3 is a map of Great Britain showing the major coal fields.

**Figure 3**
*The major coal fields of Great Britain.*

The coal in this country is mined from about 250 collieries at a cost of £12 per tonne, in 1974. Deep mines are dug, sometimes a kilometre or more deep, and tunnels or 'roads' are cut sideways into the layers, or *seams* of coal. Modern mining equipment is very sophisticated. Huge automatic cutters rip the coal from the seams as the roof of the tunnel is kept up by hydraulically operated supports which move along with the cutters. The cutter travels backwards and forwards between two parallel tunnels, eating into the coal face all the time.

**Figure 4**
*A modern coal cutter and roof supports.*

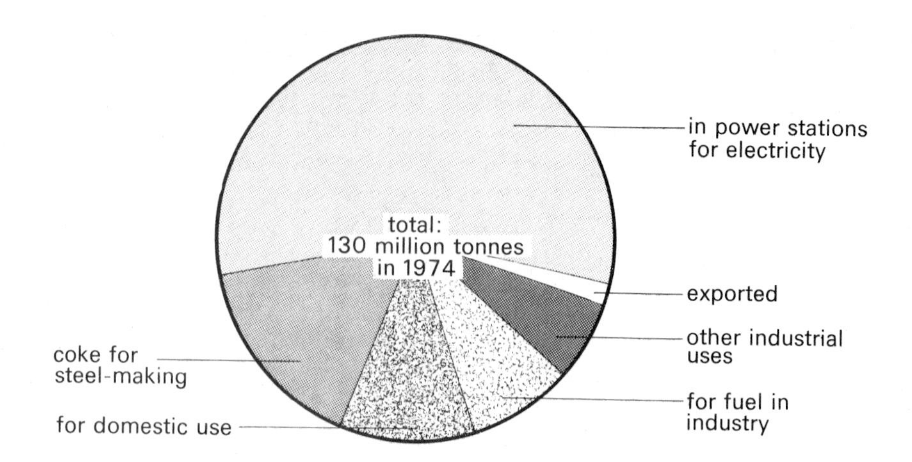


**Figure 5**
*Uses of coal in Britain in 1974.*

## 19.4 Chemicals from coal

Before large quantities of natural gas were discovered under the North Sea, ordinary gas, or Town Gas, was made from coal. The coal was heated to a high temperature in large towers called retorts. It was not allowed to burn because air was not let in. Instead, all that happened was that the volatile parts of the coal were driven out. This process is called *destructive distillation.*

The main products of the destructive distillation of coal were:

**1** Coal gas. This was a mixture of about 50% hydrogen with smaller amounts of carbon monoxide, methane and other gases.

**2** Coal tar. This was a tarry residue from which many chemicals could be extracted. These included creosote and pitch for roads and fences, and tar oil for spraying trees; naphthalene for moth balls and firelighters; phenol for plastics, soaps and disinfectants; drugs, dyes and many other compounds could also be made.

**3** Ammoniacal liquor. This was an aqueous product which was condensed from the gas. It contained ammonia from which fertilizer could be made.

**4** Coke. This is the solid residue left after everything else had boiled off. Different types of coal produce different qualities of coke. Steel making coke is made from hard bituminous coal. Other types of coke are sold as domestic smokeless fuels, like Coalite.

As coal became more and more expensive to mine, some town gas was made directly from oil by passing it over a catalyst at a high temperature. This gave a gas similar in composition to coal gas. However, by 1972, most of Britain's gas supply came from natural gas from the North Sea oil beds. Town gas is no longer made for domestic use, although coal is still destructively distilled to make coke and chemicals.

Nevertheless, coal must not be dismissed as a fuel of the past. Many of our power stations burn coal to produce electricity and as our oil stocks run out, coal could come to be of major importance once again.

## 19.5 Oil

Like coal, oil is a fossil fuel. Millions of years ago, most of the living creatures that existed were in the sea, and were animals with shells. As they died, their bodies fell to the beds of the oceans. This accumulation slowly happened over thousands of years, they became covered with silt and mud. Over millions of years, they were compressed and the organic matter was converted into oil. Changes in the structure of the Earth helped to trap the oil in porous layers of rock (rather like a sponge soaking up water). (See figure 6.)

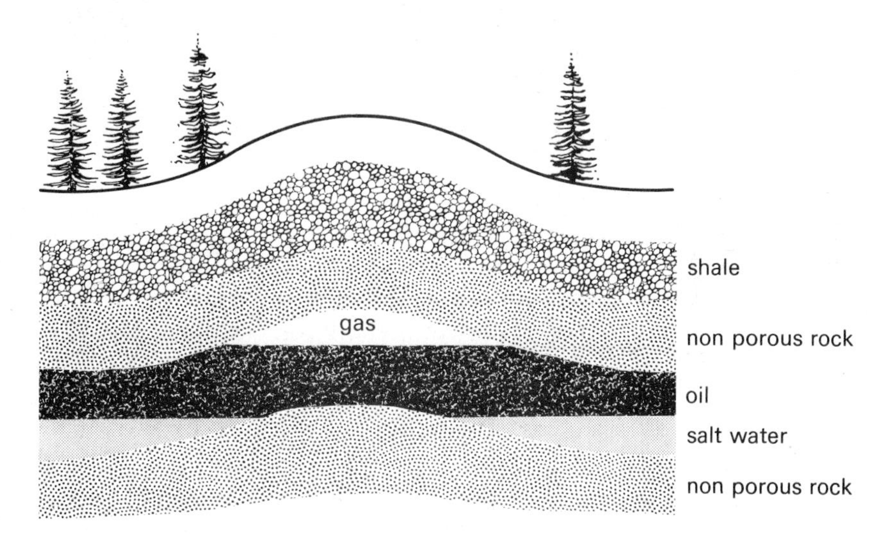


**Figure 6**
*The oil is trapped in a porous layer of rock.*

Geologists can recognize this sort of structure from aerial photographs, or from a seismic survey. This involves setting off a small explosion under the ground and recording the echoes that rebound from the different layers of rock.

Once a possible site for oil has been identified, a test drilling is made. This is a very expensive operation, especially since the well may be dry, or yield only gas. Figure 7**a** shows a land drilling rig. Different types of drill heads have to be used for the different layers of rock. 10-metre lengths of drill pipe are connected together to reach down into the Earth. As they become worn out, the drill heads have to be changed. This may be after only a few metres of drilling if the rock is hard, but the whole string of pipes has to be pulled up and disconnected, section by section. This is very hard and dangerous work. If the oil is located under the sea, the process is much more difficult. An oil drilling platform has to be used, and these alone cost millions of pounds each. (See figure 7**b**.)

**Figure 7a** *A land drilling rig.* **7b** *A North sea oil drilling rig.*

## 19.6 Natural Gas

If the oil well is a good one, the first thing that comes up is gas. Natural gas is made of about 90% methane. Many of our North Sea oil beds produce gas which is piped ashore to be used as a fuel. It is much cleaner than the old coal gas and it has a higher heat energy content. Despite the fact that it is 'free', a lot of money has been spent on finding it and getting it up.

The pressure of this gas will force the oil up to the surface. Once oil has been found, the drilling rig can be removed and replaced by a pump which forces the oil into storage tanks at oil refineries. These are usually built near the coast, so that great super-tankers bringing in imported oil can also discharge their loads. An oil refinery is shown in figure 8.

**Figure 8**
*An oil refinery.*

## 19.7 Oil refining

Oil is a mixture of hundreds of different chemicals. They can be separated initially by fractional distillation. This is the same type of process that was used to separate ethanol from water in chapter 2, but a much larger and more complicated piece of apparatus called a fractionating column is used. Figure 9**a** shows a fractionating column at an oil refinery. Figure 9**b** shows what it looks like from the outside.

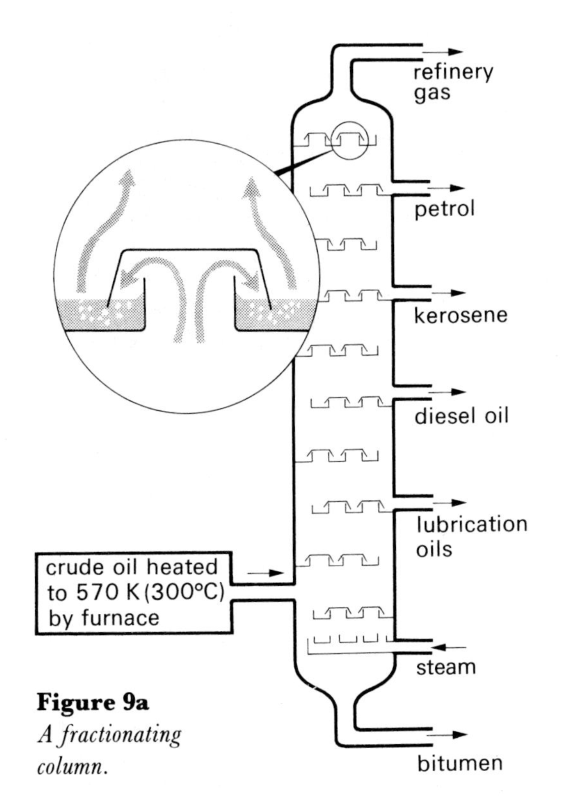


**Figure 9a**
*A fractionating column.*

**Figure 9b**
*A fractionating column.*

The insert which is shown on the left hand side shows a bubble cap. As the vapours are forced up the column they pass through bubble caps. Some of the vapour condenses; more vapour bubbles through the condensed vapour on its way up. This happens many times and the result is that only the volatile, low boiling point liquids reach the top of the column. The heavy, high boiling point liquids continually condense and run down to the bottom. All the way up the column, liquids of different boiling points are collected at different levels.

The heavy, high boiling liquids have large molecules. They are thick, viscous liquids. The light, low boiling point liquids have small molecules – they are runnier and more volatile. They are also more flammable. Figure 10 shows some of the uses of the different liquids that are obtained from the oil. The liquids are called fractions and each fraction is still a mixture of many other chemicals. The fractions are distilled again, and treated in other ways to get many more chemicals out of them.

| fraction | boiling point | size of molecule | uses |
|---|---|---|---|
| Refinery gas | less than 40 °C | 1–4 carbon atoms long | bottled gas such as Calor gas, and plastics. |
| Petrol and Naphtha | between 40 ° and 75 °C | 4–12 carbon atoms | Fuel for cars, chemicals. |
| Kerosine | between 150 ° and 240 °C | 9–16 carbon atoms | Fuel for jets, paraffin. |
| Diesel oil | between 220 ° and 250 °C | 15–25 carbon atoms | Fuel for trains, lorries and tractors. |
| Lubrication oils | between 250 ° and 350 °C | 20–70 carbon atoms | Chemicals, oils, waxes and polish. Fuels for ships and heating. |
| Bitumen | above 350 °C | all the residue left at the bottom of the column | Materials for road making, water-proofing, roofing. |

**Figure 10** *The uses of oil.*

## 19.8 Other sources of energy

Our fossil fuels are not going to last for very long at the rate we are using them.

Oil will probably run out at the beginning of the next century – natural gas may last a little longer.

Coal should last for perhaps another 200 years. What will happen then? How will we run buses and cars; heat our homes; or cook our food? Where will we get chemicals from? What will we do without plastic? (Remember that our metals will not last for ever.)

Already, nuclear power is being used to generate electricity, and a great deal of research is going on into the design of efficient and safe reactors for the future.

One source of energy that was used a great deal at one time is the wind. Ships moved by the energy of the wind which they caught in their sails; windmills converted the wind's energy into power for grinding corn. Scientists are now experimenting with giant propellors which can generate large amounts of electricity.

More recently, experiments have been started to use the power of the waves to generate electricity. Huge floats are put onto the water, and as they nod up and down in the waves, they can work a generator. Experimental buoys have been made that can generate enough electricity to light a lamp on top of them to warn ships, and it is estimated that strings of floats like this, sited round the coast of the British Isles could provide enough electrical energy for most of our present electrical needs.

Some areas of the Earth can get their energy from underground, from hot springs and geysers. This is called geothermal energy. We could get geothermal energy in this country if we drilled holes deep enough in the ground, but it would make the energy very expensive.

Of course, all the energy in our fossil fuels came originally from the sun. The sun made the trees and animals grow. People have come to realize that it may be possible to harness the sun's energy directly. This could be done with photo electric cells like the ones used in the light metres on cameras. They convert light energy into electrical energy. As yet, however, they are not very efficient, and to generate the same amount of electricity as a large power station, you would need many hectares of solar cells. Instead, many people are now considering fitting solar panels to the roofs of their houses. These contain flowing water, which heats up in the warmth of the sun. The water may then be used for domestic purposes. Even on cloudy days, the sun's energy gets through. (See figure 11.)

**Figure 11**
*A solar panel.*

Finally, experiments are going on to try to imitate the way energy is made in the sun. Every second, millions of tonnes of the hydrogen atoms that make up the sun are fused together to make molecules of helium. This gives out an immense amount of heat, so much so that we can feel it 93 million miles away. The experiments are far from complete, but some scientists think that in the middle of the next century we may get all of our energy by this method.

But what will replace plastic when the coal and oil run out?

19.9
## Flames and burning

**Flames.** Flames are chemical reactions that you can see. When a candle is lit, the wax melts and vaporises. The vapour burns in oxygen from the air. The flame is energy produced in the chemical reaction. A candle flame is yellow because much of the carbon in the wax is not burnt, and the tiny particles of carbon in the flame glow brightly. Some flames however, are almost invisible, because everything in the fuel is being burnt and turned into gases.

A flame is a mixture of different reactions, which are taking place at different temperatures and with different amounts of oxygen. This is shown in figure 12.

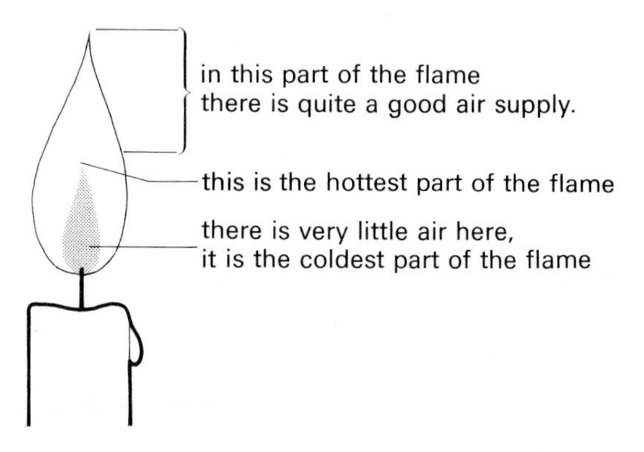


**Figure 12**
*A candle flame.*

**The bunsen burner.** This is a cleverly designed piece of apparatus that was first made 200 years ago. The jet at the bottom must be just the right size so that the speed of the gas rushing up the chimney of the burner is the same as the speed of burning of the flame at the top. The air hole in the collar at the bottom of the chimney controls the amount of air that is mixed with the gas. If the hole is closed, the gas is not mixed with air as it emerges from the chimney. The flame is luminous and sooty like a candle flame. When the air hole is opened, air mixes with the gas, and the flame is almost invisible and much noisier. The flame is a great deal hotter because the combustion in the flame is better. (See figure 13.)

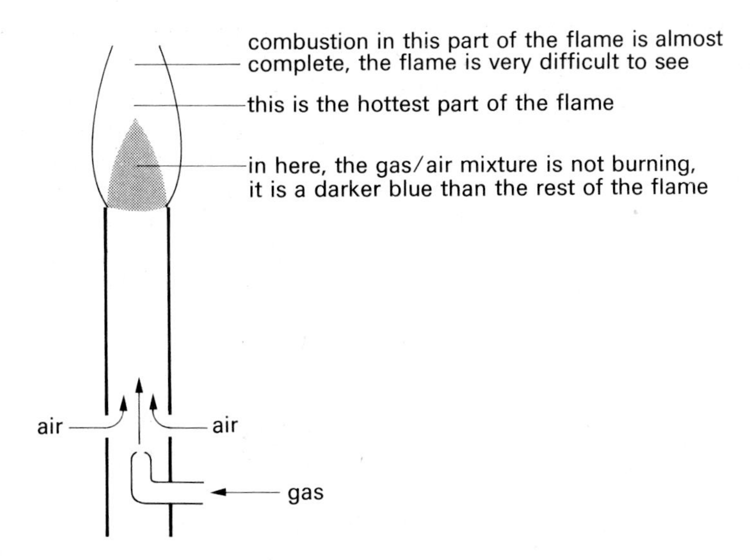


**Figure 13** *The bunsen burner flame.*

**Summary**

At the end of this chapter you should be able to:

**1** Say what a fuel is.

**2** Give some examples of common fuels, and the forms of energy that may be obtained from them.

**3** Give 'rules' for an ideal fuel.

**4** Compare two fuels by means of their heat content values.

**5** Describe how coal was formed.

**6** Give examples of the uses of coal.

**7** Describe the products of the destructive distillation of coal and give some of their uses.

**8** Describe how oil was formed, and outline the method by which it is found, and removed from the ground.

**9** Explain how oil is fractionally distilled, and describe the fractions and their uses.

**10** Discuss possible energy sources for the future.

**11** Describe the flames produced by a candle and a bunsen burner, and say why they are different.

# Fuel for thought

*Bravo cooling off.*

*The world's fuel supply has been boosted by finding new sources of old fuels. Two very different sources of fuels – and the problems of developing them – are illustrated in these two articles.*

**Blow out on Bravo 169**

On 23rd April 1977, an oil rig in the North Sea called Bravo 169 went out of control and started shooting 200 tonnes of crude oil into the sea every hour.

The first sign that something was going seriously wrong was nothing more than a small trickle of mud. The rig was undergoing normal, routine – but slightly dangerous – maintenance during which none of the normal safety valves could be in place. This seemingly rather dangerous state of affairs occurs every two or three years.

If the safety valves are removed, the 3000 metre tube reaching down beneath the sea bed is pumped full of special mud, the weight of which is normally sufficient to plug the well. It was this mud which began to surge upward. Shortly before the blow out started, a control valve had been fitted – but it was installed the wrong way up. The mud began to surge through. It shot out faster and faster, until it was rushing out at 1000 km/h The valve would not work – and nothing could be done to stop the oil flow. The risk of the oil exploding was considerable, so the 120 men on the rig were evacuated and the experts were called in.

**'Boots' Hansen to the rescue**

On April 25th, 'Boots' Hansen and Richard Hallenburger arrived at the rig. They came from 'Red Adair's Oil Well Fire and Blow out Company' – experts in the art of blow out control.

The arguments began as to whose fault the mistake had been – but in the meantime, Boots Hansen still had the job of stopping the oil. An oil slick had already built up. Ships designed like massive vacuum cleaners had been sent in to suck up the spilt oil, but the sea was too rough. One broke down, the other only managed to collect two hundred tonnes of oil.

Hampered both by high winds which forced them off the rig, and low winds which meant that all the oil came straight down onto the rig – the team tried to block off the oil. Over the next week, they tried a total of four times, each time without success. One week and thirty thousand tonnes of oil later, the world's leading experts on the plugging of blow outs had not been able to stop the oil flow. There was only one man left who could be called upon – Red Adair, the greatest of all the trouble-shooters.

**Bring in Red Adair**
The sixty-two-year-old, slightly greying red-haired Adair, admitted that he preferred to stay at home in Texas and play with his grand-children. But even though he had come all this way, he wasn't going to take over from his men. 'I would never take a job off Boots,' he said. 'I'll just go along and get dirty with the rest of them.'

'I've got to be back in Houston to meet some people on Monday, and I'm going to stay here until we finish, so we'll have it done in a couple of days or sooner.'

Within half a day after his arrival, the rig had been tamed. A small adjustment to the machinery Boots had already installed made it possible to slowly close the valve, and the flow of oil simply stopped.

**Big 'uns every time, please**
Red Adair has put out over a thousand oil rig fires. The spurting oil has a tendency to explode – at Bravo 169, special fire fighting tugs sprayed the rig continually with water to reduce the temperature of the hot oil. Red was once blown 45 metres by an oil well explosion. On landing, all he said was 'give me the real big 'uns every time'.

His job is a risky one, but he is quite well paid. It seems that he never works for less than £3000 a day. Anyone looking for a new job?

**The Petrol Tree**
A Nobel Prize-winning chemist who has discovered a tree which he believes could become a major source of petroleum is still seeking government finance for research.

Professor Melvin Calvin, of the University of California, winner of the 1961 Nobel prize for his work on photosynthesis needs £100 000 over three years, to cultivate a three-acre site in southern California on which he wants to grow a variety of the *euphorbia* family, known in the American West as the 'gopher tree'.

**Fuel does grow on trees!**
'I discovered that the gopher tree contains one-third hydrocarbons in its latex or sap. It is much like crude oil and you can do anything with it that you can do with oil – turn it into gasoline, or plastics. Now it is a question of deciding on the best variety of tree and the best method of harvesting and processing.'

The tree, the family of which he decided to investigate while waiting two hours in a petrol queue during an oil shortage, is known botanically as *euphorbia lathyrus*. It is in between bush and tree height, and has thick dark green leaves. Its common name comes from the mole-like rodent, the gopher. Because gophers dislike its oozing sap, the tree is used to ward them off.

**Cultivate semi-desert areas**
As the tree grows well on arid soil he envisages billions of largely uncultivated acres in the south-west United States and in areas such as Iran, Brazil, Australia and West Africa turned into 'Petrol plantations'.

Professor Calvin is reconciled to the difficulty of getting his idea accepted by the American bureaucracy, which he puts down partly to the usual inertia which greets any new idea. 'I thought the discovery should be public; I avoided commercial interests. So now I suffer from trivial blocks in the bureaucracy where everyone has to have approval from someone else. Things like this move so incredibly slowly.'

If research goes well, he believes that the petrol tree could produce oil at a price of 50p per 100 litres.

*The gopher 'petrol tree'.*

## Questions

**1** Explain why oil is a good example of a fuel.

How many different forms of energy may be made starting from a supply of wood? Say what pieces of apparatus, engines or machines would have to be used to make the changes.

**2** Look again at the 'rules' for an ideal fuel. Now think about coal, gas, and wood. Compare how each one satisfies the 'rules'.

**3** If you were given 1 kilogram of coal and 1 kilogram of wood, which one would give you the most heat when it was burned? How do you know?

Imagine you are going on a mountain survival course, in winter, and you were told to bring some emergency rations. Which would you take; bread and butter, sugar, or yoghurt? Explain your answer.

**4** Draw a series of pictures or diagrams to show the way in which coal was formed. Give some reasons why coal is so expensive today.

**5** Before most laboratories were converted to natural gas, town gas was always used as a good reducing agent for laboratory reactions. Why was this? How would Town gas react with hot copper(II) oxide and hot lead oxide? Draw the apparatus you would use for this reaction and write the equations for the reactions that would take place. (Try Chapters 11 and 12 for ideas.)

**6** Do a 'This is your life' story for oil. Start with the way in which it was made millions of years ago, and come up to date with its recent history. Some interesting characters in the story might be the first cave man to find a black sticky liquid oozing out of the ground; a worker on a North Sea oil rig; and a Chemist in an oil refinery. What will you say about the future of oil?

**7** Draw the apparatus the teacher would use to fractionally distil some oil in the laboratory. What precautions would he take? (This would in fact be a dangerous experiment, because some of the fractions might be carcinogenic.)

**8** Some oil was fractionally distilled in a laboratory fume cupboard and these fractions were obtained.

| Fraction | boiling point |
|---|---|
| A | 50 °C |
| B | 200 °C |
| C | 230 °C |
| D | 300 °C |
| E | 400 °C |

Which fraction do you think would:

**a** have the biggest molecules.
**b** be most flammable.
**c** be used to power a tractor.
**d** be most viscous.
**e** be used in jet engines.
**f** burn with the smokiest flame.
**g** be used as a fuel for ships.

# 20 Plastics

## 20.1 Polymers or Plastics?

Plastics are polymers. Let's sort these two words out.

A polymer is a big molecule – usually a long one – made up of one small molecule, altered slightly and then repeated many thousands of times. The small molecule is called a monomer. The big molecule is called a *polymer*. Mono means one. Poly means many. For example:

$$\text{ethene gas} \xrightarrow[200\,^{\circ}\text{C}]{\text{high pressure}} \text{polyethene}$$

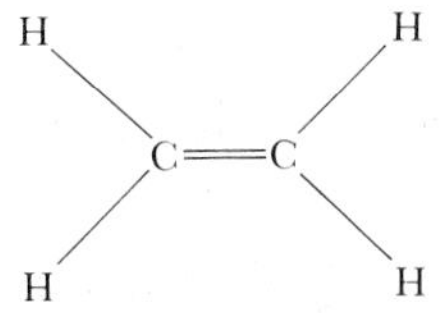

```
H         H
 \       /
  C == C
 /       \
H         H
```

ethene is the monomer

```
     H   H   H   H   H
     |   |   |   |   |
...— C — C — C — C — C —...
     |   |   |   |   |
     H   H   H   H   H
```

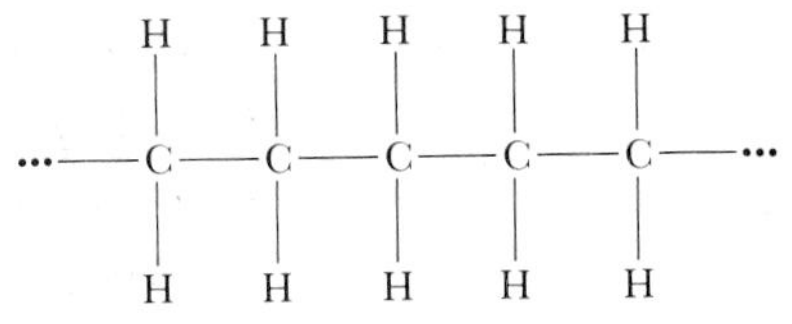

polyethene is the polymer.

A polymer chain may be 50,000 monomer units long.
There is more about polyethene later in this chapter.

## 20.2 Plastics

Plastics are polymers, but of a special sort. The word Plastic means pliable or mouldable, and this is the characteristic which all plastics have in common.

Plastics can be moulded into shape.

Many plastics are pliable and bendable.

In addition, all plastics are synthetic, (man-made) materials.

Plastics have many advantages over naturally occurring substances such as wood, stone and metals:

**1** They do not corrode in air and water.

**2** They do not rot with diseases.

**3** They do not erode with frost and rain.

**5** They are thermal insulators.

**6** They are much lighter than wood, stone or metals.

**7** Some of them can be easily bent.

**8** Some of them have very great strength. (See figure 1.)

**Figure 1**
*Mini car supported by Terylene film.*

**9** They can be coloured when they are manufactured and do not need to be painted afterwards.

**10** They are often cheaper than wood, metal or stone.

**11** They can be easily moulded into almost any shape.

But these marvellous materials do have some disadvantages:

**1** They are difficult to dispose of. They do not rot away. When they are burnt, they often produce smoke and poisonous gases.

**2** They can be a serious fire hazard. Plastics may burn easily, especially in a fire that is already burning well. Molten plastic can inflict severe burns.

**3** They are sometimes not as pleasing to the eye as wood, stone or metal.

**4** Because they are cheap, they are often used where wood, stone or metal would be better.

There are many different types of plastics and manufacturers give them various brand names. Figure 2 shows some of these names and their proper chemical names. We shall look at some of them in a later section.

| manufacturer's name | chemical name |
|---|---|
| Acrilan<br>Courtelle<br>Dralon<br>Orlon<br>Perspex | Acrilic plastics |
| BriNylon<br>Enkalon<br>Perlon<br>Tendrelle | Nylon plastics |
| Crimplene<br>Dacron<br>Terylene | Polyester plastics |

**Figure 2**
*Plastics by other names.*

Most of these names are to be found on the labels of dresses, jumpers and underwear – in fact on almost any sort of clothing. The clothing industry is one of the major users of plastics.

## 20.3 Two sorts of plastics

Plastics can be divided into two sorts depending upon the way they behave when they are heated.

## 20.4 Thermoplastics:

These get soft and runny when they are heated, but become hard again when they are cooled. You can repeat this process over and over again. This means that the plastics can be moulded into shape when they are hot and soft. When they are cooled they retain that shape.

The long polymer chains in a thermoplastic lie alongside each other, and they may be entwined, but the chains are not linked together. They can slide over each other. (See figure 3.)

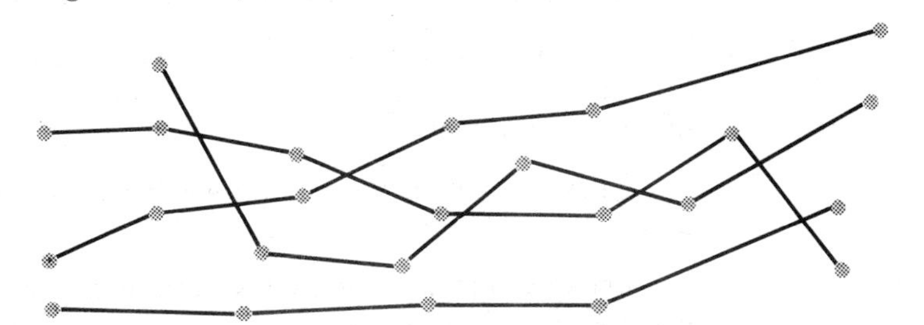

**Figure 3**
*The polymer chains in a thermoplastic.*

The next paragraphs give details of some common thermoplastics.

**Polyethene.** This is commonly called Polythene. Sometimes coloured polyethene is sold under the name of Alkathene.

Low density polyethene is made by compressing ethene and oxygen to about 2000 times atmospheric pressure at a temperature of about 200 °C:

$$\begin{array}{c}H\qquad H\\ C{=}C\\ H\qquad H\end{array} \xrightarrow[200\,^{\circ}\mathrm{C}]{2\,000\ \mathrm{Atms}} \begin{array}{c}\\ H-\\ \\ \end{array}\begin{array}{ccccc}H & & H & & H\\ C & - & C & - & C\\ H & & H & & H\end{array}-\left[\begin{array}{ccc}H & & H\\ C & - & C\\ H & & H\end{array}\right]_n-\begin{array}{ccc}H & & H\\ C & - & C\\ H & & H\end{array}\begin{array}{c}\\ -H\\ \\ \end{array}$$

This is called a *polymerisation* reaction. Polyethene has a very long polymer chain. *n* may be as big as 50 000.

This type of polyethene is soft and flexible, light enough to float on water, and when made into sheets, it is transparent. It is used to make bags for foods of all sorts, washing-up liquid containers, and sheets for water-proofing and insulation.

One disadvantage of low density polyethene however, is that at boiling water temperature, it softens so much that it becomes soft and floppy and loses its shape.

Scientists found that by using special catalysts and a much lower pressure of only 30 atmospheres, ethene would polymerise to form a much harder, stiffer polymer called high density polyethene. This does not lose its shape in hot water and can be used for washing-up bowls, buckets, toys, dustbins, and because of its rigidity, milk and beer crates. (See figure 4.)

**Polyvinyl chloride.** This name is usually shortened to P.V.C. P.V.C. is made by mixing a monomer called vinyl chloride with warm soapy water under pressure. After a catalyst has been added, the polymer slowly forms as a solid and is extracted and dried. (Vinyl chloride is a very dangerous cancer forming chemical and great care is taken when handling it. P.V.C. is harmless.)

P.V.C. is an ideal material for all sorts of jobs. It is waterproof and weather resistant, so it is made into gutters and pipes on houses, and for soft drink bottles. It is flexible when thin, so it can be made into covering for electric cables, upholstery in cars, baby pants and coats. It is rigid when thick, so toys, light switches and curtain rails are made from it.

The equation for the manufactured P.V.C. is:

$$\begin{array}{c}H\qquad H\\ C{=}C\\ Cl\qquad H\end{array} \xrightarrow[\text{water}]{\text{hot}} \begin{array}{c}\\ H-\\ \\ \end{array}\begin{array}{ccc}H & & H\\ C & - & C\\ Cl & & H\end{array}-\left[\begin{array}{ccc}H & & H\\ C & - & C\\ Cl & & H\end{array}\right]_n-\begin{array}{ccccc}H & & H & & H\\ C & - & C & - & C\\ Cl & & H & & Cl\end{array}\begin{array}{c}\\ -H\\ \\ \end{array}$$

**Figure 4**
*A milk crate made of high density polythene.*

**Polystyrene.** This is a polymer with a slightly more complicated structure. The monomer styrene is heated in hot water and it forms tiny droplets of the polymer:

$$\begin{matrix} H & & H \\ & C{=}C & \\ C_6H_5 & & H \end{matrix} \xrightarrow{\text{about } 100\,^{\circ}\mathrm{C}} \cdots\!-\!\begin{matrix} H \\ | \\ C \\ | \\ C_6H_5 \end{matrix}\!-\!\begin{matrix} H \\ | \\ C \\ | \\ H \end{matrix}\!-\!\begin{matrix} H \\ | \\ C \\ | \\ C_6H_5 \end{matrix}\!-\!\begin{matrix} H \\ | \\ C \\ | \\ H \end{matrix}\!-\!\begin{matrix} H \\ | \\ C \\ | \\ C_6H_5 \end{matrix}\!-\!\begin{matrix} H \\ | \\ C \\ | \\ H \end{matrix}\!-\!\cdots$$

Some polystyrene is used to make disposable 'plastic' cups, and casings for transistor radio sets. Plastic model kits are made of the same thing. Another sort is made into food containers, like egg boxes, imitation glass containers for jams and yoghurt, and ballpoint pens.

In yet another type, the polymer is whipped up with air and called *expanded polystyrene.* It is very light and is an excellent thermal insulator. It is used for ceiling tiles and packaging fragile items. The little trays you get meat in at supermarkets are often made of expanded polystyrene.

However, polystyrene has two big disadvantages. Firstly, it gets soft at about 90 °C and runny at 150 °C. This means that it is a fire hazard. Secondly, the clear type of polystyrene used for cups and containers becomes brittle if left in bright sunlight.

Nevertheless, more than 140 000 tonnes of this plastic are produced each year in Britain.

**Nylon** You are probably wearing something made of nylon at this moment. It is a plastic that can be made into thin fibres which may be woven into cloth, sometimes mixed with other synthetic or natural materials.

Its structure is not as simple as those we have seen already, but one type of nylon may be made when solutions of hexanedioic acid and diaminohexane are mixed.

$$H-O-\overset{\overset{O}{\|}}{C}-C_4H_8-\overset{\overset{O}{\|}}{C}-O-H \qquad \underset{H}{\overset{H}{\gt}}N-C_6H_{12}-N\underset{H}{\overset{H}{\lt}}$$

hexanedioic acid diaminohexane

A molecule of water is removed from the two molecules and they join together. This is repeated thousands of times to form a polymer chain.

$$\cdots-\overset{\overset{O}{\|}}{C}-C_4H_8-\overset{\overset{O}{\|}}{C}-\underset{\underset{H}{|}}{N}-C_6H_{12}-\underset{\underset{H}{|}}{N}-\cdots$$

This is called Nylon 6.6 because the two monomer units each contain six carbon atoms. Terylene is another polymer with a similar structure.

## 20.5 Putting the shape into plastics

Thermoplastics can be squeezed, blown, rolled or sucked into shape. The following paragraphs explain some of the methods in more detail.

**Extrusion.** Thin nylon fibres, drain pipes, curtain rails and any object that can be made in a long, continuous strip are formed by this method. Granules of the plastic are fed into a machine. Heaters make the material soft as it is squeezed or 'extruded' through a nozzle. Then the soft, shaped plastic is cooled with cold air so that it hardens. (See figure 5**a**.)
Figure 5**b** shows very fine nylon fibres being made by this method.

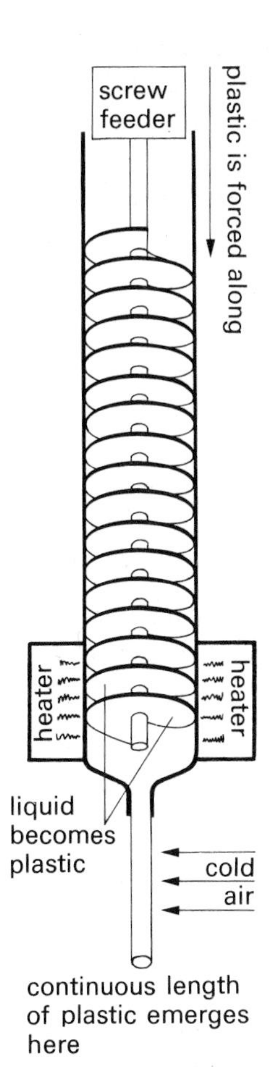


**Figure 5a**
*The formation of an item by extrusion.*

**5b**
*Nylon fibres being made by extrusion.*

**Injection moulding.** This is much the same as extrusion, but the softened plastic is forced into a mould, and after it has been cooled, the mould is opened, and the shaped plastic item removed. (See figure 6.)

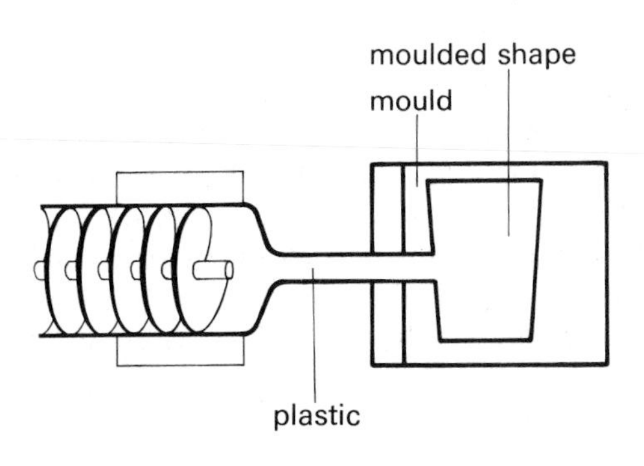


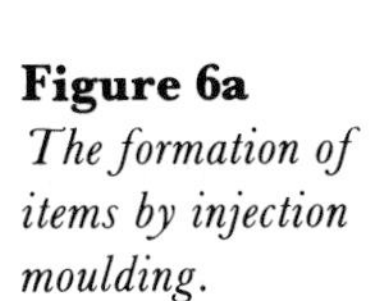

**Figure 6a**
*The formation of items by injection moulding.*

Sometimes the soft plastic is blown into the mould by compressed air. This is called blow moulding.

## 20.6 Thermosetting plastics:

These plastics become permanently hard once they are heated. After that, they cannot be softened and remoulded by heating. Their structure differs from those of thermoplastics in that the polymer chains are cross linked during the initial heating. Once these linkages are formed, the plastic remains rigid. (See figure 7.) The next three sections give some details of the most common thermosetting plastics.

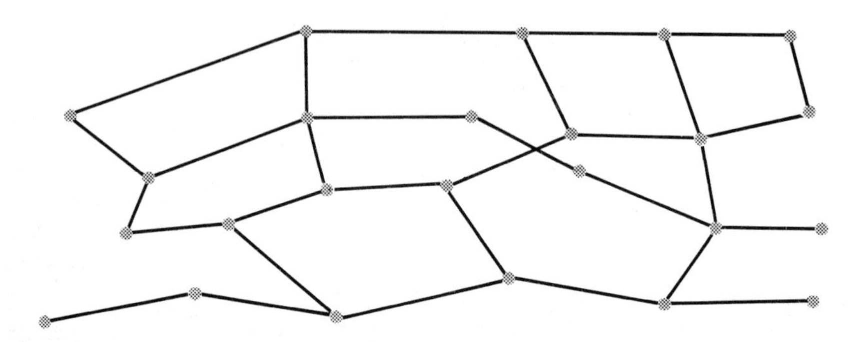

**Figure 7**
*The polymer chains in a thermosetting plastic are cross-linked.*

**Melamine.** This plastic is often used to make children's dishes and cups. It is hard and stiff and difficult to break. Extreme force, however, will snap it. It will withstand hot water without bending at all, but when it is heated in a fire, it chars.

Its structure is very complicated.

**Bakelite.** This was one of the first plastics to be made. It is made from an organic compound called *phenol*, which has carbon atoms arranged in rings.

When phenol is heated with methanol, their molecules are joined together, and a syrupy liquid called a *resin* is formed. This consists of long polymer chains, but they do not as yet have any cross-linkages. The resin is mixed with a filler, such as sawdust, or silica powder, and then put into the mould of the object to be made. When heated, the cross-links form up, and the resin sets hard into a stiff, heat resistant plastic.

Bakelite and other similar plastics are used for electric plugs, and switches, saucepan handles, insulation on electronic printed circuits, and even as heat shields on the nose cones of space rockets.

**Glass reinforced polyester – G.P.R.** Organic esters are polymerised and the resin that is formed is mixed with glass fibres before another chemical is added to make it harden. A very tough, lightweight plastic is made which has very great strength.

It is used for the bodies of boats and canoes, garden pools, piping, roofing, car repairs – in fact anywhere that strength and lightness, or resistance to corrosion are needed.

## 20.7 Shaping thermosetting plastics

**Compression moulding.** Thermosetting plastic items cannot be extruded or injected because they harden as soon as they are heated. Instead, they are moulded and heated at the same time by compressing them into shape. (See figure 8.)

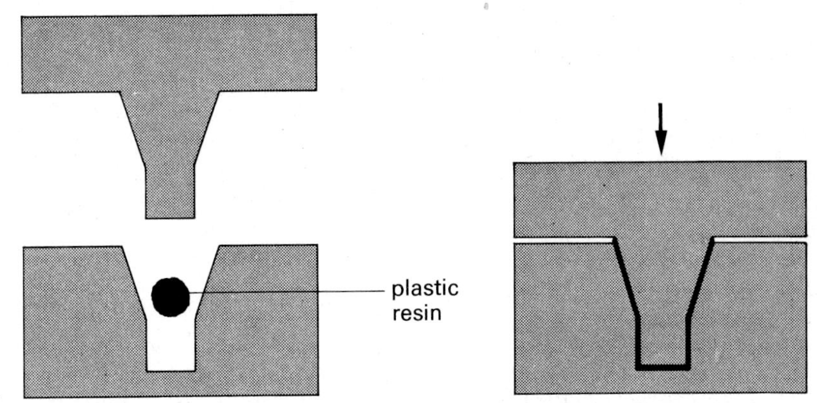


**Figure 8**
*Compression moulding an item in thermosetting plastic.*

**Lamination.** Thermosetting plastics can be sandwiched in layers between paper or cloth, and the mixture can be squashed in a powerful press. The result is a thin, tough, heat proof laminate. Formica, and other kitchen working surface materials are made in this way.

## Summary

At the end of this chapter you should be able to:

**1** Explain what the terms monomer and polymer mean.

**2** Say what is meant by the word plastic.

**3** List the advantages of plastics over natural materials.

**4** List some disadvantages of plastics.

**5** Explain the difference between thermoplastics and thermosetting plastics, in terms of their behaviour and their structures.

**6** Give examples of thermoplastics and thermosetting plastics, and their everyday uses.

**7** Describe how plastic objects may be made by extrusion, injection moulding, compression moulding and laminating.

# It's a record!

*Buying a gramophone record is very easy. You just go to the shop, and pay your money. When you put it on the turntable, and put the needle in the groove, out floods the sound of a pop group, or a full orchestra. But have you ever thought of all the things that must happen before a record reaches the shop?*

First of all the musicians get together in a recording studio, and engineers record their performance on tape. They don't usually make just one recording. They make several, and then cut the tape up, fitting all the best bits together to make a 'master' tape.

Next, the sound from the tape is transferred to an aluminium disc which is covered with cellulose acetate. The disc is put on a turntable, just like that on a record player. This revolves at the speed of the record, which may be $33\frac{1}{3}$ or 45 revolutions a minute. The tape is connected to an electronic machine that has an arm with a cutting stylus on the end of it. As the tape plays, the stylus vibrates and converts the sound into a wavy groove cut into the disc. If the notes are low, the waves are long, and if the notes are high, then the waves are short. The loudness of the notes makes the waves either wide, or narrow if it is quiet. Look at a record under a strong light and you will see these grooves.

The continuous groove slowly curves in towards the centre of the record. On a long playing record, it loops about eight hundred times before reaching the end of the record. Each groove is about one four-hundredth of a centimetre deep.

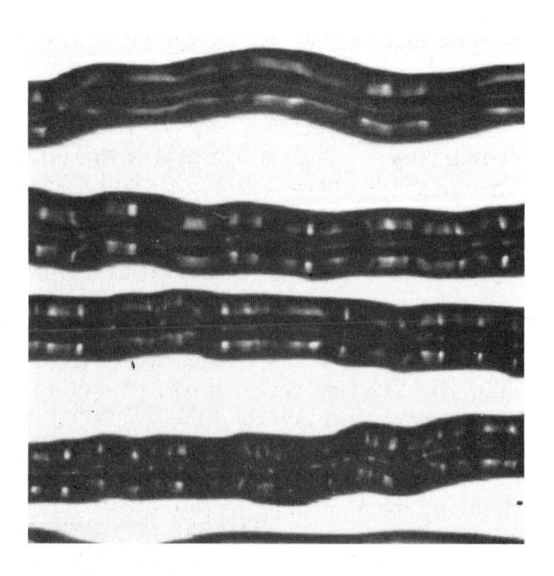

*The grooves on a gramophone record.*

*A test record is made.*

The next stage is to transfer the grooves on the very delicate aluminium disc to plastic. The disc is first carefully cleaned and then coated with a very thin layer of silver. Next it is put into a nickel plating bath. The electrodes used are the silver coated disc as cathode; the anode is pure nickel. The electrolyte contains nickel ions, and when the electricity is switched on, a layer of nickel is slowly deposited on top of the silver. After two hours, the layer of nickel and silver is carefully split away from the aluminium which was coated with cellulose acetate, and the result is a *negative* where all the grooves are back to front, rather like a negative in photography.

From the negative, a *positive* is made. This is done by putting the negative back into the electrolysis bath again, and more nickel is deposited onto it. When this is split off, the new disc has the grooves the right way round – just like the original aluminium disc.

This positive is sometimes known as a 'mother'. While the original nickel negative disc is stored in case some accident now happens, the 'mother' is inspected under a microscope to make sure that there is no damage to the grooves in any way.

For the last time, the positive 'mother' disc goes back into the electrolysis bath and another negative is made. This is called the 'stamper' because it will be used to make the plastic discs. Two stampers have to be made of course; one for each side of the record, and after inspection they have holes drilled through their centres. This hole must be drilled to an accuracy of several hundredths of a centimetre. Many stampers will be made, because they wear out quickly.

Now comes the actual pressing of the record. The two stampers are clamped in a press which can be heated by steam and cooled by water. A small piece of the plastic is put between the stampers. This plastic is mainly polyvinyl chloride mixed with carbon black, and other chemicals called *stabilisers*. The press closes, and the steam is switched on. The press exerts a pressure of many tonnes, and the plastic is forced between the stampers, to take on the shape of the record. The grooves are exactly reproduced in plastic. After 30 seconds the record is cooled with water and the press is opened. During this process, the labels have been added. All that remains is to trim off the surplus plastic around the edge on a circular cutter.

However, before the record is packed, it is inspected. Experts play it on extremely sensitive players and listen for errors. If the record is found to be faulty in any way, it is rejected.

So, as you listen to your record, think about these processes. The way in which the sound gets back out of the grooves and into your ears – is another story altogether.

## Questions

**1** Bridges are not made of plastic. Give one reason why this is so. However, if plastic could be used, in what ways might it be better than steel?

**2** Explain in your own words what a polymer is.
Suppose you had the monomer:

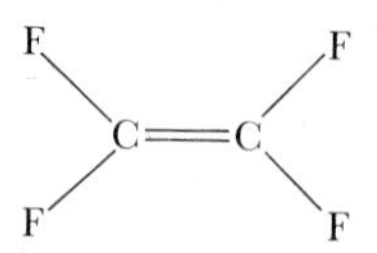


1,1,2,2-tetrafluoroethene

Draw the shape of the chain that is formed when the monomer is polymerised.

**3** What is the difference between a thermoplastic and a thermosetting plastic?
Why cannot low density polyethene be used for washing-up bowls?
Why would it be very difficult to make fibres (threads) of Bakelite for knitted garments?

**4** Polyethene and P.V.C. are made when their monomers join together by a process called *addition*. Nylon and Bakelite are made when their monomers join together by a process called *condensation*. Use the chapter to find out and explain the difference between them.

**5** Which would be the best method of plastic shaping for making:
- **a** a plastic chopping board.
- **b** a washing-up bowl.
- **c** polyethene tubing.
- **d** electric light switch covers.

**6** Make a trip around your house and list all the things that are made of plastic. Can you identify any of the plastics? Try to suggest the methods by which the different objects were made.

**7** **a** Orange squash bottles are now often made of plastic instead of glass. Suggest two ways in which this is an improvement, and two ways in which it is a nuisance.
**b** In most new houses, the floors are covered with plastic tiles instead of being left as wood or stone. In what ways is this better? Can you think of a big disadvantage?
**c** A large part of your clothing is probably made of nylon, or another man-made fibre instead of wool or cotton. Is it better or worse than natural materials? Different people will have different answers to this question. How does the price compare?

**8** Why is polystyrene a good insulator for fridges and 'cold' boxes, but no good for table mats?

# Up to date names

The names of most of the chemicals mentioned throughout this book have been kept deliberately simple in order that you will not be confused. Some of them however, should have more up to date names based upon the I.U.P.A.C. (International Union of Pure and Applied Chemistry) system of naming, and you may well see them in other books.

**Metals**

The I.U.P.A.C. system has been used for naming metal compounds whenever a metal forms more than one ion. In such a case, the charge of the ion is put into brackets after the metal, as a Roman numeral. Iron(II) sulphate, $FeSO_4$, means iron sulphate containing the $Fe^{2+}$ ion. Iron(III) chloride, $FeCl_3$, means iron chloride containing the $Fe^{3+}$ ion. Copper(I) oxide, $Cu_2O$, means copper oxide containing the $Cu^{+}$ ion. Copper(II) nitrate, $Cu(NO_3)_2$, means copper nitrate containing the $Cu^{2+}$ ion.

Sometimes more complicated metal compounds which do not contain simple ions are named in this way.
Manganese(IV) oxide means manganese oxide that might contain the $Mn^{4+}$ ion, if it existed in this simple form.

**Acids**

In Chapter 15 you met sulphuric acid and sulphurous acid. More correctly, these should be called:

sulphuric(VI) acid, $H_2SO_4$ (sulphuric acid)

and

sulphuric(IV) acid, $H_2SO_3$ (sulphurous acid).

The numbers in brackets refer to the ions that sulphur might form in the compound, if it could.
This means that a sulphate should be called a sulphate(VI), and a sulphite should be called a sulphate(IV).

Similarly, in Chapter 16, nitric acid ($HNO_3$) should be called nitric(V) acid, and its salt called a nitrate(V), and nitrous acid ($HNO_2$) should be called Nitric(III) acid, and its salt a nitrate(III), instead of a nitrite.

| group I | group II | | | | | | | | | | |
|---|---|---|---|---|---|---|---|---|---|---|---|
| | | | | 1 1 H hydrogen | | | | | | | |
| 7 3 Li lithium | 9 4 Be beryllium | | | | | | | | | | |
| 23 11 Na sodium | 24 12 Mg magnesium | | | | | | | | | | |
| 39 19 K potassium | 40 20 Ca calcium | 45 21 Sc scandium | 48 22 Ti titanium | 51 23 V vanadium | 52 24 Cr chromium | 55 25 Mn manganese | 56 26 Fe iron | 59 27 Co cobalt | 59 28 Ni nickel | 64 29 Cu copper | 65 30 Zn zinc |
| 85 37 Rb rubidium | 88 38 Sr strontium | 89 39 Y yttrium | 91 40 Zr zirconium | 93 41 Nb niobium | 96 42 Mo molybdenum | 98 43 Tc technetium | 101 44 Ru ruthenium | 103 45 Rh rhodium | 106 46 Pd palladium | 108 47 Ag silver | 112 48 Cd cadmium |
| 133 55 Cs caesium | 137 56 Ba barium | 139 57 La lanthanum | 178·5 72 Hf hafnium | 181 73 Ta tantalum | 184 74 W tungsten | 186 75 Re rhenium | 190 76 Os osmium | 192 77 Ir iridium | 195 78 Pt platinum | 197 79 Au gold | 201 80 Hg mercury |
| 223 87 Fr francium | 226 88 Ra radium | 227 89 Ac actinium | | | | | | | | | |
| | | | 140 58 Ce cerium | 141 59 Pr prae-sodimium | 144 60 Nd neodimium | 147 61 Pm promethium | 150 62 Sm samarium | 152 63 Eu europium | 157 64 Gd gadolinium | 159 65 Tb terbium | 162 66 Dy dysprosium |
| | | | 232 90 Th thorium | 231 91 Pa prot-actinium | 238 92 U uranium | 237 93 Np neptunium | 242 94 Pu plutonium | 243 95 Am americium | 247 96 Cm curium | 247 97 Bk berkelium | 251 98 Cf californium |

**Relative atomic masses based on internationally agreed figures.**

| Element | Symbol | Atomic number | Relative atomic mass |
|---|---|---|---|
| Actinium | Ac | 89 | |
| Aluminium | Al | 13 | 26·9815 |
| Americium | Am | 95 | |
| Antimony | Sb | 51 | 121·75 |
| Argon | Ar | 18 | 39·948 |
| Arsenic | As | 33 | 74·9216 |
| Astatine | At | 85 | |
| Barium | Ba | 56 | 137·34 |
| Berkelium | Bk | 97 | |
| Beryllium | Be | 4 | 9·0122 |
| Bismuth | Bi | 83 | 208·980 |
| Boron | B | 5 | 10·811 |
| Bromine | Br | 35 | 79·909 |
| Cadmium | Cd | 48 | 112·40 |
| Caesium | Cs | 55 | 132·905 |
| Calcium | Ca | 20 | 40·08 |
| Californium | Cf | 98 | |
| Carbon | C | 6 | 12·01115 |
| Cerium | Ce | 58 | 140·12 |
| Chlorine | Cl | 17 | 35·453 |
| Chromium | Cr | 24 | 51·996 |
| Cobalt | Co | 27 | 58·9332 |
| Copper | Cu | 29 | 63·54 |
| Curium | Cm | 96 | |
| Dysprosium | Dy | 66 | 162·50 |
| Einsteinium | Es | 99 | |
| Erbium | Er | 68 | 167·26 |
| Europium | Eu | 63 | 151·96 |
| Fermium | Fm | 100 | |
| Fluorine | F | 9 | 18·9984 |
| Francium | Fr | 87 | |
| Gadolinium | Gd | 64 | 157·25 |
| Gallium | Ga | 31 | 69·72 |
| Germanium | Ge | 32 | 72·59 |
| Gold | Au | 79 | 196·967 |
| Hafnium | Hf | 72 | 178·49 |
| Helium | He | 2 | 4·0026 |
| Holmium | Ho | 67 | 164·930 |
| Hydrogen | H | 1 | 1·00797 |
| Indium | In | 49 | 114·82 |
| Iodine | I | 53 | 126·9044 |
| Iridium | Ir | 77 | 192·2 |
| Iron | Fe | 26 | 55·847 |
| Krypton | Kr | 36 | 83·80 |
| Lanthanum | La | 57 | 138·91 |
| Lawrencium | Lw | 103 | |
| Lead | Pb | 82 | 207·19 |
| Lithium | Li | 3 | 6·939 |
| Lutetium | Lu | 71 | 174·97 |
| Magnesium | Mg | 12 | 24·312 |
| Manganese | Mn | 25 | 54·9380 |
| Mendelevium | Md | 101 | |

| group III | group IV | group V | group VI | group VII | group 0 |
|---|---|---|---|---|---|
| | | | | | $^{4}_{2}$He helium |
| $^{11}_{5}$B boron | $^{12}_{6}$C carbon | $^{14}_{7}$N nitrogen | $^{16}_{8}$O oxygen | $^{19}_{9}$F fluorine | $^{20}_{10}$Ne neon |
| $^{27}_{13}$Al aluminium | $^{28}_{14}$Si silicon | $^{31}_{15}$P phosphorus | $^{32}_{16}$S sulphur | $^{35·5}_{17}$Cl chlorine | $^{40}_{18}$Ar argon |
| $^{70}_{31}$Ga gallium | $^{73}_{32}$Ge germanium | $^{75}_{33}$As arsenic | $^{79}_{34}$Se selenium | $^{80}_{35}$Br bromine | $^{84}_{36}$Kr krypton |
| $^{115}_{49}$In indium | $^{119}_{50}$Sn tin | $^{122}_{51}$Sb antimony | $^{128}_{52}$Te tellurium | $^{127}_{53}$I iodine | $^{131}_{54}$Xe xenon |
| $^{204}_{81}$Tl thallium | $^{207}_{82}$Pb lead | $^{209}_{83}$Bi bismuth | $^{210}_{84}$Po polonium | $^{210}_{85}$At astatine | $^{222}_{86}$Rn radon |

| | | | | |
|---|---|---|---|---|
| $^{165}_{67}$Ho holmium | $^{167}_{68}$Er erbium | $^{169}_{69}$Tm thulium | $^{173}_{70}$Yb ytterbium | $^{175}_{71}$Lu lutecium |
| $^{254}_{99}$Es einsteinium | $^{253}_{100}$Fm fermium | $^{256}_{101}$Md mendelevium | $^{254}_{102}$No nobelium | $^{257}_{103}$Lw lawrencium |

**Approximate atomic masses for calculations.**

| Element | Symbol | Atomic mass for calculations |
|---|---|---|
| Aluminium | Al | 27 |
| Bromine | Br | 80 |
| Calcium | Ca | 40 |
| Carbon | C | 12 |
| Chlorine | Cl | 35·5 |
| Copper | Cu | 64 |
| Helium | He | 4 |
| Hydrogen | H | 1 |
| Iodine | I | 127 |
| Iron | Fe | 56 |
| Lead | Pb | 207 |
| Lithium | Li | 7 |
| Magnesium | Mg | 24 |
| Manganese | Mn | 55 |
| Nitrogen | N | 14 |
| Oxygen | O | 16 |
| Phosphorus | P | 31 |
| Potassium | K | 39 |
| Silicon | Si | 28 |
| Silver | Ag | 108 |
| Sodium | Na | 23 |
| Sulphur | S | 32 |
| Zinc | Zn | 65 |

| Element | Symbol | Atomic number | Relative atomic mass |
|---|---|---|---|
| Mercury | Hg | 80 | 200·59 |
| Molybdenum | Mo | 42 | 95·94 |
| Neodymium | Nd | 60 | 144·24 |
| Neon | Ne | 10 | 20·179 |
| Neptunium | Np | 93 | |
| Nickel | Ni | 28 | 58·71 |
| Niobium | Nb | 41 | 92·906 |
| Nitrogen | N | 7 | 14·0067 |
| Nobelium | No | 102 | |
| Osmium | Os | 76 | 190·2 |
| Oxygen | O | 8 | 15·9994 |
| Palladium | Pd | 46 | 106·4 |
| Phosphorus | P | 15 | 30·9738 |
| Platinum | Pt | 78 | 195·09 |
| Plutonium | Pu | 94 | |
| Polonium | Po | 84 | |
| Potassium | K | 19 | 39·102 |
| Praseodymium | Pr | 59 | 140·907 |
| Promethium | Pm | 61 | |
| Protactinium | Pa | 91 | |
| Radium | Ra | 88 | |
| Radon | Rn | 86 | |
| Rhenium | Re | 75 | 186·2 |
| Rhodium | Rh | 45 | 102·905 |
| Rubidium | Rb | 37 | 85·47 |
| Ruthenium | Ru | 44 | 101·07 |

| Element | Symbol | Atomic number | Relative atomic mass |
|---|---|---|---|
| Samarium | Sm | 62 | 150·35 |
| Scandium | Sc | 21 | 44·956 |
| Selenium | Se | 34 | 78·96 |
| Silicon | Si | 14 | 28·086 |
| Silver | Ag | 47 | 107·868 |
| Sodium | Na | 11 | 22·9898 |
| Strontium | Sr | 38 | 87·62 |
| Sulphur | S | 16 | 32·064 |
| Tantalum | Ta | 73 | 180·948 |
| Technetium | Tc | 43 | |
| Tellurium | Te | 52 | 127·60 |
| Terbium | Tb | 65 | 158·924 |
| Thallium | Tl | 81 | 204·37 |
| Thorium | Th | 90 | 232·038 |
| Thulium | Tm | 69 | 168·934 |
| Tin | Sn | 50 | 118·69 |
| Titanium | Ti | 22 | 47·90 |
| Tungsten | W | 74 | 183·85 |
| Uranium | U | 92 | 238·03 |
| Vanadium | V | 23 | 50·942 |
| Xenon | Xe | 54 | 131·30 |
| Ytterbium | Yb | 70 | 173·04 |
| Yttrium | Y | 39 | 88·905 |
| Zinc | Zn | 30 | 65·37 |
| Zirconium | Zr | 40 | 91·22 |

# Index

## Answers to numerical questions

**Chapter 1** page 19 7. **a** 60 °C **b** 2 minutes **c** 220 °C

**Chapter 2** page 32 4. **a** lead nitrate **b** lead nitrate
**c** approx. 80 g per 100 g of water **d** 8 °C **e** 50 °C
**f** approx. 80 g per 100 g of water
5. 238 g per 100 g of water.

**Chapter 6** page 91 4. **a** 23 g, 27 g, 39 g, 32 g, 32 g
**b** 130 g, 112 g, 48 g, 110 g, 56 g
**c** 6 g **d** 33 g **e** 4 g **f** 2 g **g** 4 g
**h** 325 g **i** 70 g **j** 168 g

5. **a** 1 mole, 1 mole, 1 mole of atoms
**b** 2 moles, 4 moles, 2 moles, 0·5 moles

6. **a** 18 g **b** 88 g **c** 50 g **d** 710 g **e** 153 g

7. 50 g

8. 36 g

9. $FeCl_3$

10. ZnO

**Chapter 7** page 106 7. **a** 16 g **b** 36 g **c** 44 g

**Chapter 8** page 117 3. 80%

**Chapter 9** page 134 7. **a** 213 g **b** 18 g **c** $n = 9$